Gene
Therapy

Principles
and
Applications

Edited by
Thomas Blankenstein

Birkhäuser Verlag
Basel • Boston • Berlin

Editor

Prof. Dr. Thomas Blankenstein
Max-Delbrück-Centrum für molekulare Medizin
Robert-Rössle-Strasse 10
D-13122 Berlin-Buch
Germany

Library of Congress Cataloging-in-Publication Data
　　Gene therapy: principles and applications / edited by Thomas Blankenstein
　　　　　p. cm.
　　Includes bibliographical references and index.
　　ISBN 0-8176-5972-2 (hardcover: alk. paper).-- ISBN 3-7643-5972-2 (hardcover: alk. paper)
　　1. Gene therapy.　I. Blankenstein, Thomas.
　　RB155.8.G464 1999
　　616'.042--dc21　　　　　　　　　　　　99-37011
　　　　　　　　　　　　　　　　　　　　　　CIP

Deutsche Bibliothek Cataloging-in-Publication Data
　　Gene therapy: principles and applications / ed. by Thomas Blankenstein
　　Basel; Boston; Berlin; Birkhäuser, 1999
　　ISBN 3-7643-5972-2 (Basel...)
　　ISBN 0-8176-5972-2 (Boston)

The publisher and editor can give no guarantee for the information on drug dosage and administration contained in this publication. The respective user must check its accuracy by consulting other sources of reference in each individual case.

The use of registered names, trademarks etc. in this publication, even if not identified as such, does not imply that they are exempt from the relevant protective laws and regulations or free for general use.

© 1999 Birkhäuser Verlag, PO Box 133, CH-4010 Basel, Switzerland
Printed on acid-free paper produced from chlorine-free pulp. TCF ∞
Printed in Germany
Cover design: gröflin. Graphic Design, Basel

ISBN 3-7643-5972-2
ISBN 0-8176-5972-2

9 8 7 6 5 4 3 2 1

Contents

List of Contributors

ALESSANDRO AIUTI, Telethon Institute for Gene Therapy, Milano, Italy

MUHAMMAD AMJAD, Dept. of Medicine, Jefferson Medical College, Thomas Jefferson University, Jefferson Alumni Hall, Philadelphia, USA

OMAR BAGASRA, Division of Infectious Diseases, Department of Medcine, Jefferson Medical College, Thomas Jefferson University, Piladelphia, USA

JOHN A. BARRANGER, Dept. of Human Genetics, Graduate School of Public Health, University of Pittsburgh, Pittsburgh, USA

THOMAS BLANKENSTEIN, Max-Delbrück-Centrum für Molekulare Medizin, Berlin, Germany

CHIARA BONINI, Telethon Institute for Gene Therapy, Milano, Italy

CLAUDIO BORDIGNON, Telethon Institute for Gene Therapy, Milano, Italy

XUETAO CAO, Dept. of Immunology, Second Military Medical University, Shanghai, P.R. China

GUILIA CASORATI, Unita d`Immunochimica, DIBIT, Istituto San Raffaele, Milano, Italy

SOPHIE CAYEUX, Robert-Rössle-Klinik, Virchow Klinikum, Berlin, Germany

FREDERICA CAVALLO, Dipartimento di Scienze Cliniche e Biologiche, University of Torino, Orbassano, Italy

GÜNTHER CICHON, Max-Delbrück-Centrum für molekulare Medizin, Berlin, Germany

KENNETH CORNETTA, Section of Hematology/Oncology, Indianapolis, USA

PAOLO DELLABONA, Unita d'Immunochimica, DIBIT, Istituto San Raffaele, Milano, Italy

BERND DÖRKEN, Robert-Rössle-Klinik, Virchow Klinikum, Berlin, Germany

GUIDO FORNI, Dipartimento di Scienze Cliniche e Biologiche, University of Torino, Orbassano, Italy

ALAN M. GEWIRTZ, Hematology/Oncology and Hematology/Molecular Diagnosis Sections, Depts. of Internal Medicine and Pathology, University of Pensylvania, School of Medicine, Philadelphia, USA

MICHAEL M. GOTTESMAN, Laboratory of Biology, National Cancer Institute, National Institutes of Health, Bethesda, Maryland, USA

MICHAEL HALLEK, Laboratorium für Molekulare Biologie, Genzentrum, and Medizinische Klinik III, Klinikum Großhadern, Ludwig-Maximilian-Universität München, München, Germany

ALASTAIR INNES, Scottish Adult Cystic Fibrosis Service, MRC Human Genetics Unit, Western General Hospital, Edingburgh, UK

TAKASHI IWAZAWA, Depts. Surgery, Mol. Genetics and Biochemistry, Pittsburgh Cancer Institute, University of Pittsburgh, School of Medicine, Pittsburgh, USA

TORU KITAGAWA, Depts. Surgery, Mol. Genetics and Biochemistry, Pittsburgh Cancer Institute, University of Pittsburgh, School of Medicine, Pittsburgh, USA

ROBERT KOTIN, NIH, NHLBI, Bethesda, Maryland, USA

THOMAS LICHT, Laboratory of Molecular Biology, National Cancer Institute, National Institutes of Health, Bethesda, Maryland, USA

PIER-LUIGI LOLLINI, Istituto di Cancerologia, University of Bologna, Bologna, Italy

MICHAEL T. LOTZE, Divisions of Surgical Oncology and Biologic Therapy, University of Pittsburgh, Cancer Institute, Pittsburgh, USA

CORNELIS MELIEF, Department of Immunohaematology and Blood Bank, Leiden University Medical Center, Leiden, Netherlands

ERICA MENGEDÉ, Department of Immunohaematology and Blood Bank, University Hospital Leiden, Leiden, Netherlands

DORIS MICHL, Laboratorium für Molekulare Biologie, Genzentrum, Medizinische Klinik, Ludwig-Maximilians-Universität München, München, Germany

MUHAMMAD MUKHTAR, Dept. of Medicine, Jefferson Medical College, Thomas Jefferson University, Jefferson Alumni Hall, Philadelphia, USA

PATRIZIA NANNI, Istituto di Cancerologia, University of Bologna, Bologna, Italy

ELFRIEDE NÖSSNER, GSF-Forschungszentrum für Umwelt und Gesundheit GmbH, Institut für Molekulare Immunologie, München, Germany

RIENK OFFRINGA, Department of Immunohaematology and Blood Bank, Leiden University Medical Center, Leiden, Netherlands

FERRY OSSENDORP, Department of Immunohaematology and Blood Bank, Leiden University Medical Center, Leiden, Netherlands

IRA PASTAN, Laboratory of Molecular Biology, National Cancer Institute, National Institutes of Health, Bethesda, Maryland, USA

DAVID J. PORTEOUS, Scottish Adult Cystic Fibrosis Service, MRC Human Genetics Unit, Western General Hospital, Edingburgh, UK

ZHIHAI QIN, Max-Delbrück-Zentrum für Molekulare Medizin, Berlin, Germany

PAUL D. ROBBINS, Department of Molecular Genetics and Biochemistry, University of Pittsburgh, School of Medicine, Pittsburgh, USA

DOLORES J. SCHENDEL, GSF-Forschungszentrum für Umwelt und Gesundheit GmbH, Institut für Molekulare Immunologie, München, Germany

KAROL SIKORA, Chief, WHO Cancer Programme and Professor of Cancer Medicine, Imperial College School of Medicine, Hammersmith Hospital, London, UK

MICHAEL STRAUSS, Max-Delbrück-Centrum für Molekulare Medizin, Berlin, Germany

HIDEAKI TAHARA, Departments of Surgery, Molecular Genetics and Biochemistry, University of Pittsburgh, Pittsburgh, USA

René Toes, Department of Immunohaematology and Blood Bank, Leiden University Medical Center, Leiden, Netherlands

Michael J. Vallor, Dept. of Human Genetics, Graduate School of Public Health, University of Pittsburgh, Pittsburgh, USA

Richard G. Vile, Molecular Medicine Program, Mayo Clinic, Rochester, Minnesota, USA

Ellen van der Voort, Department of Immunohaematology and Blood Bank, Leiden University Medical Center, Leiden, Netherlands

Ernst Wagner, Boehringer Ingelheim Austria, Forschung & Entwicklung, Wien, Austria

Clemens-Martin Wendtner, Laboratorium für Molekulare Biologie, Genzentrum, and Medizinische Klinik III, Klinikum Großhadern, Ludwig-Maximilians-Universität München, München, Germany

Ernst-Ludwig Winnacker, Laboratorium für Molekulare Biologie, Genzentrum, Ludwig-Maximilians-Universität München, München, Germany

Th. Wölfel, I. Medizinische Klinik und Poliklinik, Klinikum der Johannes-Gutenberg-Universität, Mainz, Germany

Gerhard Wolff, Robert-Rössle-Klinik, Virchow Klinikum, Berlin, Germany

Introduction

K. Sikora

Gene therapy is one of the fastest developing areas in modern medical research. Transcending the classical preclinical and clinical disciplines, it is likely to have far-reaching consequences in the practice of medicine, as we enter the next millennium. Currently, there are over 200 seperate active clinical trials with over 2,500 patients entered. These studies involve over 20 countries and include patients with a wide range of diseases, including cancer, HIV infection, cystic fibrosis (CF), haemophilia, diabetes, immune deficiencies, metabolic disorders, ischaemic heart disease and arthritis.

Gene therapy can be defined as the deliberate transfer of DNA for therapeutic purposes. There is a further implication that only specific sequences containing relevant genetic information are used; otherwise, transplantation procedures involving bone marrow, kidney or liver could be considered a form of gene therapy. The concept of transfer of genetic information as a practical clinical tool arose from the gene-cloning technology, developed during the 1970s. Without the ability to isolate and replicate defined genetic sequences, it would be impossible to produce purified material for clinical use. The drive for the practical application of this technology came from the biotechnology industry with its quest for complex human biomolecules produced by recombinant techniques in bacteria. Within a decade, pharmaceutical-grade insulin, interferon, interleukin 2 and tumour necrosis factor were all involved in clinical trials. The next step was to obtain gene expression *in vivo*.

Suitable Target Diseases

Genetic disorders were the obvious first target for such therapies. Abortive attempts were made in the early 1980s to treat two patients with thalassaemia. These experiments were surrounded by controversy, as the preclinical evidence of effectiveness was not adequate and full ethical approval had not been given. We now know that thalassaemia – a disorder in which there is transcriptional dysregulation of the globin chains of haemoglobin – may not be an ideal target for gene therapy, as we still do not have good methods of regulating the expression level of inserted constructs. The features of suitable target diseases are listed in Table 1.

T. Blankenstein (ed.) Gene Therapy
©1999, Birkhäuser Verlag Basel

Table 1 Features of a disease suitable for gene therapy

- life threatening
- gene cloned
- efficient gene transfer to relevant tissue possible
- precise regulation of gene not required
- proper processing of protein product
- correct subcellular localisation of protein
- persistence of gene expression to avoid repetitive dosing
- measurable surrogate end points
- effects of disease must be potentially reversible

One common strand linking all clinical gene-therapy studies is the difficulty in delivering genes to the right place and effectively controlling their expression. These problems are outlined in the first section of this book, which considers both viral and physical methods of gene delivery. Selective targeting may also be used to enhance the tissue specificity of delivery.

The aim of gene manipulation varies with different diseases. In patients with single-gene defects, such as haemophilia, all that may be required is a suitable *protein factory* to produce enough circulating coagulation factor to be effective, physiologically. In CF, enough CFTR product must be selectively expressed by those cells where its lack causes pathological damage, such as in the lung and gastrointestinal epithelia. A variety of metabolic disorders result from genetic abnormalities in liver proteins. Some form of tissue targeting may be necessary before effective therapeutic strategies can be developed. With cancer, the problem is the requirement to target every single malignant cell. Although a variety of ingenious methods are being examined, currently, it would seem more realistic to utilise systems that do not require the correction of the somatic defect resulting in malignancy.

Over the last seven years, a growing number of protocols have been approved by regulatory authorities throughout the world – the majority in the USA. Protocols for cancer gene therapy are now much more successful than all others. This reflects the difficulty in treating patients with advanced cancer by conventional chemotherapy as well as the low risk-benefit potential of such treatments and the relatively high level of research funding in this area of biomedical research. There are 205 trials currently active: 158 are for cancer research, and can be located on the NIH Office for Recombinant DNA Technology (ORDA) and the European Working Group on Gene Therapy (EWGGT) databases.

Enabling Technology

As with any mission-orientated project, certain milestones can be identified, without which further progress would have been impossible (Tab. 2). There are many future hurdles to be overcome before the successful clinical application of this technology is achieved on a routine basis. Most of the problems include vector development.

Table 2 Enabling technology

Year	Discovery
1928	Transformation of *D. pneumoniae*
1944	DNA is transforming substance
1952	DNA is genetic material
1953	Double helix structure
1963	Genetic code elucidated
1968	Transformation by DNA viruses
1976	Cloning of globin genes
1979	Cloned genes expressed
1981	Clinical thalassaemia studies
1982	Recombinant proteins for clinical use
1985	Retroviral vectors developed
1990	First adenoise deaminase (ADA) trial
1990	Gene marking studies
1991	Cystic fibrosis protocol
1992	Familial hypercholesterolaemia studies
1992	Large number of cancer trials

Perhaps the most significant development yet to come is the development of systemically administered highly selective targeting vectors with high efficiency of stable incorporation into all cells of a chosen population. The delivery problem currently pervades the whole of this exciting field.

Ethical and Safety Considerations

Perhaps the biggest risk from gene manipulation *in vivo* is the possibility of insertional mutagenesis and the activation of oncogenes, leading to neoplasia. Clearly,

3

such risks are important factors in the consideration of the ethical basis for gene therapy of disorders such as CF, haemophilia and the haemoglobinopathies; for patients with metastatic cancer, the risks are low. Such patients are often desperate for some form of treatment and are already searching for the gene therapy programmes described in the media. Therapies with even minimal a potential benefit will be avidly considered. In this situation, the biggest problem is offering false hope. It is unrealistic to expect such new strategies to be effective immediately. The first patients entering trials will provide much information for little personal benefit. This must be recognised by both the investigator and the patient to reduce the breakthrough mentality that surrounds novel treatments.

Various countries have now established regulatory bodies for gene therapy. Most are modelled on the US Recombinant DNA Advisory Committee, working closely with the country's existing drug regulatory body – the Food and Drug Administration (FDA). The success of the FDA in taking over much of the paperwork has recently led to the disbanding of the recombinant DNA advisory commitee (RAC). A parallel system has been established in the UK, where the Gene Therapy Advisory Committee (GTAC) together with the Medicines Control Agency (MCA) evaluate proposals from a scientific, ethical and safety standpoint. The creation the European Medicines Evaluation Agency (EMEA) – a single agency for drug and biotechnology product evaluation in Europe, based in London – provides an opportunity to standardise approaches across an increasing number of countries. The potential risks are detailed in Table 3.

Table 3   Potential risks of gene therapy

- insertional mutagenesis leading to cancer
- recombination of disabled vector resulting in environmental pollution by infectious recombinant virus
- toxic shock caused by viraemia
- transfer of non-viral exogenous genetic material
- contamination with other deleterious viruses or organisms
- physiological effects of over-expression

Germ-line Gene Therapy

So far no government has seriously considered germ-line gene therapy. The technology is relatively straightforward, as the problem of targeting can be solved by direct manipulation. The very sophisticated developments in *in vitro* fertilisation together with the growing experience in the generation of transgenic and knockout animals and plants means that, sooner or later, we will have to discuss the ethics of such approaches. Techniques for homologous recombination – the targeted replacement of old genes with new – are now possible, at least in mice. Furthermore, the likely explosion in genome information coming both from brute force sequencing of the human genome and from the genetic dissection of functionally related genes in simpler organisms as diverse as the nematode, zebrafish and yeast, will almost certainly open new possibilities for genetic intervention, which could transcend generations. It is this permanency which is most frightening to the ethicists, especially if unforeseen problems arise. At some time in the future, a policy of *genetic cleansing* may become a serious option for governments trying to contain spiralling health-care costs. Ethics, like beliefs, values and culture, change with time. So, perhaps today's heresy will be tomorrow's routine.

Severe Combined Immunodeficiency

The first clinical trial to use a therapeutic gene began on the 14th of September, 1990 at the National Institutes of Health in Washington, USA and involved a young girl suffering from severe combined immunodeficiency. Bone marrow cells were removed and transduced *ex vivo* with a retroviral vector carrying a cDNA copy of the adenosine deaminase (ADA) gene. ADA converts deoxyadenosine to its metabolites. If the enzyme is not present, deoxyadenosine accumulates and is toxic to some cells, especially T lymphocytes. Since these cells are intimately involved in the correct functioning of the immune system, ADA deficiency is usually fatal, due to the development of severe combined immunodeficiency of both T and B cell lineages.

This disorder was an attractive early target for gene therapy. The gene had been cloned in 1983 and was well characterised. Bone marrow transplantation was an accepted treatment if a suitable matching donor could be found. Bone marrow cells could be manipulated in the laboratory, and a check for successful transduction could be carried out after appropriate selection procedures. Furthermore, the child could be maintained on exogenous ADA, administered with polyethylene glycol, so that a sudden decline in immune function was unlikely.

Experience with the first two patients at the NIH answered a number of important questions. Safe *ex vivo* manipulation was shown to be possible, although the expression of the transduced ADA gene was only transient. This meant that periodic infusions of T lymphocytes were necessary. The first patient showed a dramatic increase in T cell numbers and function. She showed improved delayed hypersensitivity tests and increased isohaemagglutinins – useful short-term surrogate endpoints. Since this pioneering study, other groups, in different countries, have set out to repeat and improve the procedure. Several groups have attempted to obtain more long-lasting effects by gene transfer to CD34 + bone marrow stem cells. The disease is extremely rare and very variable in its clinical course, making the assessment of new technology very difficult. However the principle of successful gene transfer, with at least transient functional improvement, has now been verified several times.

Cystic Fibrosis

There is a vast range of disorders associated with the lung, including malignant disease, asthma and infection. Two hereditary disorders, CF and alpha-1 antitrypsin deficiency, are the most obvious targets for gene therapy. CF is an autosomal recessive disorder, resulting in abnormal electrolyte transport of epithelial cells. It is typically characterised by a collection of sticky mucus in the lung, pancreas and liver, which results in chronic inflammation and obstruction. The most common cause of death is respiratory failure, caused by sequential bouts of severe chest infection. CF is caused by a mutation in the *CFTR* gene. This encodes a membrane protein that acts as a chloride channel, pumping Cl out of cells in response to an increase in cAMP. It has been shown that when normal *CFTR* cDNA is transferred into epithelial cells isolated from CF patients, the cells are able to respond to increasing levels of cAMP and can secrete Cl. Thus, the abnormalities associated with CF can essentially be reversed in the laboratory. Can this be achieved *in vivo* in patients?

Two main approaches have been used with encouraging, but not dramatic results. The first used an adenovirus-based vector, which had the epithelial cells lining the brocheoalveolar tree as its natural target. Here, the *CFTR* gene was driven by the adenovirus type-2 major late promoter. The second strategy was to use cationic liposomes, which are positively charged and, therefore, can bind negatively charged DNA on their outer surface. Initial experiments have used electrical conductance changes in the nasal membranes as a surrogate endpoint. Data from two phase-I studies show 20% correction of conductance abnormalities with good patient tolerance. Current studies involve the use of aerosol inhalers to saturate at least the upper part of the respiratory system. Calculations have shown that, if the correct form of the *CFTR* gene is expressed in only one in ten cells, then reversal of the pathophysiology is possible. Clearly optimising gene delivery, uptake and stable expression are the main goals of investigators. The future therapy of CF may well involve the regular use of genetic aerosols from an early age.

Familial Hypercholesterolaemia

Familial hypercholesterolaemia (FH) is an autosomal dominant disorder in which there is a defect in the low-density lipoprotein *(LDL)* receptor gene. The effect is a reduced ability to breakdown LDL and intermediate-density lipoproteins. Individuals who are heterozygous for FH have increased serum levels of LDL and are more prone to premature coronary-artery disease. Homozygous individuals have very high serum LDL levels and severe atherosclerosis. Current therapeutic approaches include LDL apheresis, ileal bypass and liver transplantation. The success of tissue replacement therapy makes gene replacement the obvious next step, if the technological hurdles can be overcome. Furthermore, such genetic intervention represents a good paradigm for many other metabolic disorders involving the liver.

The existence of an excellent animal model – the Watanabe heritable hyperlipidaemic rabbit – led to a series of preclinical experiments to verify the principles. Newborn-rabbit hepatocytes were isolated and transduced with a recombinant retrovirus, containing the normal LDL receptor. On returning the transduced cells to the animal, near-normal levels of LDL were achieved. The clinical protocol involves the resection of a small wedge of patient's liver and the preparation of a cell suspension using collagenases. The hepatocytes are transduced *ex vivo* with appropriate retrovirus and returned to the remaining liver via the portal vein. Suggestions of transient clinical benefit have now been obtained and published.

Gene Marking

There are several situations where the use of a genetic marker to tag tumour cells may help in making decisions regarding the optimal treatment for an individual patient. The insertion of a foreign marker gene into cells from a tumour biopsy and replacing the marked cells into the patient prior to treatment can provide a sensitive new indicator of minimal residual disease after chemotherapy. The most common marker is the gene (neo R gene), for neomycin phosphotransferase an enzyme which metabolises the aminoglycoside antibiotic G418. This gene, when inserted into an appropriate retroviral vector, can be stably incorporated into the host cell's genome. Originally detected by antibiotic resistance, it can now be picked up more sensitively by means of the polymerase chain reaction (PCR). In this way, as few as one tumour cell amongst one million normal cells can be identified. This procedure can help in the design of aggressive chemotherapy protocols.

Cancer

The key problem in the effective treatment of patients with solid tumours is the similarity between tumour and normal cells. Local therapies such as surgery and radiotherapy can succeed, but only if the malignant cells are confined to the area treated. This is the case in approximately one third of cancer patients. For the majority, some form of systemic selective therapy is required. Whilst there are many cytotoxic drugs available, only a small proportion of patients are actually cured by them. The success stories of Hodgkin's disease, non-Hodgkin's lymphoma, childhood leukaemia, choriocarcinoma and germ-cell tumours have just not materialised for the common cancers, such as those of the lung, breast or colon. Despite enormous efforts in new drug development, clinical trials of novel drug combinations, the addition of cytokines, high-dose regimens and even bone marrow rescue procedures, the gains have been marginal.

Against this disappointing clinical backdrop, we have seen an explosion of information on the molecular biology of cancer. Although our knowledge of growth control is still rudimentary, we have at last had the first glimpse of its complexity. This has brought a new vision with which to develop novel selective mechanisms to destroy tumours.

The main problem facing the gene therapist is how to get new genes into every tumour cell. If this cannot be achieved, then any malignant cells that remain unaffected will emerge as a resistant clone. Presently, we do not have ideal vectors. Despite this drawback, there are, already, over 150 protocols accepted for clinical trials involving cancer patients world-wide: the majority are in the USA. The ethical issues are fairly straightforward, with oncology providing some of the highest possible benefit-risk ratios. There are several strategies currently under investigation that will be considered in detail later in this volume.

Gene correction

The downregulation of abnormal oncogene expression has been shown to revert the malignant phenotype in a variety of *in vitro* tumour lines. It is possible to develop *in vivo* systems, such as the insertion of genes encoding complementary (antisense) mRNA to that produced by the oncogene. Such anti-genes specifically switch off the production of the abnormal protein product. Another mechanism to selectively switch off a gene is through the use of ribozymes which cleave specific RNA sequences.

Mutant forms of the *c-ras* oncogene are an obvious target for this approach. Up to 85% of human pancreatic cancers contain a mutation in the twelfth amino acid of this protein and reversal of this change in cell lines leads to the restoration of normal growth control. Clearly, the major problem is to ensure that every single tumour cell gets infected. Any cell which escapes will have a survival advantage and produce a clone of resistant tumour cells. For this reason, it may be that future treatment schedules will require the repetitive administration of vectors in a similar way to fractionated radiotherapy or chemotherapy.

In cell culture, malignant properties can often be reversed by the insertion of normal tumour suppressor genes, such as *RB-1*, *TP53* and *DCC*. Although tumour suppressor genes were often identified in rare tumour types, abnormalities in their expression and function are abundant in common human cancers. As with antigene therapy, the difficulty in this approach lies in the delivery of actively expressed vectors to every single tumour cell *in vivo*. Nevertheless, clinical experiments are in progress in lung cancer patients, where retroviruses which encode *TP53* genes are being administered bronchoscopically.

Suicide Gene Therapy

The main problem with existing chemotherapy is its lack of selectivity. If drug-activating genes could be inserted, which would only be expressed in cancer cells, then the administration of an appropriate pro-drug could be highly selective. There are now many examples of genes, preferentially expressed in tumours. In some cases, their promoters have been isolated and coupled to drug-activating enzymes. Examples include alphafoetoprotein in hepatoma, prostate-specific antigen in prostate cancer, and *c-erbB2* in breast cancer. Such promoters can be coupled to enzymes such as cytosine deaminase or thymidine kinase, thereby producing unique retroviral vectors which are able to infect all cells, but can only be expressed in tumour cells. These suicide (or Trojan horse) vectors may not have absolute tumour specificity, but this may not be essential; it may be possible to perform a genetic prostatectomy or breast ductectomy, effectively destroying all tumour cells.

Immunogene Therapy

The presence of an immune response to cancer has been recognised for many years. The problem is that human tumours seem to be predominantly weakly immunogenic. If ways could be found to elicit a more powerful immune stimulus, then effective immunotherapy could become a reality. Several observations from murine tumours indicate that one reason for weak immunogenicity of certain tumours is the failure to elicit a T helper-cell response. This, in turn, releases the necessary cytokines to stimulate the production of cytolytic T cells which can destroy tumours. The expression of cytokine genes, such as interleukin 2 (IL2), tumour necrosis factor (TNF) and interferon, in tumour cells has been shown to bypass the need for T helper cells in mice. Similar clinical experiments are now in progress. Melanoma cells have been prepared from biopsies and infected with a retrovirus containing the *IL2* gene. These cells are being used as a vaccine to elicit a more powerful immune response.

Conclusion

There are many problems yet to be solved before the routine application of gene therapy is achieved inclinical practice. Despite the negative views often expressed by critics, the principles of clinical gene transfer have essentially been proven valid.

Patient benefit, in terms of permanent disease reversion, will clearly take much longer to achieve, but the pace of the genetic age is accelerating: the human genome project; a greater understanding of transcription control; our understanding of gene function from the study of simpler organisms; the use of increasingly sophisticated subtraction techniques to compare DNA and RNA from different cells and tissues; the development of precisely targeting, highly selective vectors; and the uncovering of novel cellular control processes will make this an exciting and promising area for the next decade. Furthermore the dissection of the genetic basis of different diseases will almost certainly result in a much better understanding of their molecular pathogenesis, giving new targets for rational drug design. Classical pharmacology and molecular genetics will become essential partners in the future of therapeutic development. This book provides a very timely summary of current technology and an excellent glimpse into the future.

I. Gene Transfer Methods

1 Retroviral Vectors

P. D. Robbins

Summary

Retroviral vectors represent the most widely used vector system for gene therapy. They are currently in the clinic for treatment of a wide variety of diseases, from cancer to virus infection to arthritis. Although the current vectors have worked efficiently for gene transfer in the initial trials, it is likely that modifications of both the vector and packaging lines should result in even higher titer, targetable viruses with elevated and/or regulated gene expression. Thus, it is likely that retroviral vectors will continue to be the vector of choice for many gene therapy applications.

Over the last five years, gene therapy has gone from the laboratory to phase 1 clinical trials (for reviews see Crystal, 1995; Mulligan, 1993). In the future it is likely that gene therapy will be accepted as one, if not the, form of treatment for many lethal and non-lethal diseases, both genetic and acquired. The rate limiting step for successful gene therapy, however, is the efficiency of gene transfer and the subsequent level and duration of expression of the therapeutic gene. The vector systems currently being used for gene transfer include both RNA and DNA viruses as well as non-viral systems such as liposomes. Of the vectors currently available for gene therapy applications, retroviral vectors are the most widely used and the most developed. Indeed, retroviral vectors are being used in approximately 80% of the gene therapy clinical trials conducted to date.

The advantages of retroviral vectors, most of which are derived from Moloney murine leukemia virus (MoMuLV), are that they are easily manipulated *in vitro*, the vector stably integrates into the host DNA without any apparent pathogenicity, moderate to high titers can be achieved, a wide range of target cells can be infected and reasonably large genes can be inserted into the vector. The disadvantages of the current retroviral vectors are that they require cell division for integration, their titers are low compared to DNA virus vectors such as adenovirus and herpes simplex virus and that they insert randomly into the host genome causing mutations and theoretically, cancer. However, no form of cancer has been directly associated with integration of replication-defective retroviral vectors. Moreover, recent developments in packaging lines, methods for concentrating virus and modification of viral envelope have led to better infection efficiencies of a variety of target cells and viral stocks with higher titers. It is also possible that the ability of the lenti family of retroviruses, such as HIV, to infect non-dividing cells (Bukrinsky et al., 1993; Galley et al., 1995) could be transferred to the current vectors, thus allowing the application of retroviral vectors for *in vivo* gene delivery to non-dividing cells. This chapter will summarize the development and application of retroviral vectors to gene therapy.

1.1 Retroviral Life Cycle

Retroviruses are RNA viruses which have a DNA intermediate that is stably integrated into the host genome. Three genes are carried by the virus, *gag*, *pol*, and *env*, encoding proteins that are processed post-translationally into multiple polypeptides required for viral replication and packaging. The proteins encoded by *gag*, *pol*, and *env* can be supplied *in trans* for the virus to replicate. All that is required for viral replication *in cis* are the 5' and 3' long-terminal repeats (LTRs) which contain promoter, poly-adenylation and integration sequences, a packaging site termed *psi*, a tRNA binding site as well as several additional sequences involved in reverse transcription. Retroviruses bind to a target cell through an interaction with a receptor specific for the viral *env* protein. For the ecotropic envelope, which allows for infection of murine cells, the receptor is a cationic amino acid transporter (Wang et al., 1991). The receptor target for the MuLV amphotropic envelope, which allows for infection of cells from a wide variety of species, is the phosphate transporter, Ram-1 (Miller and Miller, 1994). Other viral envelopes also recognize specific cell-surface proteins, thereby limiting the ability of the virus to infect a wide range of cell types.

Following binding of the viral envelope to its receptor, there is a conformational change in the envelope protein, resulting in fusion of the virus to the cell. The virus is imported into the cells in endosomes where it is released by a viral protein-associated process. The viral RNA is then reverse transcribed into DNA by the virally-encoded, virion-associated reverse transcriptase (*pol*). The resulting double-stranded viral DNA is then transported to the nucleus. For many types of retroviruses, mitosis and thus breakdown of the nuclear envelope is required for the viral RNA to make it to the nucleus (Lewis and Emerman, 1994). The exception to the requirement of cell division for infection are lentiviruses such as HIV which appear to have acquired both *cis* and *trans*-acting factors that allow for active transport of the DNA to the nucleus of non-dividing cells (Bukrinsky et al., 1993; Galley et al., 1995). Once the proviral DNA makes it to the nucleus, it integrates into the host DNA using a virally-encoded integrase. The integrated provirus is then transcribed, producing RNAs encoding *gag*, *pol*, and *env* protein, which allow for packaging of the full-length unspliced viral RNA containing the *psi* site: the spliced message does not contain the *psi* sites and thus is not packaged. At least one of the *gag* polypeptides binds directly to the viral RNA to facilitate packaging into the virion which is released from the cell by budding from the *env*-coated cell membrane. An important feature of the virus life cycle is that the infected cell stably produces viruses for the life of the cell without affecting the growth properties of the cell. Thus, stable producer lines can be generated which allow for continued recombinant virus production.

The LTRs in the retrovirus are important for viral transcription, polyadenylation, replication and integration of the provirus into the genome. The LTR in the provirus has three regions that are defined functionally, U3, R, and U5. The U3 region contains sequences which are important for transcription including the enhancer and promoter and sequences for polyadenylation. Transcription is initiated at the start of the R region. Thus, in the viral RNA, the 5' end contains the R and U5 regions derived from the 5' LTR in the provirus, whereas, the U3 region, derived from the 3' LTR, is present at the 3' end. During viral replication, the R-U5 and U3 regions are duplicated so that two full copies of the LTR are present at both the 5' and 3' ends of the provirus. For the construction of retroviral vectors, one of the important features is that modifications in the U3 region of the 3' LTR are duplicated into the 5' LTR in the provirus during reverse transcription in infected cells. Moreover, modifications can be made in the U3 region, including large deletions and/or insertions of gene expression cassettes, without affecting the viral titer. Indeed, selectable marker, suicide, or antisense genes have been inserted into the U3 region of the 3' LTR just upstream from the MuLV enhancer where they are expressed after integration into the host genome.

1.2 Types of Vectors

1.2.1 Murine Leukemia Virus

The majority of retroviral vectors currently being used are derived from Moloney murine leukemia virus (MoMuLV) (Boris-Lawrie and Temin, 1993; Dranoff et al., 1993; Gilboa et al., 1986; Guild et al., 1988; Miller et al., 1993; Robbins et al., 1993; Salmons and Gunzberg, 1993). As described above, MoMuLV contains three genes, *gag*, *pol*, and *env* that are expressed from two transcripts, one spliced (*env*) and one unspliced (*gag, pol*). The *gag*, *pol*, and *env*, proteins are then processed into smaller polypetides that are used for generation of infectious virus. The three genes can be removed from the virus without affecting the ability of the viral RNA to be packaged into infectious virions as long as they are provided *in trans*. Indeed, all that is required *in cis* are the viral LTRs, the packaging site, the tRNA binding site and several additional sequences important for reverse transcription. Thus, recombinant viruses can be produced carrying therapeutic gene or genes which, following infection and stable integration into the host genome, are replication-defective due to the absence of *gag*, *pol*, and *env* proteins.

There are many different types of retroviral vectors which have been developed for gene transfer. The current MoMuLV-based retroviral vectors can be classified into three general classes by the type and position of the enhancer/promoter driving expression of the therapeutic gene: internal promoter vectors, the LTR-based vector and gene cassette vectors (Fig. 1-1). Each of these vector systems have certain advantages that makes it suitable for certain gene transfer/therapy applications.

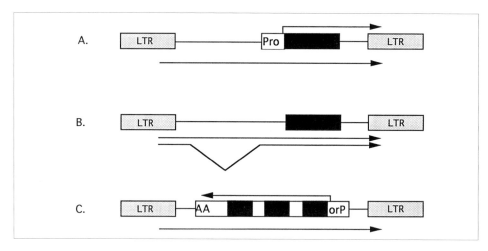

Figure 1-1. Diagram of the three general types of retroviral vectors, internal promoter, LTR-based, and gene cassette, classified by position and type of the enhancer/promoter driving expression of the therapeutic gene. The black boxes represent coding sequences for a therapeutic protein. The empty boxes represent introns or 3' non-coding sequences. The position of the polyadenylation site in gene cassette vector is indicated by AA. The position and direction of the promoter driving expression of the therapeutic gene is indicated by Pro, which is in the reverse orientation in the gene cassette vector (orP). The promoter driving gene expression in the LTR-based vector is present within the 5' LTR.

1.2.2   Internal Promoter Vectors

One standard type of vector that has been utilized extensively is the internal promoter vector (Fig. 1-1a). This type of vector has an exogenous promoter inserted between the viral LTRs to drive expression of either a therapeutic gene and/or a selectable marker gene. In certain internal promoter vectors the enhancer and/or promoter in the U3 region of the 3' LTR is mutated so that it is no longer active. Inactivation of the 3' U3 region does not affect viral replication, but does reduce expression after infection when the 3' U3 region is copied in the 5' U3 region during reverse transcription. In these crippled vectors, expression is mediated predominantly by the internal promoter. The inactivation of viral transcription allows for the appropriate expression from cell type specific promoters or higher levels of expression from constitutive promoters. Alternatively, the 3' LTR can be used to drive expression of a second gene in the vector. However, the presence of two active promoters in the context of the provirus appears to reduce the levels of expression from the two promoters or block expression from one of the promoters.

1.2.3   LTR-based Vectors

A second type of vector which has been used extensively is the LTR-based vector where the gene of interest is expressed from the LTR: there is no internal promoter (Fig. 1-1b). In this type of vector, the therapeutic gene can be expressed from the unspliced message or from the spliced envelope message. One example of an LTR-based vector is MFG where the gene of interest is inserted at the normal position of *env* in MuLV (Dranoff et al., 1993; Ohashi et al., 1992; Robbins et al., 1993). The initiation codon of the gene of interest is actually fused to the start codon of *env* (ATG) and is expressed from a spliced transcript whose 5' leader sequence is identical to the normal env message in MoMLV infected cells. It has been demonstrated that a region of *gag* is required *in cis* to increase viral titer by enhancing the packaging of the RNA, possibly by stabilizing the RNA and/or affecting its secondary structure (Bender et al., 1987). Thus, MFG and other high titer LTR-based and internal promoter vectors contain the 5' portion of *gag*. The start codon of the *gag* protein is mutated in both the LTR-based and internal promoter vectors so that it will not be expressed from the unspliced full length RNA. MFG and the related N_2-based vectors, such as G2, have been shown to express genes efficiently in hematopoietic cells, fibroblasts, myoblasts, endothelial and epithelial cells, and hepatocytes as well as other cell types. However, expression usually is not regulated in a tight tissue-specific manner since the viral promoter is utilized.

The *cis*-acting elements in LTR-based vectors such as MFG also can be altered to increase expression in certain cell types (Couture et al., 1994). The enhancer/promoter element within the U3 region of myeloid proliferative sarcoma virus (MPSV) functions more efficiently in hematopoietic cells than does the enhancer/promoter within the MuLV LTR. The introduction of the MPSV U3 region into the 3' LTR in the MFG provirus results in a virus able to express more efficiently in hematopoietic stem cells than the MuLV LTR (Reviere et al., 1995). In addition, the LTRs of other murine viruses also exhibit tissue tropism. It is also possible to insert an enhancer/promoter from an unrelated virus, such as human cytomegelovirus (HCMV), or from a cellular gene into the 3' LTR, resulting in enhanced expression in certain cell types compared with the normal MuLV enhancer/promoter.

Another example of a *cis*-acting alteration which can enhance a retroviral vector performance is the B2 mutation. MoMuLV-based vectors do not express following infection of embryonic carcinoma cell lines, such as F9. The screening for mutants which can express in F9 cells identified a single base change (B2) in the sequence in the tRNA binding site which appears to affect the binding of a transcriptional repressor (Barklis et al., 1986). Insertion of the B2 mutation into MFG as well as other vectors has allowed for enhanced expression in certain cell types *in vivo* and in culture. In particular, the combination of the B2 mutation and the MPSV enhancer in the MFG vector increased expression six fold in hematopoietic cells following long-term reconstitution of mice transplanted with MFG-infected hemtopoietic stem cells (Reviere et al., 1995).

In order to express multiple genes from a vector, such as MFG, it is possible to insert the second gene at the position of *gag* where it would be expressed from the unspliced *gag-pol* message. However, a more effective way to express multiple genes is to use sequences first identified in picornaviruses, termed internal ribosome entry sites (IRES), to facilitate translation of multiple proteins from a single, polycistronic message (Morgan et al., 1992). Indeed, multiple IRES elements have been used to efficiently express three genes from the single spliced message in MFG (Zitvogel et al., 1994). It appears that the use of IRES elements allows for expression of a second gene at almost the equivalent level as the first gene and does not dramatically reduce expression of the first gene compared with a vector carrying only one gene. Thus, the use of an IRES element(s) insures that multiple genes will be expressed at similar levels. It is also possible to use IRES elements in the context of internal promoter based vectors where the internal promoter drives expression of a polycistronic message.

1.2.4   Gene Cassette Vectors

A third type of retroviral vector which is used for specific gene therapy applications is the "gene cassette" vector where the entire gene with its own regulatory sequences is inserted into the viral vector (Fig. 1-1c). In this type of vector, the retroviral vector acts solely as a carrier of the therapeutic gene cassette into the cell, playing no role for expression of the gene after stable integration. In order to obtain appropriate expression of the gene cassette it has to be inserted in reverse orientation so that the introns are not removed during viral replication and/or the viral RNA is not inappropriately polyadenylated. A good example of how the reverse orientation vectors may be used for is gene therapy of β-thalassemia, where tissue specific, position independent β-globin expression is required for treatment (Dzierzak et al., 1988; Sadelain et al., 1995).

The β-globin gene with its promoter and enhancer located with an intron as well as 3' to the gene is inserted in the reverse orientation into a simple retroviral vector with a crippled 3' LTR. To improve the position independent expression level, locus control region elements (LCRs) also have been inserted into the vector (Sadelain et al., 1995). The appropriate combination of β-globin regulatory elements allows for tissue specific globin expression in erythroid cells. The limitation to these vectors is that there is a size constraint to the gene cassette, given that it includes both introns and exons, that can be inserted. In addition, cryptic polyadenylation and splice sites in the gene cassette in the reverse orientation can lead to reduced viral titer.

1.2.5   Other Types of Retroviruses for Use as Vectors

There are several different types of avian viruses which are being used for gene transfer and which, with refinements, could be used for gene therapy applications. One type includes the reticuloendotheliosis viruses (REVs) which includes spleen necrosis virus (SNV) and reticuloendotheliosis virus strain A (REV-A). SNV, in particular has been developed as a vector for gene transfer with the appropriate packaging lines generated. Another type of avian retrovirus, Rous sarcoma virus (RSV), also has been used for gene transfer. One potential advantage to RSV is that it is possible to insert a therapeutic gene in place of its non-essential *src* gene and still have a replication-competent virus. The ability of a virus to replicate in certain situations, such as for infection of tumor cells *in vivo,* may be an important feature for efficient delivery of a therapeutic gene such as the suicide *tk* gene to tumors.

Another class of retroviruses which theoretically could be used as vectors for gene therapy are the lentiviruses. Although lentiviruses, such as HIV, SIV, or EIAV, are currently not being used extensively as vectors, there are certain features of these viruses which make them appropriate for gene therapy applications. In particular, lentiviruses have the advantage of being able to infect non-diving cells due, in part, to the ability of the viral genome to target the nucleus without mitosis, as is required for MoMuLV-base vectors (Lewis and Emerman, 1994) At least part of the targeting of the genome to the nucleus is mediated by the viral *gag*-derived matrix protein (Bukrinsky et al., 1993; Galley et al., 1995). Thus, it is possible that the HIV matrix protein and other *cis* and *trans*-acting regions of lentiviruses could be inserted into MoMLV vectors to make the vectors capable of infecting non-dividing cells, although the initital HIV-MoMuLV chimerics have not given cell-division independent infection. (Deminie and Emerman, 1994). Alternatively, it is possible that lentiviruses themselves could be modified to work as vectors for gene therapy applications. In fact, the use of crippled HIV vectors, expressing HIV inhibitory proteins, may be well suited for gene therapy for AIDs since they should infect the appropriate cell types and express only in cells infected with wild type (wt) HIV.

1.3   Packaging Lines

Since the viral proteins are only required *in trans* for packaging of RNAs containing a *psi* site, it is possible to make packaging lines by stably expressing *gag, pol,* and *env* proteins (Fig. 1-2). The initial packaging line for MuLV-derived vectors was generated simply by deleting the packaging sequence from the MoMuLV provirus and stably introducing the crippled construct into 3T3 cells. The resulting cell lines expressed high levels of *gag, pol,* and *env* proteins, but did not produce infectious virus (Cone and Mulligan, 1984; Mann et al., 1983). Introduction of retroviral vectors stably into these cells by transfection resulted in moderate viral titers, but also a in high frequency of replication competent virus arising due to recombination between the packaging-site deleted virus and the retroviral vector with a functional packaging site. To reduce the frequency of recombination a packaging cell line was generated which carried a deletion of the packaging site and had the 3' LTR replaced with SV40 polyadenylation sequences (Miller and Buttimore, 1986). In this type of packaging line, at least two recombination events are required for wildtype virus to arise. To further reduce the possibility of replication competent virus from arising in stable producers, the proteins required for packaging were separated on to two plasmids, one expressing *gag* and *pol* and the other *env* (Danos and Mulligan, 1988; Markowitz et al., 1988; see Fig. 1-2) In addition, other safety factors such as not us-

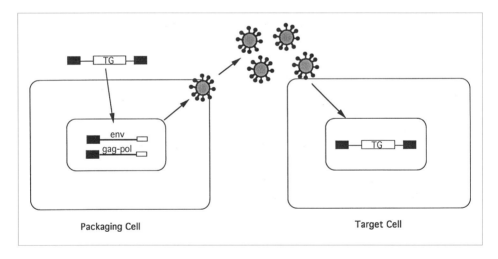

Figure 1-2. Generation of a "safe" packaging line for production of replication-defective retroviral vectors (Adapted from Salmons and Gunzberg, 1993). Plasmids expressing either gag-pol and env are stably transfected into a recipient cell line. Subsequently, the retroviral vector is either stably or transiently transfected into the packaging line. The virus produced by the transfected cells can then be used to infect a target cell where, following stable integration, the virus is unable to replicate due to the absence of the viral proteins.

ing the 3' and/or 5' LTRs were also incorporated into the new generation of packaging lines. In these new "safe" packaging lines multiple recombination events are necessary for replication competent virus to arise.

It also has been shown that the use of 3T3 cells to generate packaging lines results in co-packaging of endogenous murine retroviruses, such as VL30, along with the recombinant virus at a certain frequency. Thus, a high percentage of virions produced from 3T3-based packaging lines have other viral RNAs. In addition, 3T3 based lines appear to produce an inhibitory factor which, following concentration of the virus, reduces infection by the recombinant virus. The inhibitory factor(s) may be defective viral particles, viral proteins and/or other factors. Thus, the use of other non-murine cell types, such as human or canine cell lines, for the generation of packaging lines has led to producers with increased titers (Pear et al., 1993). More importantly, the virus produced by these cells appears able to infect certain cell types, such as human T-cells, more efficiently than viruses produced from 3T3-based packaging lines. In addition, the use of cell lines which are more "transfectable" than 3T3 cells such as 293 cells has allowed for the rapid production of high viral titers by transient transfection. In particular, 293-based packaging cells have given transient retroviral titers of greater than 106 per ml (Pear et al., 1993)

In addition to using normal ecotropic and amphotropic MuLV-derived envelope proteins, it is also possible to use other viral envelopes to either expand or limit the viral host range. A packaging line derived using MoMLV *gag* and *pol* and the envelope from gibbon ape leukemia virus (GaLV), which by use of a different receptor appears to infect hematopoietic cells more efficiently (Miller et al., 1991). An alternate way of producing virus with an expanded host range is through the generation of pseudotyped virus. Virus particles can be generated with the envelope G protein for Vesicular Stomatitis Virus in place of the normal MuLV *env*. The VSV-G containing particles are more stable and can be effectively concentrated by centrifugation achieving viral titers of greater than 10^9 (Yee et al., 1994). The pseudotyped particles can infect cells from a wide range of species from zebra fish to humans (Lin et al., 1994). In addition, it is possible to produce virus by transient transfection of 293 cells which contain both MLV envelope and VSV G proteins. This mixed envelope virus is able to infect non-murine cells due to the presence of the VSV-G protein, but appears to be more infectious because of the MLV envelope (Robbins, unpublished).

In addition to using certain envelope proteins which exhibit cell tropism, it is also possible to use modified envelopes which may allow for targeted infection (reviewed in Salmons and Gunsberg, 1993). One method is to conjugate an antibody to a specific cell surface marker to the virion. This can be accomplished using an antibody to the viral *env* protein coupled by streptavidin to the specific antibody. However, this approach appears to be target protein specific and/or antibody specific. Alternatively, the envelope can be modified directly by making chimeras with specific proteins (Kasahara et al., 1994; Somia et al., 1995). For instance, the ecotropic envelope has been fused to erythropoietin (epo), allowing for binding to cells expressing the epo receptor (Kasahara et al., 1994). The titers of these viruses, although partially specific, is low. However, the titer of the epo-env chimeric virus was shown to be higher when a normal ecotropic env was also incorporated into the virion along with the chimeric receptor (Kasahara et al., 1994). Although the ecotropic receptor does not bind to human cells, it appears to increase virus entry upon epo-directed binding to epo receptor-expressing human cells. It is also possible to incorporate into the virion a ligand for a specific cell surface receptor at a low frequency by over-expressing the ligand in the producer cell line. However, titers of virus prepared in this manner are low and the ability to incorporate a ligand into the virion is protein specific. For a specific application, such as targeting viral infection to the liver, it is possible to asialate *env* protein so that it can bind to the liver-specific asialoglycoprotein receptor. These approaches to targeting viral infections to specific cell types, as they improve in specificity and efficiency, should allow for greater applications of retroviral vectors to *in vivo* gene delivery.

1.4   Retroviral Vector Clinical Applications

Retroviral vectors have been used preferentially relative to other viral vector systems and have been used for both *in vivo* and *ex vivo* applications for both acquired and genetic diseases. The ability of retroviruses to stably infect dividing cells without inducing an immune response allows retroviruses to be used for long-term therapies for inherited diseases. Because of the limitation of retroviral vectors to infect only dividing cells and because of their relatively low titers, their use has been restricted mainly to *ex vivo* gene therapy approaches. However, in some cases, either retroviruses or producer cells have been used for gene delivery *in vivo*.

1.4.1   *Ex vivo* Clinical Applications

The first clinical application of retroviral vectors was for the marking of tumor infiltrating lymphocytes (TILs), isolated from primary tumors and propagated in culture, to determine if the manipulated cells actually target to the tumor *in vivo*. Similarly, retroviral vectors were used to infect bone marrow from cancer patients to determine whether or not the relapse was due to contamination of transplanted cells (Brenner, 1994; Brenner et al., 1994). Interestingly, the use of retroviral vectors as markers has provided substantial clinical information, more so than phase I trials using retroviral vectors carrying therapeutic genes. In addition, these experiments help to demonstrate the safety of using replication-defective retroviral vectors to deliver genes to human cells, even in immuno-compromised individuals.

Retroviral vectors also have been used for treatment of ADA deficiency using an *ex vivo* approach. ADA deficiency or severe combined immunodeficiency (SCID) is the result of loss of expression of the gene for adenine deaminase which affects the viability the function of T-cells. Reconstitution of ADA activity in T-cells by gene transfer results in normal T-cell function. Transduction of primary T-cells from two SCID patients with a retroviral vector expressing ADA, followed by reinfusion of the genetically-modified cells, has shown therapeutic efficacy (Blaese et al., 1995) Indeed, the T-cells in the patients receiving gene theapy with exogenous ADA activity have normal immune function at least a year post-infusion. However, the fact that the patients have also been maintained on systemic PEG-ADA has prevented firm conclusion as to the efficacy of gene therapy for ADA-deficiency using T-cells as the targets. Recently, stem cell gene therapy for ADA-deficiency using cord blood has been attempted in several infants, with promising results in that a certain percentage of ADA-positive T-cells from the treated individuals has been detected 2 years post-transplant.

Recently, several protocols for treatment of Gaucher disease, a lysosomal storage disease, have been initiated. It has been demonstated previously that hematopoietic stem cells from mice can be efficiently infected in culture with a retroviral vector expressing glucocerebrosidase, the enzyme deficient in Gaucher disease. Following transplantation and reconstitution of lethally-irradiated mouse recipients, a high level of expression of glucocerebrosidase in all hematopoietic lineages was detected (Ohashi et al., 1992). For the clinical application of gene therapy for treatment of Gaucher disease, CD34+ cells have been isolated from peripheral blood and infected with retroviral vectors expressing the therapeutic enzyme. The genetically-modified CD34+ cells, which contain a sub-population of hematopoietic stem cells, were then re-infused back into non-ablated patients who were then monitored for the extent of repopulation with the genetically modified hematopoietic cells. A similar appoach either with or without ablation of the endogenous bone marrow could be used for treatment of a variety of hematopoietic genetic diseases.

An alternative approach for certain lysosomal storage diseases, which have proven effective in animal models, has been the use of organoids. In particular, primary fibroblasts have been gentically modified to over-express a therapeutic protein by retroviral infection and then implanted in a collagen matrix (Mollier et al., 1993). The transplanted cells remain viable for extended periods and secrete high levels of the therapeutic protein such as β-glucuronidase or glucocerebrosidase. Similar approaches have also been taken using organoids comprised of myoblasts instead of fibroblasts. Interestingly, the efficiency of retroviral infection of myoblasts and the subsequent level of retroviral-mediated gene expression appears to be higher than in fibroblasts. Clinical protocols using genetically-modified fibroblasts and/or myoblast implants will soon be initiated.

The particular application where retroviral vectors have been used the most extensively is for *ex vivo* cancer gene therapy. Tumor cells have been isolated from patients, modified by retroviral infection to express an immunostimulatory agent such as interleukin 2 or GM-CSF and then injected back into the patient to enhance the immune response (Berns et al., 1995). Alternatively, autologous fibroblasts, modified in culture by retroviral infection and then injected into the primary tumor, have been used as a delivery system for specific cytokines. A number of phase 1 trials are in progress or have recently been completed with extensive analysis showing the safety of using retroviral vectors in the clinic. In addition, efficient infection of a number of different tumor types has been demonstrated using retroviral vectors derived from amphotropic packaging lines. Moreover, partial and complete responses have been observed in a small percentage of the treated patients suggesting that *ex vivo* gene therapy for cancer, using immunostimulatory agents, may be an important armament in the battle against cancer, although further development is needed.

Retroviral vectors are also are being used for treatment of autoimmune diseases such as arthritis. Although arthritis is not necessarily a genetic disease, it and other connective tissue diseases are good candidates for treatment by gene therapy. An *ex vivo* method has been developed to express therapeutic genes intra-articularly using transplantation of genetically-modified synoviocytes (Bandara et al., 1993). Using the rabbit knee as a model joint, transgene expression following transplantation of genetically-modified rabbit synoviocytes was shown to persist for up to six weeks. The expression of interleukin-1 receptor antagonist protein (IL-1Ra) in the rabbit knee, using this methodology, was able to block the intra-articular pathophysiological changes in a rabbit antigen-induced model of arthritis. These promising results have led to initiation of a clinical protocol to assess the safety and efficacy of using transplantation of genetically-modified synovial fibroblasts into arthritic joints for treatment of rheumatoid arthritis.

1.4.2 *In vivo* Clinical Applications

Retroviral vectors have also been used for specific *in vivo* gene delivery applications. The fact that the MoMuLV-based vectors infect only dividing cells allows for partial specificity for infection of tumor cells *in vivo*. This specificity can be used to deliver to tumors, *in vivo*, "suicide genes" such as those encoded by the HSV thymidine kinase or cytosine deaminase genes which convert the prodrugs ganciclovir and 5-fluoro-cytosine respectively to cytotoxic drugs. One example of how retroviral-mediated delivery of suicide genes to tumors is being used *in vivo* for cancer gene therapy is in the treatment of brain tumors. An amphotropic packaging cell line, producing a defective retroviral vector carrying the HSV *tk* gene, has been injected into human brain tumors followed by treatment with ganciclovir (Culver et al., 1992). In murine studies, this form of therapy efficiently eradicated established brain tumors. The efficacy of this approach, however, in the clinic has been limited by the relatively poor infection of tumor cells *in vivo* following injection of the producer line. Higher titer producers, modified envelope proteins, and/or the use of replication-competent viruses may allow for better infection *in vivo*.

Retroviral vectors have also been used for *in vivo* delivery of antigenic proteins to vaccinate against viral infections such as HIV. A retroviral vector expressing HIV *env* and *rev* proteins was injected intra-muscularly into HIV-infected individuals and a subsequent immune response was observed similar to that observed in mice and non-human primates (Irwin et al., 1994). However, it is not clear if the injected virus induces the immune response by infecting muscle or if other targets, such as in the draining lymph nodes, could be involved. Clearly, though, these experiments and others demonstrate the feasibility of using retroviral vectors *in vivo*.

1.5 Ackowledgements

This work was supported in part by Public Health Service grants CA59371, CA55227, and DK44935. The author would like to thank Drs. Hideaki Tahara, Steve Ghivizzani, Laurence Zitvogel, John Barranger, Mike Lotze, and Chris Evans for helpful discussion.

References

Bandara, G., Mueller, G. M., Galea-Lauri, J., Tyndall, M. H., Georgescu, H. I., Suchanek, M. K., Hung, G. L., Glorioso, J. C., Robbins, P. D. and Evans, C. H. (1993) Intraarticular expression of the interleukin-1 receptor antagonist protein by *ex-vivo* gene transfer. *Proc. Natl. Acad. Sci. USA* 90: 10764–10768.

Barklis, E., Mulligan, R. C. and Jaenisch, R. (1986) Chromosomal position of virus mutation permits retrovirus expression in embryonal caricinoma cells. *Cell* 47: 391–399.

Bender, M. A., Palmer, T. D., Gelinas, R. E. and Miller, A. D. (1987) Evidence that the packaging signal of Moloney murine leukemia virus extends into the gag region. *J. Virol.* 61: 1639–1646.

Berns, A. J., Clift, S., Cohen, L. K., Donehower, R. C., Dranoff, G., Hauda, K. M., Jaffee, E. M., Lazenby, A. J., Levitsky, H. I., Marshall, F. F. and al., e. (1995) Phase I study of non-replicating autologous tumor cells injections using cells prepared with or without GM-CSF gene transduction in patients with metastatic renal cell carcinoma. *Hum. Gene Ther.* 6: 347–368.

Blaese, R. M., Culver, K. W., Miller, A. D., Carter, C. S., Fleisher, T., Clerici, M., Shearer, G., Chang, L., Chiang, Y., Tolstoshev, P., Greenblatt, J. J., Rosenberg, S. A., Klein, H., Berger, M., Mullen, C. A., Ramsey, W. J., Morgan, R. A. and Anderson, W. F. (1995) T lynphocytes-directed gne therapy for ADA-SCID: initial trial results after 4 years. *Science* 270: 475–480.

Boris-Lawrie, K. A. and Temin, H. M. (1993) Recent advances in retrovirus vector technology *Curr. Opin. Genet. Develop.* 3: 102–109.

Brenner, M. K. (1994) Genetic marking and manipulation of hematopoietic progenitor cells using retroviral vectors. *Immunomethods* 5: 204–210.

Brenner, M. K., Rill, D. R., Heslop, H. E., Rooney, C. M., Roberts, W. M., Li, C., Nelson, T.

and Krance, R. A. (1994) Gene marking after bone marrow transplantation. *Eur. J. Cancer.* 30A: 1171–1176.

Bukrinsky, M. I., Haggerty, S., Dempsey, M. P., Sharova, N., Adzhubel, A., Spitz, L., Lewis, P., Goldfarb, D., Emerman, M. and Stevenson, M. (1993) A nuclear localization signal within HIV-1 matrix protein that governs infections of non-dividing cells. *Nature* 365: 666–669.

Cone, R. D. and Mulligan, R. C. (1984) High efficiency gene transfer into mammalian cells: generation of helper free recombinant retrovirus with broad mammalian host range. *Proc. Natl. Acad. Sci. USA* 81: 6349–6353.

Couture, L. A., Mullen, C. A. and Morgan, R. A. (1994) Retroviral vectors containing promoter/enhancer elements exhibit cell-type-specific gene expression *Hum. Gene Ther.* 5: 667–677.

Crystal, R. G. (1995) Transfer of genes to humans: early lessons and obstacles to success *Science* 270: 404–409.

Culver, K. W., Ram, Z., Wallbridge, S., Oldfield, E. H. and Blaese, R. M. (1992) *In vivo* gene transfer with retroviral vector-producer cells for treatment of experimental brain tumors *Science* 256: 1550–1552.

Danos, O. and Mulligan, R. C. (1988) Safe and efficient generation of recombinant retroviruses with amphotropic and ecotropic host ranges. *Proc. Natl. Acad. Sci. USA* 85: 6460–6464.

Deminie, C. A. and Emerman, M. (1994) Functional exchange of an oncoretrovirus and a lentivirus matrix protein. *J. Virol.* 68: 4442–4449.

Dranoff, G., Jaffee, E., Lazenby, A., Golumbek, P., Levitsky, H., Brose, K., Jackson, V., Hamada, H., Pardoll, D. and Mulligan, R. C. (1993) Vaccination with irradiated tumor cells engineered to secrete murine granulocyte-macrophage volony-stimulating factor stimulates potent, specific, and long-lasting anti-tumor immunity. *Proc. Natl. Acad. Sci. USA* 90: 3539–3543.

Dzierzak, E. A., Papayannopoulou, T. and Mulligan, R. C. (1988) Lineage-specific expression of a human beta-globin gene in murine bone marrow transplant recipients reconstituted with retrovirus-transduced stem cells. *Nature* 331: 35–41.

Galley, P., Swingler, S., Aiken, C. and Trono, D. (1995) HIV-1 infection of nondividing cells: C-terminal tyrosine phosphorylation of the viral matrix protein is a key regulator. *Cell* 80: 379–388.

Gilboa, E., Eglitis, M. A., Kantoff, P. W. and Anderson, W. F. (1986) Transfer and expression of cloned genes using retroviral vectors. *Biotechiques.* 4: 504–512.

Guild, B. C., Finer, M. H., Houseman, D. E. and Mulligan, R. C. (1988) Development of retrovirus vectors usefull for expressing genes in cultured murine embryonal cells and hematopoietic cells *in vivo. J. Virol.* 62: 3795–3801.

Irwin, M. J., Laube, L. S., Lee, V., Austin, M., Chada, S., Anderson, C. G., Townsend, K., Jolly, D. J. and Warner, J. F. (1994) Direct injection of a recombinant retroviral vector induces human immunodeficiency virus-specific immune responses in mice and non-human primates. *J. Virol.* 68: 5036–5044.

Kasahara, N., Dozy, A. M. and Kan, Y. W. (1994) Tissue-specific targeting of retroviral vectors through ligand-receptor interactions. *Science* 266: 1373–1376.

Lewis, P. F. and Emerman, M. (1994) Passage through mitosis is required for oncoretroviruses but not for the human immunodeficiency. *J. Virol.* 68: 510–516.

Lin, S., Gaiano, N., Culp, P., Burns, J. C., Friedmann, T., Yee, J. K. and Hopkins, N. (1994)

Integration and germ-line transmission of a pseudotyped retroviral vector in zebrafish. *Science* 265: 666–669.

Mann, R., Mulligan, R. C. and Baltimore, D. (1983) Construction of a retrovirus packaging mutant and its use to produce helper-free defective retrovirus. *Cell* 33: 153–159.

Markowitz, D. G., Goff, S. P. and Bank, A. (1988) Safe and efficient ecotropic and amphotropic packaging lines for use in gene transfer experiments. *Trans. Assoc. Amer. Physician.* 101: 212–218.

Miller, A. D. and Buttimore, C. (1986) Redesign of retrovirus packaging lines to avoid recombination leading to helper virus production. *Mol. Cell. Biol.* 6: 2895–2902.

Miller, A. D., Garcia, J. V., Suhr, N. V., Lynch, C. M., Wilson, C. and Eiden, M. V. (1991) Construction and properties of retrovirus packaging lines based on gibbon ape leukemia virus. *J. Virol.* 65: 2220–2224.

Miller, A. D., Miller, D. G., Garcia, J. V. and Lynch, C. M. (1993) Use of retroviral vectors for gene transfer and expression. *Methods Enzymol.* 217: 581–599.

Miller, D. G. and Miller, A. D. (1994) A family of retroviruses that utilize related phosphate transporters for cell entry. *J. Virol.* 68: 8270–8276.

Mollier, P., Bohl, D., Heard, J. M. and Danos, O. (1993) Correction of lysosomal storage in the liver and spleen of MPS VII mice by manipulation of genetically modified skin fibroblasts. *Nat. Genet.* 4: 154–159.

Morgan, R. A., Couture, L., Elroy-Stein, O., Ragheb, J., Moss, B. and Anderson, W. F. (1992) Retroviral vectors containing putative internal ribosome entry sites: development of a polycistronic gene transfer system and applications to human gene therapy. *Nucl. Acid. Res.* 20: 1293–1299.

Mulligan, R. C. (1993) The basic science of gene therapy *Science* 260: 926–932.

Ohashi, T., Boggs, S., Robbins, P., Bahnson, A., Patrene, K., Wei, F. -S., Wei, J. -F., Fei, Y., Li, J., Lucht, L., Clark, S., Kimak, M., He, H., Mowery, P. and Barranger, J. A. (1992) Efficient transfer and sustained high expression of the human glucocerebrosidase gene in mice and their functional macrophages following transplantation of bone marrow transduced by a retroviral vector. *Proc. Natl. Acad. Sci. USA* 89: 11332–11336.

Pear, W. S., Nolan, G. P., Scott, M. L. and Baltimore, D. (1993) Production of high-titer helper free retroviruses by transient transfection. *Proc. Natl. Acad. Sci. USA* 90: 8392–8396.

Reviere, I., Brose, K. and Mulligan, R. C. (1995) Effects of retroviral design on expression of human adenosine deaminase in murine bone marrow transplant recipients engrafted with genetically modified cells. *Proc. Natl. Acad. Sci. USA* 92: 6733–6737.

Robbins, P. D., Tahara, H., Mueller, G. M., Hung, G., Bahnson, A., Zitvogel, L., J. Galea-Lauri, Ohashi, T., Patrene, K., Boggs, S. S., Evans, C. H., Barranger, J. A. and Lotze., M. T. (1993) *Ann. New York Acad. Sci.* 716: 72–81.

Sadelain, M., Wang, C. H., Antoniou, M., Grosveld, F. and Mulligan, R. C. (1995) Generation of a high titer retroviral vector capable of expressing high levels of the human beta-globin gene. *Proc. Natl. Acad. Sci. USA* 92: 6728–6732.

Salmons, B. and Gunzberg, W. H. (1993) Targeting of retroviral vectors for gene therapy *Hum. Gene Ther.* 4: 129–141.

Somia, N. V., Zoppe, M. and Verma, I. M. (1995) Generation of targeted retroviral vectors by using sisngle-chain variable fragment: an approach to *in vivo* gene delivery. *Proc. Natl. Acad. Sci. USA* 92: 7570–7574.

Wang, H., Kavanough, M. P., North, R. A. and Kabat, D. (1991) Cell surface receptor for ecotropic murine retroviruses is a basic amino acid transporter. *Nature* 352: 729–731.

Yee, J. K., Miyanohara, A., LaPorte, P., Bouic, K., Burns, J. C. and Friedmann, T. (1994) A general method for the generation of high-titer, pantropic retroviral vectors: highly efficient infection of primary heptocytes. *Proc. Natl. Acad. Sci. USA* 91: 9564–9568.

Zitvogel, L., Tahara, H., Mueller, G., Wolf, S. F., Gubler, U., Gately, M., Robbins, P. D. and Lotze, M. T. (1994) Construction and characterization of retroviral vectors expressing biologically active human interleukin 12. *Hum. Gene Ther.* 5: 1493–1506.

2 Adenovirus Vectors for Gene Therapy

G. Wolff

2.1   Introduction

In the fast growing area of vector systems, one of the most advanced is the replication deficient-recombinant adenovirus (Ad vector). It is remarkably efficient at transferring genes to a wide variety of target cells *in vitro* and *in vivo* and at expressing exogenous genes at a high level [Crystal, 1995; Horwitz, 1996]. After infection, the Ad vector expresses the transgene in the nucleus of replicating and non-replicating cells in an epi-chromosomal fashion, where the duration of gene expression is transient over time [Rosenfeld et al., 1991; Kass-Eissler et al., 1994; Smith et al., 1993; Engelhard et al., 1994; Yang et al., 1994a, b].

In 1953 Rowe *et al.*, and Hilleman and Werner independently discovered a new cytopathic agent in cultures of tonsils and adenoids from humans with acute respiratory illness. Following the nomenclature of 1956, this group of respiratory tract viruses was termed adenoviruses (Ad) [Enders et al., 1956]. Based on their immunological, biological, and biochemical characteristics, there are 49 adenovirus serotypes (Ad1 to Ad49), and based on their hemagglutination properties, there are six distinguishable subgroups (subgroup A to subgroup F) [for review, Shenk, 1996]. Adenoviruses are responsible for a variety of clinical manifestations in humans such as respiratory infection, conjunctivitis, hemorrhagic cystitis and gastroenteritis. The Ad types 12, 18 and 31 are able to induce tumors in rodents [Trentin et al., 1962]. However, there are no known associations between adenoviruses and human malignancies [Green et al., 1980]. During the 1970s and 1980s, an intensive effort was made to characterize the molecular biology of the adenoviruses. Thus, it could be demonstrated that the genome of all subgroups has the same structural organization. The best characterized types are Ad type 2 (Ad2) and Ad type 5 (Ad5) of the subgroup C. Both have been sequenced completely and are identical in 95% of the genome [Robert et al., 1984; Chroboczek et al., 1992]. Since the early 1980s Ad vectors have been used as expression systems in mammalian cells [Thummel et al., 1983; Berkner and Sharp, 1983; Berkner, 1988]. Because of the high tropism of Ad2 and Ad5 for epithelial cells from the respiratory tract, in the early 1990s, Crystal and co-workers established the adenovirus-mediated gene transfer *in vivo* as vector system for gene therapy of the lung [Rosenfeld et al., 1991,

T. Blankenstein (ed.) Gene Therapy
©1999, Birkhäuser Verlag Basel

1992]. Using this strategy, on April 17, 1993, a patient suffering from cystic fibrosis received the first *in vivo* administration of an adenovirus vector carrying the cDNA of the cystic fibrosis transmembrane conductance regulator (CFTR) protein.

2.2 Molecular Biology of Adenovirus

The adenovirus is a non-enveloped, 35.9 kb, linear, double-stranded DNA virus. The protein coat comprises 240 hexons and 12 pentons forming 252 capsomers in an icosahedral structure (Fig. 2-1). Each penton is composed of a penton base and

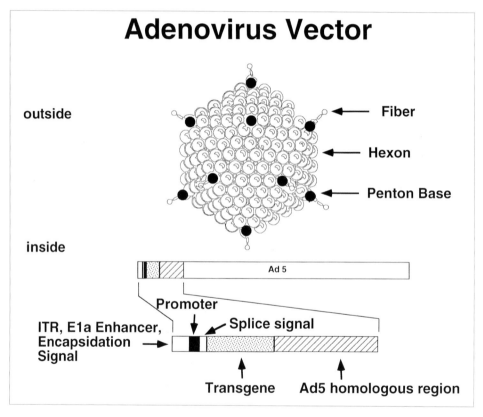

Figure 2-1. Structure of the Adenovirus Vector. The protein coat (outside) comprises 240 hexons and 12 pentons forming 252 capsomers in an icosahedral structure. Each penton is composed of a penton base and a fiber. The viral genome (inside) is inside of the capsid. The genome of Ad 5 consists of 35.9 kb. To increase the space for insertions, the Ad vector of first generation contains different deletions in its genome, E1A–, partial E1B– and partial E3–. The expression cassette is placed in the position of E1 at the left side of the adenovirus genome.

a fiber. The viral genome is inside of the capsid and has a terminal polypeptide (TP) covalently linked, at the 5' end, to a dCMP residue. It carries two distinguishable groups of genes, the early and the late transcription units. The early genes consist of E1A, E1B, E2, E3, E4 and two delayed early units, IX and IVa2. The group of late genes consist of L1 to L5 [for review, Shenk, 1996].

The adenovirus attacks the cell by binding to a coxsackievirus and adenovirus receptor protein (CAR) and internalizes via the integrins $\alpha_v\beta_3$ and $\alpha_v\beta_5$ [Bergelson et al., 1997; Wickham et al., 1993]. This way, the adenovirus reaches the endosomal pathway and avoids lysosomal degradation. Inside the endosome, a stepwise disassembly programm takes place allowing the adenovirus to release its genome into the nucleus. During this process, the pH of the endosome decreases leading to the release of the fiber from the virion, and the dissociation of the penton base [Seth et al., 1984]. The resulting rupture of the endosome allows the DNA to escape from inside the degraded capsid and to enter the nucleus. This uncoating process of the adenovirus starts immediately after internalization and ends 40 min after infection with translocation of the adenovirus DNA into the nucleus. As early as 60 min after infection, the adenovirus begins to transcribe its genome in the host cell [Greber et al., 1993].

To complete the life cycle after infection, the adenovirus has to replicate its DNA and assemble the virion within the host cell. At first, the viral replication is initiated by E1A. For that, E1A and E1B-55 kDa activate the cell cycle progression. In parallel, apoptosis of the infected cell is blocked by inactivation of p53 through E1B-19 kDa [Kao et al., 1990; Yew and Berk, 1992]. Viral replication is continued by transcription of the early genes, initiation of DNA replication by E2A and E2B, and by the transcription of the late genes. In favor of the adenovirus replication, transport and translation of host mRNA is inhibited by different viral genes (E1B-55 kDa, E4, VA RNA, Tripartite Leader, L4-100kDA and L3-23 kDa). Finally, the viral replication is completed by virion assembly, mediated by L1-L5 and pIX [for review, Shenk, 1996; Horwitz, 1996].

To avoid destruction of the infected cell by the immune system, the adenovirus interferes with the host defence system at several points. The E1A and VA RNA antagonize α and β interferons [Anderson et al., 1987; Kitajewski et al., 1986]. The adenovirus proteins E3-14.7 kDa or a protein complex formed between E3-14.5 kDa and E3-10.4 kDa blocks TNF-α mediated cytolysis of the infected cell [Gooding et al., 1991]. The arachidonic acid release mediated by TNF-α is antagonized by the E3-14.7 kDa protein [Zilli et al., 1992]. The E3-gp19 blocks transport of adenoviral proteins to the cell surface, thereby inhibiting the recognition by cytotoxic T-lymphocytes [Burgert et al., 1987].

2.3 Generation of Recombinant Adenovirus Vectors

The genome of Ad 5 consists of 35.9 kb. In general, the adenovirus can package about 105% of its genome length. This capacity allows insertion of up to 1.8 kb of foreign DNA. To increase the space for insertions, the Ad vector of first generation contains different deletions in its genome, E1A–, partial E1B– and partial E3–. Therefore, it can carry up to 7.5 kb of foreign DNA. As there is about 1.5 kb needed for promotor/enhancer elements and RNA processing sequences, an effective space for cDNA inserts of about 5 kb is available. In contrast to the E3 region of the adenovirus genome, E1A is essential for adenovirus replication. However, this function can be delivered *in trans* by a human embryonic kidney cell line, 293, where E1A is inserted stably [Graham et al., 1977]. In general, an Ad vector can be constructed with three tools: 293 cells, the viral backbone DNA, and a plasmid carrying the expression cassette and viral elements that complete the viral backbone ("shuttle-vector"). Cotransfection into 293 cells allows homologous recombination between overlapping regions of the viral backbone DNA and the "shuttle-vector". Because this takes place with E1A supplied *in trans*, the viral infectious cycle is terminated with the generation of the replication-deficient recombinant adenovirus carrying the transgene of choice [Graham and Prevec, 1991; Bett et al., 1994].

The new Ad vector is generated by cotransfection of the viral backbone and the "shuttle-vector" into 293 cells. The 293 cell monolayer is overlayed by agar, incubated at 37 °C and checked for the development of plaques indicating successful homologous recombination. This can be identified after 7 to 10 days by a cytopathic effect (CPE) as indicated by the rounding-up of the cells and, therefore, the resulting formation of single plaques. Once the CPE occurs, single plaques are harvested and amplified individually by another round of infection on the next 293 cell monolayer. Because this time the monolayer is infected without an agar overlay, the Ad vector can be harvested by collecting the cytopathic cells. Typically, CPE-formation starts after 24 to 30 h and is fully developed after 36 h. When the CPE occurs before 24 h, other causes than virus production have to be assumed and the preparation needs to be discarded. After harvesting the cells containing the Ad vector produced ("crude virus stock" = CVS), the vector is separated by breaking up the cells by five freeze and thaw cycles resulting in the "crude virus lysate" (CVL). The Ad vector is further separated from the cell debris and purified by ultracentrifugation. At the end of the procedure, the Ad vector is aliquoted and stored in a stabilization buffer [Graham and Prevec, 1991]. To verify the correct function and activity of the construct generated, the bioactivity of the adenovirus vector requires to be evaluated.

2.4   Evaluation of the Bioactivity of Adenovirus Vectors

In the context of Ad vectors as potential therapeutic pharmaceuticals in gene therapy, the quality and the yield of Ad vector production needs to be evaluated. There
are no widely accepted international standard methods for evaluation. However,
there are at least four recommended methods that should be used in a validated
manner: (I) quantification of the adenovirus vector by optical density; (II) quantification of the adenovirus vector by detection of the formation of visible plaques in a
monolayer of 293 cells; (III) verification of the absence of wild type contamination
and adenovirus replication; and (IV) quantification of transgene expression
[Mittereder et al., 1996].

The quantification of the adenovirus vector by optical density is determined by
measuring the optical density (260 nm) of a diluted aliquot of the purified Ad vector. The concentration of adenovirus vector is calculated by multiplying the ab-
sorbance by the appropriate dilution factor and then dividing by the extinction co-
efficient determined for purified wild-type adenovirus ($\varepsilon_{260} = 9.09 \times 10^{-13}$ OD ml
cm virion^{-1}) [Maizel et al., 1968]. The value is based on the fact that 1.1×10^{12} virions of wild-type Ad 5 correspond to an OD of 1.0 at 260 nm. The resulting concentration of vector particle is reported as optical particle units (OPU). The quantification of adenovirus vector using optical density measurements demonstrates a
physical means to determine the concentration of virions. It does not distinguish between functional and non-functional virions. In addition, it is influenced by a variety of factors like the purity of the preparation, the actual behavior of small particles in fluid or the accuracy of the instrument used. However, it is a fast estimation
of the yield, and together with the quantification of the functional virions in the
preparation, it is important for the assessment of the Ad vector preparation.

The quantification of adenovirus vector by detection of the formation of visible
plaques in a monolayer of 293 cells determines the number of biologically active
virions in the batch of the Ad vector. The plaque assay is performed with different
dilutions of the purified Ad vector (10^{-5} to 10^{-10}) on 293 cells. The resulting concentration of the vector is expressed as plaque forming units (PFU) per ml. In an ideal situation, the absolute number of virions present in the preparation is equal to the
number of biologically active virions (particle/PFU ratio). However, this ratio is, in
practice, between 50 and 100. This ratio is of importance for the quality of «clinical grade» preparations of Ad vectors for clinical application.

The proof for the absence of wild type contamination and adenovirus replication
is of special importance. In every round of Ad vector amplification, there is a theoretical risk of homologous recombination in favor of the wild-type adenovirus. Indeed,
experimental studies have demonstrated that this risk increases with the number of
amplifications [Lochmuller et al., 1994; Hehir et al., 1996]. Therefore, each batch
of Ad vector produced needs to be analyzed for the absence of replication-competent adenovirus. A fast routine method is the polymerase chain reaction (PCR) de-

tecting the presence of E1A contamination in the preparation using a sequence of E4 as control [Eissa, 1994]. However, the presence of E1A in the actual preparation does present the assembly of replication competent virons. The appropriate test is the detection of replication competent adenovirus (RCA) which is carried out on the lung carcinoma cell line A549. This cell line is free of the E1A sequence and free of any "E1a like activity" [Imperiale et al., 1984; Spergel and Cheng-Kiang, 1991]. Using A549 cells in parallel to 293 cells in the "plaque assay" allows the simultaneous quantification of replication competent and biologically active virions.

The ability of the Ad vector to express sufficient amounts of the transgene is tested on HeLa cells. This human cervix carcinoma cell line is very receptive to adenovirus infection and has been used in a variety of classical studies in the past. Based on the type of transgene, an appropriate quantitative assay should be employed to quantify the amount of transgene after infection with a standard dose of the Ad vector. In parallel cytopathic effects can also be evaluated on Hela cells. On cells that are receptive to adenovirus infection, only doses above 100 PFU per cell lead to CPEs as indicated by the rounding up and detaching of the cells from the plate. If CPEs arise below this dose, then the preparation needs to be discarded.

2.5   Immunology of Adenovirus Vectors

Although the Ad vectors are efficient in entry and translocation, the inital robust expression of the foreign transgene is transient over time. At least in part, the expression is limited by a variety of anti-adenovirus defence mechanisms. These defence mechanisms arrayed against the Ad vector are complex and include both the innate (unspecific) and the adaptive (specific) immune response.

Based on the observations that Ad vector-mediated gene expression is transient over time, extensive studies were carried out to elucidate the underlying mechanisms [Rosenfeld et al., 1991; Kass-Eissler et al., 1994; Smith et al., 1993; Engelhard et al., 1994; Yang et al., 1994a]. As a result of these early studies, the loss of expression over time could be explained, at least in part, by the host adaptive immune response. Cells infected with the Ad vector are eliminated by a T-cell-mediated response including cytotoxic T-cells [Yang et al., 1994a, b; Zabner et al., 1994; Dai et al., 1995; Zsengeler et al., 1995]. This paradigm is similar to the classical description of the host immune response to wild-type viral infection, where viral proteins are presented on the cell surface. These foreign proteins are recognized by antigen-specific cytotoxic T-cells in the context of HLA class I antigens and are eliminated [Mullbacher et al., 1989; Zinkernagel, 1993, 1996]. Consistent with this concept, the duration of expression of a gene transferred by an Ad vector is significantly extended in time in experimental animals that are either deficient in various components of this T-cell mediated defence system or rendered immunodeficient by pharmacological means [Dai et al., 1995; Kay et al., 1995; Yang and Wilson, 1995; Yang et al., 1995a, b, c].

The second arm of the adaptive immune system arrayed against the adenovirus is the humoral immune response. After administration of Ad vectors to the lung of experimental animals, antibodies against the Ad vector were found in the local millieu (bronchoalveolar lavage) and systemically in the blood stream [Mastrangeli et al., 1996]. Thus, repeated administration of Ad vectors into experimental animals resulted in a dramatic decline of transgene expression. These antibodies are specific to the serotype of Ad vector administered. Therefore, the use of Ad vectors of different serotypes (see "Designer and Chimeric Adenovirus Vectors") could re-establish an effective adenovirus-mediated gene transfer *in vivo*.

Although adaptive immune mechanisms undoubtedly play a significant role in the elimination of the Ad vector, they do not explain events shortly after the Ad vector was administered. After 24 to 48 h of infection with different doses of Ad vector *in vivo*, there was cellular degeneration and non-lymphocytic infiltration. In addition, depending on the animal model, different cytokines were activated. Therefore, it was assumed that mechanisms other than the T-cell mediated immune response must be involved. Another line of evidence came from efficiency studies for Ad vector-mediated gene transfer *in vivo*. First, using different routes of applications, it was demonstrated that different sites of administration lead to different patterns of organ distribution and intensity of the transgene expression [Kass-Eissler et al., 1994]. Second, systemic suppression of the immune system improved the transgene expression at a time point where lymphocytic infiltration in the infected organ was still absent. Third, there was an early phase of vector loss after *in vivo* administration [Worgall et al., 1997]. These data suggest that the innate immune system plays a significant role in the elimination of Ad vectors.

To date the best studied *in vivo* model for the concerted action of host defence mechanisms against adenovirus is the adenovirus-mediated gene transfer to the liver. The liver is one of the most important target organs for gene therapy. This organ consists mainly of parenchymal cells, and hepatocytes (about 80% of the organ volume and 60% of the cell number). Nonparenchymal cells (20% of the organ volume and 40% of the cell number) include Kupffer cells, endothelial cells, pit cells and fat storing cells, fibroblasts, and cells from the bilary duct system. The Kupffer cells are the tissue-specific macrophages of the liver and are essential for eliminating pathogens entering the system from the gut. Independent of the mode of application (direct injection via the portal vein or the systemic circulation), the Ad vector comes in contact with the nonparenchymal cells building a barrier against infection around the hepatocytes. Here the Kupffer cells are involved in the elimination of adenoviruses, amounting to 50% of the dose administered [Wolff et al., 1997]. In parallel, there is a dramatic decline in Ad vector DNA within the liver after *in vivo* administration that is not demonstrated in primary cultures of hepatocytes [Worgall et al., 1997]. The mechanism for this still remains unclear. In the following 2 to 5 days, the defence against the adenovirus is dominated by a nonlymphocytic infiltration and a ballooning degeneration of hepatocytes. These effects are closely dependent on the dose of vector [Yang et al., 1994a, b; Yang and Wilson, 1995]. At the end of the first week after infection, the nonlymphocytic infiltration disappears and a lymphocytic infiltration follows that is dominated by cytotoxic T-cells. Whereas the T-cell infiltration occurs only in immunocompetent animals, the unspecific, early inflammation is present in both, immunocompromized and immunocompetent animals [Barr et al., 1995; Dai et al., 1995; Yang et al., 1994a, b, 1995a, b, c; Zsengeller et al., 1995). Despite these multilayered defence mechanisms, the amount of surviving Ad vectors is still enough to allow a transient but strong transgene expression making the liver the most effective organ for adenovirus-mediated gene transfer.

2.6   Designer and Chimeric Adenovirus Vectors

The replication-deficient recombinant Ad vectors of the first generation are E1A–, partial E1B-, and partial E3-based on the Ad5 genome where the expression cassette is placed in the position of the E1 deletion (Fig. 2-1). Soon it became evident that even in the absence of E1A–, there was still a low transcription rate of the early viral gene E2A and that its deletion resulted in an Ad vector with prolonged transgene persistence and a decreased inflammatory response. Because of the improved features, this construct was called second generation Ad vector [Engelhardt et al., 1994]. Other deletions, like E4, demonstrated that these vectors were more stable *in vivo* than Ad vectors from the first generation and, hence, more efficent for use in possible applications in human gene therapy [Krougliak and Graham, 1995; Gao et al., 1996]. The deletion of different parts of the adenovirus genome is aimed to make the Ad vector "stealth" to the immune system. In the context of the MHC class I restricted immunity against the Ad vector, fewer adenoviral (foreign) proteins are presented on the surface of the infected cell. Here, the ultimate Ad vector is the one that is completely free of any adenoviral gene [Kochanek et al., 1996; Krishna et al., 1996].

In addition, there are other factors like epichromosomal localization, counting for transient expression. Theoretically, this limitation could be circumvented by repeated administration of the Ad vector. However, there is also a humoral immune response against the Ad capsid mediated by induction of neutralizing antibodies leading to a dramatic decline of gene transfer and expression after reapeated administration *in vivo* [Wohlfahrt et al., 1985; Wohlfahrt, 1988]. One strategy to circumvent the Ad vector-specific immunity by neutralizing antibodies is to switch the serotype of the Ad vector with subsequent administrations. Studies have demonstrated that Ad5 vectors can bypass the neutralizing immune response resulting from prior administration of wild-type adenoviruses from a different serotype [Mastrangeli et al., 1996]. This concept was recently verified with the use of Ad5 and Ad2 based vectors which have been used in clinical gene therapy trials. Despite the homology of these vectors for the penton base (98%), hexon (86%), and fiber (70%), there is no significant crossover in the humoral immunity [Adam et al., 1995]. Consequently, it was demonstrated that the alternate use of different sero-types of Ad vectors (within the same subgroup or from different subgroups) can by-pass the humoral immunity against the Ad vector [Mack et al., 1997].

In contrast to replication-deficient recombinant adenovirus vectors, replication-competent recombinant adenoviruses are designed especially only for cancer gene therapy. This recently-developed new generation uses the capabiltiy of an E1B-55 kDa deleted wild type Ad to replicate selectively in p53 mutated tumor cells. The E1B-55 kDa protein binds to and inactivates p53 in the infected host cell [Kao et al., 1990; Yew and Berk, 1992]. This activity is essential for adenovirus replication be-cause E1A induces p53 dependend apoptosis. If an adenovirus (E1A+) is mutated in E1B-55 kDa, the replication is blocked by p53. In turn, if the mutated adenovirus infects p53 mutated cells, replication takes place resulting in cell death by the lytic infection cycle. In addition, the newly generated replication-competent virions infect neighboring cells leading to the next round of lytic infection. This process continues until cells reach a state where a functional p53 terminates the replication. Therefore, the difference in the p53 status between tumors carrying a p53 mutation or deletion and the surrounding non-tumorigenic tissue carrying the wild type p53 can be used for the selective killing of tumor cells [Bischoff et al., 1996].

In contrast to the strategy of making "stealth" the Ad vector to the immune sys-tem by deletions in the genome, the development of targeted Ad vectors focuses on the modification of the protein coat. The protein coat of the adenovirus is assem-bled from 240 hexons and 12 pentons forming 252 capsomers in an icosahedral structure (Fig. 2-1). Each penton is composed of a penton base and a fiber. The widespread distribution of receptors binding the pentons of the adenovirus limits the use of Ad vector for a targeted gene transfer. Therefore, it is necessary to ablate the endogenous viral tropism and introduce a new, novel tropism. This can be achieved either by exchanging parts of the protein coat or by placing new molecules on the surface of the Ad vector. The penton base of the protein coat contains a bind-ing motif, arg-gly-asp (RGD). This motif is recognized by the integrins $\alpha_v\beta_3$ and $\alpha_v\beta_5$ which mediate the internalization of the Ad vector [Wickham et al., 1993]. However, cells of other than epithelial origin express different types of integrins re-sulting in decreased or no infectibility by the Ad vector. Consequently, replacement of the RGD motif by other peptide motifs can target the Ad vector to other integrin receptors than $\alpha_v\beta_3$ and $\alpha_v\beta_5$ [Wickham et al., 1995]. In addition, to target the bind-ing of the Ad vector to cell-type specific receptors the fiber protein can be modifed either by the introduction of chimeric fiber-fusionproteins or by placing specific monoclonal antibodies on top of the fiber protein [Michael et al., 1995, Douglas et al., 1996; Krasnykh et al., 1996; Wickham et al., 1996].

The attemps of vector targeting and to "stealth" the vector to the immune sys-tem represent new and promising strategies to improve adenovirus-mediated gene transfer *in vivo*. Additionally, pharmacological means are under investigation which increase the *in vivo* efficiency and transgene persistence of the Ad vector [Dai et al., 1995; Kay et al., 1995; Yang et al., 1995a, b, c; Wolff et al., 1997]. However, for clinical applications, further studies are needed to demonstrate their *in vivo* efficiency and safety in human gene therapy.

2.7 Clinical Applications

The concept of human gene therapy is to use a gene as a therapeutical drug in somatic cells [Anderson, 1993; Mulligan, 1993; Friedmann, 1996]. In general, the transgene can be delivered by an *ex vivo* or *in vivo* strategy. In both, the expression cassette is usually transferred by a vector delivering the gene to an intracellular site where it can function appropriately in a local and/or systemic fashion. For these approaches, the adenovirus-mediated gene transfer is an attractive strategy. Ad vectors can be produced in high titers resulting in an absolute yield of up to 10^{13} viral particles per ml. The vector expresses the transgene easily in the nucleus of proliferating and non-proliferating cells in an epichromosomal fashion, thus avoiding the alteration of the genome of the target cell or insertional mutagenesis. However, the immune response and the epichromosomal position of the transgene limit the duration of expression. Therefore, the replication-deficient recombinant adenovirus is a well suited vector system for *in vivo* gene therapy where a high but transient gene transfer and expression is needed [Munaf et al., 1994; Crystal, 1995]. Until 1996, there are 22 gene therapy protocols published using Ad vectors as the gene transfer system [*Hum. Gene Ther.* 1996, 18: 2287–2315].

The first disease where an Ad vector was applied in human gene therapy was cystic fibrosis. Cystic fibrosis (CF) is the most frequent genetic cause of death in the caucasian population [Collins, 1993]. CF is based on mutations in the gene of the cystic fibrosis transmembrane conductance regulator (CFTR) protein. Because the CFTR protein acts as a chloride channel on the apical surface of epithelial cells, dysfunction of the protein leads to a defect in chloride transportation. Although multiple organs are affected, the life threatening organ is the lung. The CFTR gene was identified in 1989 [Kerem et al., 1989; Rommens et al., 1989; Riordan et al., 1989]. Mutants of Ad2 and Ad5 vectors carrying a CFTR-cDNA were constructed and used in different clinical phase I trials. These studies demonstrated the feasibility of adenovirus-mediated *in vivo* gene transfer to humans. However, the clinical benefit remains to be demonstrated [Zabner et al., 1993; Crystal et al., 1994; Knowles et al., 1995].

The limitations for long term expression changed the focus of the application of Ad vectors in human gene therapy. Early experiments demonstrated that Ad vectors could express transgenes in tumor cells in amounts, so far unreachable for other vector systems. In addition, taking into account that the goal of cancer gene therapy is the killing of tumor cells, there is only a temporary gene expression necessary. However, assuming that sufficient amounts of the Ad vector infect all cells, the expression of the transgene in an epichromosomal fashion results in a loss over time when the cells are dividing. Theoretically, this will lead to tumor growth out of non-infected cells. To overcome this conflict, there are two strategies under intensive investigation: induction of a "bystander effect" and development of replication-competent Ad vectors.

The "bystander effect" is characterized by the generation of a therapeutic product inside the transfected cells that is able to move from the infected to non-infected cells and, therefore, leads to the same therapeutic events in uninfected as well as in infected cells. At first, a target cell will be infected by a vector carrying the gene coding for a prodrug converting enzyme. Once the enzyme is expressed, it transforms a seperately administered non-toxic prodrug to its toxic form leading to the death of the infected cell ("suicide strategy") [Culver et al., 1992]. As the toxic drug leaves the cell, it enters non-infected cells, where it causes cell death again. For this strategy, a variety of prodrug converting enzymes can be used [Moolten, 1994; Deonarain et al., 1995]. Ad vectors carrying the herpes simplex thymidine kinase or the cytosine deaminase gene are under investigation and are employed for clinical phase I trials. The bystander effect is not restricted to the use of prodrug converting enzymes. The tumor suppressor gene p53 induces growth arrest in p53 defective tumor cells after direct gene transfer of a p53 cDNA into the cell. However, following gene transfer, p53 inhibits the growth of human tumor cells also through a bystander mechanism without evidence of toxicity [Cai et al., 1993; Xu et al., 1997]. These results suggest that, together with a strong transgene expression in most of the cells, the "bystander effect" can overcome the limitations caused by epichromosomal location. Taking advantage of these effects, a phase I trials are ongoing with the use of an Ad vector carrying the wild-type p53 gene in patients with non-small cell lung cancer [Roth, 1996].

The development of replication-competent Ad vectors is another strategy to circumvent the limitations of non-integration of the transgene into tumor cells. This concept uses the capabiltity of an E1B-55 kDa deleted wild type Ad to replicate selectively in p53 mutated tumor cells [Bischoff et al., 1996]. Replication of the Ad vector in the infected target cell leads to tumor cell death by cell lysis. Once the cell is lyzed, the Ad vector infects the next cell and induces cell lysis again. Because the replication of this mutant is stopped in cells containing wild-type p53, the lytic infection of the Ad vector is terminated when it reaches the normal tissue surrounding the infected tumor (see "Designer and Chimeric Adenovirus Vectors"). Therefore, the difference in the p53 status between tumors carrying a p53 defect and the surrounding non-tumorigenic tissue carrying the wild type p53 leads to a selective killing of tumor cells. Using this concept, a clinical phase I trial for patients with p53$^-$ squamous cell carcinoma of the head and neck was started in April 1996 [Bischoff et al., 1996 and references therein].

Taken together, the feasibility of Ad vectors to transfer genes to humans has been proven. Whereas in the beginning it was applied for heriditary disorders like CF, the current applications focus more on aquired diseases. At present, it is too early to draw conclusions about the final clinical benefit of this strategy. The technology is just "out of the bottle" and still has a long way to go. However, in the fast growing area of vector systems adenovirus vectors are so far the most effective one for *in vitro* and *in vivo* applications. Further work needs to be carried out to diminish side effects like immune response. As demonstrated by the clinical application of new generations of Ad vector, the creative combination of the knowledge about the molecular biology of the adenovirus and the disease to be treated holds a lot of promise for molecular medicine.

References

Adam, E., Nasz, I. and Lengyel, A. (1995) Antigenic homogenicity among the adenovirus hexon types of the subgroup C. *Arch. Virol.* 140: 1297–1301.

Anderson, K. and Fennie, E. (1987) Adenovirus early region 1a modulation of interferon antiviral activity. *J. Virol.* 61: 787–795.

Anderson, W. F. (1992) Human gene therapy. *Science* 256: 808–813.

Barr, D., Tubb, J., Ferguson, D., Scaria, A., Lieber, A., Wilson, C., Perkins, J. and Kay, M. A. (1995) Strain related variations in adenovirally mediated transgene expression from mouse hepatocytes *in vivo*: comparisons between immunocompetent and immunodeficient inbred strains. *Gene Ther.* 2: 151–155.

Bergelson, J. M., Cunningham, J. A., Droguett, G., Kurt-Jones, E. A., Krithivas, A., Hong,.J.S., Horwitz, M. S., Crowell, R. L. and Finberg R.W. (1997) Isolation of a common receptor for coxsackie B virus and adenovirus 2 and 5. *Science* 275: 1320–1323.

Berkner, K. L. and Sharp, P. A. (1983) Generation of adenovirus by transfection of plasmids. *Nucl. Acid. Res.* 10: 6003–6020.

Berkner, K. L. (1988) Development of adenovirus vectors for the expression of heterologous genes. *Biotechniques* 6: 616–629.

Bett, A. J., Haddara, W., Prevec, L. and Graham, F. L. (1994) An efficient and flexible system for construction of adenovirus vectors with insertions or deletions in early regions 1 and 3. *Proc. Natl. Acad. Sci. USA* 91: 8802–8806.

Bischoff, J. R., Kirn, D. H., Williams, A., Heise, C., Horn, S., Muna, M., Ng, L., Nye, J. A., Sampson-Johannes, A., Fattaey, A., McCormick, F. (1996) An adenovirus mutant that replicates selectively in p53-deficient human tumor cells. *Science* 274: 373–376.

Burgert, H. G. and Kvist, S. (1987) The E3/19K protein of adenovirus type 2 binds to the domains of histocompatibility antigens required for CTL recognition. *EMBO J.* 6: 2019–26.

Cai, D. W., Mukhopadhyay, T. Liu, Y., Fujiwara, T. and Roth, J. (1993) Stable expression of the wild-type p53 gene in human lung cancer cells after retrovirus-mediated gene transfer. *Hum. Gene Ther.* 4: 617–624.

Chroboczek, J., Bieber, F., Jacrot B. (1992) The sequence of the genome of adenovirus type 5 and its comparison with the genome of the adenovirus type 2. *Virology* 186: 280–285.

Collins, F. S. (1992) Cystic fibrosis: Molecular biology and therapeutic implications. *Science*

256: 774–779.

Crystal, R. G. (1995) Transfer of genes to humans: Early lessons and obstacles to success. *Science* 270: 404–410.

Crystal, R. G., McElvaney, N. G., Rosenfeld, M. A., Chu, C. S., Mastrangeli, A., Hay, J. G., Brody, S. L., Jaffe, H. A., Eissa, N. T. and Danel, C. (1994) Administration of an adenovirus containing the human CFTR cDNA to the respiratory tract of individuals with cystic fibrosis. *Nat. Genet.* 8: 42–51.

Culver, K. M., Ram, Z., Wallbridge, S., Ishii, H., Oldfield, E. H. and Blase, R. M. (1992) *In vivo* gene transfer with retroviral vector-producer cells for treatment of experimental brain tumors. *Science* 256: 1550–1552.

Dai, Y., Schwarz, E. M., Gu, D., Zhang, W. -W., Sarvetnik, M. and Verma, I. M. (1995) Cellular and humoral immune responses to adenoviral vectors containing factor IX gene: Tolerization of factor IX and vector antigens allows for long term expression. *Proc. Natl. Acad. Sci. USA* 92: 1401–1405.

Deonarain, M. P., Spooner, R.A and Epenetos, A. A. (1995) Genetic delivery of enzymes for cancer therapy. *Gene Ther.* 2: 235–244.

Douglas, J. T., Rogers, B. E., Rosenfeld, M. E., Michael, S. I., Feng, M. and Curiel, T., (1996) Targeted gene delivery by tropism-modified adenoviral vectors. *Nat. Biotechnol.* 14: 1574–1578.

Eissa, N. T., Chu, C. S., Danel, C. and Crystal, R. G., (1994) Evaluation of the respiratory epithelium of normals and individuals with cystic fibrosis for the presence of adenovirus E1a sequences relevant to the use of E1a-adenovirus vectors for gene therapy for the respiratory manifestations of cystic fibrosis. *Hum. Gene Ther.* 5: 1105–1114.

Engelhardt, J. F., Ye, X., Doranz, B. and Wilson, J. M. (1994) Ablation of E2A in recombinant adenovirus improves transgene persistence and decreases inflammatory response in mouse liver. *Proc. Natl. Acad. Sci. USA* 91: 6196–6200.

Enders, J. F., Bell, J. A., Dingl,e J.H. (1956) "Adenoviruses" group name proposed for new respiratory-tract viruses. *Science* 124: 119–120.

Fisher, K.J., Mikheeva, G. V., Douglas, J. T. and Curiel, D. T.(1996) Generation of recombinant adenovirus vectors with modified fibers for altering viral tropism. *Virology* 70: 6839–6846.

Friedmann, T. (1996) Human gene therapy – an immature genie, but certainly out of the bottle. *Nat. Med.* 2: 144–147.

Gao, G. -P., Yang, Y. and Wilson, J.M: (1996) Biology of adenovirus vectors with E1A and E4 deletions for liver-directed gene therapy. *J. Virol.* 70: 8934–8943.

Gooding, L. R., Ranheim, T. S., Tollefson, A. E., Aquino, L., Duerksen-Hughes, P., Horton, T. M., Wold, W. S. (1991) The 10,400- and 14,500-dalton proteins encoded by region E3 of adenovirus function together to protect many but not all mouse cell lines against lysis by tumor necrosis factor. *J. Virol.* 65: 4114–23.

Graham, F. L. and Prevec, L. (1991) Manipulation of adenovirus vectors. *In*: E. J. Murray (ed.), *Methods in Molecular Biology*. The Humana Press, Clifton, NY, pp. 109–128.

Graham, F. L., Smiley, J., Russell, W. C. and Nairn, R. (1977) Characteristics of a human cell line transformed by DNA from human adenovirus type 5. *J. Gen. Virol.* 36: 59–74.

Greber, U. F., Willetts, M., Webster, P. and Helenius, A. (1993) Stepwise dismantling of adenovirus 2 during entry into cells. *Cell* 75: 477–486.

Green, M., Wold, W. S. M. and Brachmann, K. H. (1980) Human adenovirus transforming genes: group relationships, integration, expression in transformed cells and analysis of human cancers and tonsils. *In*: M. Essex, G. Todaro, H. zur Hausen H (eds), *Cold Spring Habor conference on cell proliferation viruses in naturally ocuring tumors*. Cold Spring Habor Laboratory, Cold Spring Habor, NY., pp. 373–397.

Hehir, K. M., Armentano, D., Cardoza, L. M., Choquette, T. L., Berthelette, P. B., White, G. A., Courture, L. A., Everton, M. B., Keegan, J., Mertin, J. M., Pratt, D. A., Smith, M. P., Smith, A. E. and Wadsworth, S. C. (1996) Molecular Characterization of replication competent variants of adenovirus vectors and genome modifications to prevent their occurrence. *J. Virol.* 70: 8459–8467.

Hillemann, M. R. and Werner, J. R. (1953) Recovery of a new agent from patients with acute respiratory illness. *Proc. Soc. Exp. Biol. Med.* 85: 574–580.

Human gene marker/therapy clinical protocols (completed and uptated listing) (1996) *Hum. Gene Ther.* 18: 2287–2313.

Horwitz, M. S. (1996) Adenoviridae and their replication. *In*: B. N. Fields and D. M. Knipe (eds), *Virology*. Raven Press, New York, pp 1679–1740.

Horwitz, M. S. (1996) Adenoviruses. *In*: B. N. Fields and D. M. Knipe (eds), *Virology*. Raven Press, New York, pp. 2149–2171.

Imperiale, M. J., Kao, H. -T. Feldmann, L. T., Nevins, J. R. and Strickland, S. (1984) Common control of the heat shock gene and early adenovirus genes for a cellular E1a-like activity. *Mol. Cell. Biol.* 4: 867–874.

Kao, C. C., Yew, P. R., Berk, A. J. (1990) Domains required for *in vitro* association between the cellular p53 and the adenovirus 2 E1B 55K proteins.*Virology* 179: 806–14.

Kass-Eisler, A., Falck-Pedersen, E., Elfenbein, D. H., Alvira, M., Buttrick, P. M. and Leinwand, L. A. (1994) The impact of developmental stage, route of administration and the immune system on adenovirus-mediated gene transfer. *Gene Ther.* 1: 395–402.

Kay, M. A., Holterman, A. X., Meuse, L., Gown, A., Ochs, H. D., Linsley, P. S. and Wilson, C. B. (1995) Long-term hepatic adenovirus-mediated gene expression in mice following CTLA4Ig administration. *Nat. Genet.* 11: 191–197.

Kerem, B., Rommens, J. M., Buchanan, J. A., Markiewicz, D., Cox, T. K., Chakravarti, A., Buchwald, M. and Tsui, L. C. (1989) Identification of the cystic fibrosis gene: genetic analysis. *Science* 245: 1073–1080.

Kitajewski, J., Schneider, R. J., Safer, B., Munemitsu, S. M., Samuel, C. E., Thimmappaya, B. and Shenk, T. (1986) Adenovirus VAI RNA antagonizes the antiviral action of interferon by preventing activation of the interferon-induced eIF-2 alpha kinase. *Cell* 45: 195–200.

Knowles, M. R., Hohneker, K. W., Zhou, Z., Olsen, J. C., Noah, T. L., Hu, P. C., Leigh, M. W., Engelhardt, J. F., Edwards, L. J., Jones, K. R., et-al (1995) A controlled study of adenoviral-vector-mediated gene transfer in the nasal epithelium of patients with cystic fibrosis. *N. Engl. J. Med.* 333: 823–831.

Kochanek, St., Clemens, P. R., Mitani, K., Chen, H. -H., Chan, S. and Caskey, C. T. (1996) A new adenoviral vector: Replacement of all viral coding sequences with 28 kb of DNA independently expressing both full-length dystrophin and β-Galactosidase. *Proc. Natl. Acad. Sci. USA* 93: 5731–5736.

Krishna, J., Choi, H., Burda, J., Chen, S. -J. and Wilson, J. M. (1996) Recombinant Adenovirus deleted of all viral genes for gene therapy of cystic fibrosis. *J. Virol.* 217: 11–22.

Krougliak, V. and Graham, F. L. (1995) Development of cell lines capable of complementing E1A, and protein IX defective adenovirus type 5 mutants. *Hum. Gene Ther.* 6: 1575–1586.

Lochmuller, H., Jani, A., Huard, J., Prescott, S., Simoneau, M., Massie, B., Karpati, G., Acsadi, G. (1994) Emergence of early region 1-containing replication-competent adenovirus in stocks of replication-defective adenovirus recombinants (delta E1 + delta E3) during multiple passages in 293 cells. *Hum. Gene Ther.* 5: 1485–1491.

Mack, C. A., Song, W. R. Carpenther, H., Wickham, T. J., Kovesdi, I., Harvey, B. G., Macgovern, C. J., Isom, O. W., Rosengart, T., Falck-Pedersen, E., Hackett, N. R., Crystal, R. G. and Mastrangeli, A. (1997) Circumvention of anti adenovirus neutralizing immunity by administration of an adenoviral vector of an alternative serotype. *Hum. Gene Ther.* 8: 99–109.

Maizel, J. V., White, J. D. and Scharff, M. D. (1968) The polypeptides of adenovirus. I. Evidence for multiple protein components in the virion and a comparison of types 2, 7a and 12. *Virology* 36: 115–125.

Mastrangeli, A., Harvey, B. G., Yao, J., Wolff, G., Kovesdi, I., Crystal, R. G. and Falck-Pedersen, E. (1996) "Sero-switch" adenovirus-mediated *in vivo* gene transfer: Circumvention of anti-adenovirus humoral immune defence against repeat adenovirus vector administration by changing the adenovirus serotype. *Hum. Gene Ther.* 7: 79–87.

Michael, S. I., Hong, J. S., Curiel, D. T. and Engler, J. A. (1995) Addition of a short peptide ligand to the adenovirus fiber protein. *Gene Ther.* 2: 660–668.

Mittereder, N., March, K. L. and Trapnell, B. C. (1996) Evaluation of the concentration and bioactivity of adenovirus vectors for gene therapy. *J. Virol.* 70: 7498–7509.

Moolten, F. L. (1994) Drug sensitivity ("suicide") genes for selective cancer therapy. Cancer *Gene Ther.* 1: 279–287.

Mullbacher, A., Bellett, A. J. and HLA, R.T (1989) The murine cellular immune response to adenovirus type 5. *Immunol. Cell. Biol.* 67: 31–39.

Mulligan, R. C. (1993) The basic science of gene therapy. *Science* 260: 926–32.

Munaf, A., Lemoine, N. R. and Ring, J. A. (1994) The use of DNA viruses as vectors for gene therapy. *Gene Ther.* 1: 367–384.

Riordan, J. R., Rommens, J. M., Kerem, B., Alon, N., Rozmahel, R., Grzelczak, Z., Zielenski, J., Lok, S., Plavsic, N., Chou, J. L., Drumm, M. L., Iannuzzi, M. C., Collins, F. S. and Tsui, L. P. (1989) Identification of the cystic fibrosis gene: cloning and characterization of complementary DNA. *Science* 245: 1066–1073.

Roberts, RJ, O´Neill, KE, Yen, CT. (1984) DNA sequences from the adenovirus 2 genome. *J. Biol. Chem.* 259: 13968–13975.

Rommens, J. M., Iannuzzi, M. C., Kerem, B., Drumm, M. L., Melmer, G., Dean, M., Rozmahel, R., Cole, J. L., Kennedy, D., Hidaka, N.,Zsiga, M., Buchwald, M., Riordan, J. R., Tsui, L. C. and Collins, F. S. (1989) Identification of the cystic fibrosis gene: chromosome walking and jumping. *Science* 245: 1059–1065.

Rosenfeld, M. A., Siegfried, W., Yoshimura, K., Yoneyama, K., Fukayama, M., Stier, L. E., Paakko, P. K., Gilardi, P., Stratford-Pericaudet, L. D., Pericaudet, M., Jallat, S., Pavirani, A., Lecocq, J. -P. and Crystal, R. G. (1991) Adenovirus-mediated gene transfer of a recombinant alpha 1-antitrypsin gene to the lung epithelium *in vivo*. *Science* 252: 431–434.

Rosenfeld, M. A., Yoshimura, K., Trapnell,B.C., Yoneyama, K., Rosenthal, E. R., Dalemans, W., Fukayama, M., Bargon, J., Stier, L. E., Stratford-Pericaudet, L. D., Pericaudet, M., Guggino, W. B., Pavirani, A., Lecocq, J. -P. and Crystal, R. G. (1992) *In vivo* transfer of the human cystic fibrosis transmembrane conductance regulator gene to the airway epithelium. *Cell* 68: 143–155.

Roth, J. A. (1996) Modification of tumor suppressor gene expression and induction of apoptosis in non-small cell lung cancer (NSCLC) with an adenovirus vector expressing wild-type p53 and cisplatin. *Hum. Gene Ther.* 7: 1013–1030.

Rowe, W. P., Huebner, R. J., Gilmore, R. J., Parrott, R. N., Ward, T. G. (1953) Isolation of a cytpathogenic agent from human adenoids undergoing spontaneous degeneration in tissue culture. *Proc. Soc. Exp. Biol. Med.* 84: 570–573.

Seth, P., Fitzgerald, D. J., Willingham, M. C., Pastan, I. (1984) Role of a low-pH environment in adenovirus enhancement of the toxicity of a Pseudomonas exotoxin-epidermal growth factor conjugate. *J. Virol.* 51: 650–655.

Shenk, T. (1996) Adenoviridiae: The viruses and their replication. *In*: B. N. Fields and D. M. Knipe (eds), *Virology*. Raven Press, New York, pp. 2149–2171.

Smith, T. A. G., Mehaffey, M. G., Kayda D.B., Saunders, J. M., Yei, S., Trapnell, B. C., McClelland, A. and Kaleko, M. (1993) Adenovirus mediated expression of therapeutic plasma levels of human factor IX in mice. *Nat. Genet.* 5: 397–402.

Spergel, J. and Cheng-Kiang, S. (1991) Interleukin 6 enhances a cellular activity that functionally substitutes for E1a protein in transactivation. *Proc. Natl. Acad. Sci. USA* 88: 6472–6476.

Trentin, J. J., Yabe, Y, Taylor, G. (1962) The quest for human cancer viruses. *Science* 137: 835–849.

Thummel, C, Tjian, R., Hu, S. L. and Grodzicker T. (1983) Translation control of SV40 T antigen expressed from the adenovirus late promotor. *Cell* 33: 455–464.

Wohlfahrt, C. E., Svenssson, U. K. and Everitt, E. (1985) Interaction between HeLa cells and adenovirus type 2 virions neutralized by different antisera. *J. Virol.* 56: 896–903.

Wohlfahrt, C. E. (1988) Neutralization of adenoviruses: kinetics,stoichiometry, and mechanisms. *J. Virol.* 62: 2321–2328.

Wickham, T. J., Mathias, P., Cheresh, D. A. and Nemerow, G. R., (1993) Integrins alpha v beta 3 and alpha v beta 5 promote adenovirus internalization but not virus attachment. *Cell* 73: 309–319.

Wickham, T. J. Carrion, M. E. and Kovesdi, I. (1995) Targeting of adenovirus penton base to new receptors through replacement of its RGD motif with other receptor-specific peptide motifs. *Gene Ther.* 2: 750–756.

Wickham, T. J., Roelvink, P. W., Brough, D. E. and Kovesdi, I. (1996a) Adenovirus targeted to heparan containing receptors increase its gene delivery efficiency to multiple cell types. *Nature-Biotechnol.* 14: 1570–1573.

Wickham, T. J., Segal, D. M., Roelvink, P. W., Carrion, M. E., Lizonova, A., Lee, G. M. and Kovesdi, I. (1996b) Targeted adenovirus gene transfer to endothelial and smooth muscle cells by using bispecific antibodies. *J. Virol.* 70: 6831–8.

Wolff, G., Worgall, S., Rooijen, N. van, Song, W. -R., Harvey, B. -G. and Crystal, R. G. (1997) Enhancement of *in vivo* adenovirus-mediated gene transfer and expression by prior depletion of tissue macrophages in the target organ. *J. Virol.* 71: 624–629.

Worgall, S., Wolff, G., Falck-Pedersen, E. and Crystal, R. G. (1997) Innate immune mechanisms dominate elimination of adenoviral vectors following *in vivo* administration. *Hum. Gene Ther.* 8: 37–44.

Xu, M., Kumar, D., Srinivas, S., Detolla, L. J., Yu, S. F., Stass, S. A. and Mixson, A. J. (1997) Parenteral Gene therapy with p53 inhibits human breast cancer tumors *in vivo* through a bystander mechanism without evidence of toxicity. *Hum. Gene Ther.* 8: 177–185.

Yang, Y. and Wilson, J. M. (1995) Clearence of adenovirus-infected hepatocytes by MHC class-I restricted CD4+ CTLs *in vivo*. *J. Immunol.* 155: 2564–2570.

Yang, Y., Ertl, H. C. and Wilson, J. M. (1994a) MHC class-I restricted cytotoxic T lympho-cytes to viral antigens destroy hepatocytes in mice infected with E1-deleted recombinant adenoviruses. *Immunity* 1: 433–442.

Yang, Y., Nunes, F. A., Berencsi, K., Furth, E. E., Gonczol, E. and Wilson, J. M. (1994b) Cellular Immunity to viral antigens limits E1 deleted adenoviruses for gene therapy. *Proc. Natl. Acad. Sci. USA* 91: 4407–4411.

Yang, Y., Li, Q., Ertl. H.C.J. and Wilson, J. M. (1995a) Cellular and humoral immune re-sponses to viral antigens create barriers to lung-directed gene therapy with recombinant adenoviruses. *J. Virol.* 69: 2004–2015.

Yang, Y., Trinchieri, G. and Wilson J.M. (1995b)Recombinant IL-12 prevents formation of blocking IgA antibodies to recombinant adenovirus and allows repeated gene therapy to mouse lung. *Nat. Med.* 1: 890–893.

Yang, Y. Xiang, Z., Ertl, H. C. and Wilson, J. M. (1995c) Upregulation of class I major his-tocompatibility complex antigens by interferon gamma is necessary for T-cell-mediated elimination of recombinant adenovirus-infected hepatocytes *in vivo. Proc. Natl. Acad. Sci. USA* 92: 7257–7261.

Yew, P. R. and Berk, A. J. (1992) Inhibition of p53 transactivation required for transforma-tion by adenovirus early 1B protein. *Nature* 357: 82–5.

Zabner, J., Couture, L. A., Gregory, R. J., Graham, S. M., Smith, A. E. and Welsh, M. J. (1993) Adenovirus-mediated gene transfer transiently corrects the chloride transport de-fect in nasal epithelia of patients with cystic fibrosis. *Cell* 75: 207–216.

Zabner, J., Petersen, D. M., Puga, A. P., Graham, S. M., Couture, L. A., Keyes, L. D., Lukason, M. J., St.George, J. A., Gregory, R. J., Smith, A. E. and Welsh, M. J. (1994) Safety and efficiacy of repetitive adenovirus-mediated transfer of CFTR cDNA to airway epithelia of primates and cotton rats. *Nat. Genet.* 6: 75–83.

Zilli, D., Voelkel-Johnson, C., Skinner, T;. and Laster, S. M. (1992) The adenovirus E3 region 14.7 kDa protein, heat and sodium arsenite inhibit the TNF-induced release of arachi-donic acid. *Biochem. Biophys. Res. Commun.* 188: 177–183.

Zinkernagel, R. M. (1993) Immunity to viruses. *In:* W. E. Paul (ed.), *Fundamental Immunology.* Raven Press, New York, pp. 1211–1250.

Zinkernagel, R. M. (1996) Immunology taught by viruses. *Science* 271: 173–178.

Zsengeller, Z. K., Wert, S. E., Hull, W. M., Hu, X., Yei, S., Trapnell, B. C. and Whitsett, J. A. (1995) Peristence of replication-deficient adenovirus-mediated gene transfer in lungs of immune-deficient (nu/nu) mice. *Hum. Gene Ther.* 6: 457–467.

3 Receptor Mediated Gene Transfer

E. Wagner

3.1 Nonviral Transfection Systems

At present, the deciding factors for progress in somatic gene therapy are the development of efficient and safe delivery systems. Nonviral gene transfer formulations [1–3], although not reaching the efficiencies of viral vectors, have been introduced already into clinical *in vivo* gene therapies. Nonviral systems are attractive for the following reasons: (i) they may be very flexible with respect to the size of DNA to be transported; (ii) in contrast to biological systems, the entire diversity of chemical reactions and physical interactions under non-physiological conditions may be utilized for the synthesis and assembly of transfection material; (iii) plasmid DNA and synthetic transfection reagents can be produced at large scale with low costs; and (iv) safety testing of synthetic material can be far less laborious than testing of recombinant material.

Nonviral strategies include physical methods, such as microinjection, particle bombardment, electroporation or osmotic shock, which actually force the DNA into cells. Most other methods involve active cellular internalization of suitably complexed nucleic acid. Transfection agents include calcium phosphate precipitates, DEAE-dextran, liposomes of mono- and polycationic lipids [4–5], polycationic macromolecules like polyethyleneimine [6], dendrimers [7] and polypeptides such as polylysine, all of which exploit the anionic character of nucleic acids to form a transfection complex. These reagents share several functions that are essential for transfection: They condense the nucleic acid into sizes that can be taken up by cells and they protect the nucleic acid against degradation by cellular enzymes. Often the cationic carrier mediates binding of the transfection complex to the cell and also the transfer of the nucleic acid to the cytoplasm, either by fusion with the cell surface or vesicular membranes, or by disruption of vesicular membranes.

Viral vectors also utilize these types of functions, although in much more refined methods. Importantly, they utilize specific cellular receptors for uptake into specific target cells. It has been an attractive goal to incorporate these properties into synthetic vectors. This communication reviews some of the recent efforts to incorporate virus-like entry functions, such as ligands binding to specific receptors or domains responsible for the delivery across cellular membranes.

T. Blankenstein (ed.) Gene Therapy
©1999, Birkhäuser Verlag Basel

3.2   Receptor Mediated Uptake of DNA Complexes

Apart from viruses many other compounds are internalized into cells by receptor-mediated endocytosis. Cells use this process for the uptake of protein such as asialo-glycoproteins, transferrin, insulin, epidermal growth factor and small vitamins such as folic acid. The first steps in this process include binding of the ligand to specific cell-surface receptors and (in most cases) internalization through coated vesicles into endosomal acidic compartments. The subsequent pathways are dependent on the type of ligand/receptor pair: The low endosomal pH may or may not trigger dissociation of receptor and ligand and sorting processes may lead to degradative lysosomal compartments but also to the nucleus.

For targeting DNA to a given cell population, a suitable receptor-ligand has to be attached covalently or non-covalently by conjugation to a carrier. Pastan *et al.* suggested the delivery of DNA to cells by direct linkage to α_2-macroglobulin [8]. Liposomes coated with ligands for cellular binding have been successfully used for gene delivery [9–12]. Most of the described lipid-free receptor-mediated systems contain the cell-binding ligand conjugated to the polycationic carrier, polylysine, to provide a DNA-binding domain. Wu and colleagues employed complexes of either asialoorosomucoid-poly(L)lysine conjugates with DNA plasmids encoding marker genes or therapeutically relevant genes which were shown to result in gene expression, both *in vitro* in cultured HepG2 hepatoma cells [13] and *in vivo* in the liver of rats or rabbits [14–18]. Other ligands tested already include the iron-transport protein transferrin, asialoglycoproteins or synthetic galactose-containing ligands for the targeting of hepatocytes, insulin, antibodies targeted to cell surface proteins and other molecules (Tab. 3-1).

Table 3-1. Ligands used in receptor-mediated gene transfer

Ligand	Receptor	Target Cells	Selected References
Asialoglycoproteins	ASGP-R	Hepatocytes	[13–19]
Transferrin	Transferrin R	Ubiquitous, uptake of iron	[20–23]
Galactose-containing ligands	ASGP-R	Hepatocytes	[24–29]
Insulin	Insulin R	Wide range	[30–31]
Lectins	Carbohydrates	Epithelial cells, tumor cells	[32–33]
Anti-EGF-R	EGF-R	Squamous cell carcinomas	[34]
Anti-thrombomodulin	Thrombomodulin	Lung endothelial cells	[35–36]
Anti-secretory component F_{ab}	Polymeric Ig R	Lung epithelial cells	[37–38]
Anti-CD3	CD3	T-lymphocytes	[39]
Rat IgG	F_c R	Alveolar macrophages	[40]
Surfactant A and B	Variety	Pulmonary cell types	[41–42]

Receptor-mediated uptake was tested by binding studies of the DNA complex to the target receptor; enhanced gene expression after upregulation of the receptor; reduced transfection efficiency in cells without receptor; reduced expression with DNA complexes lacking the ligand, or in competition experiments with free ligand. Endocytosis has been demonstrated by the time-course and temperature-dependence of internalization into endocytic vesicles as well as by the effect of lysosomatropic agents. In several, but not all cases, a correlation between receptor density and transfection efficiency was found. For example, intensive uptake of transferrin-coated DNA complexes into the transferrin receptor-rich cell line K562 correlates with high expression levels [20]. Transferrin-mediated gene transfer into T-cell lines is not efficient, although these cells type do express the transferrin receptor. However, targeting DNA complexes to the CD3 T-cell receptor complex [39] results in vigorous internalization and high gene expression under appropriate conditions. As well as choosing the proper ligand, packaging of the DNA by a proper carrier and (active or passive) delivery across cellular membranes are crucial for successful transfection (see Fig. 3-1 and below).

3.3   Nucleic Acid Binding and Condensing Carriers

Poly(L)lysines of different sizes (approx. 20 to more than 600 lysines per polymer) have been successfully used for binding various ligands to the DNA. Neutralization of the charges of DNA with concomitant condensation [43] and stabilization of the DNA against nucleases, facilitates the endocytotic uptake and survival of DNA on the way to the nucleus of the cell. It is not known whether polylysine is replaced by cellular basic proteins (such as histones) from the DNA, either before or after nuclear entry. The use of the metabolically more stable poly(D)lysine had no effect on transfection levels [43, 44]. Obviously, release of polylysine by enzymatic degradation does not play a major role in the assembly of transfected DNA into functional chromatin. Polylysines cannot be considered biologically inert molecules. Polylysine is able to enhance pinocytosis [45], as well as phagocytosis [46]. It can activate membrane phospholipase and hence may affect membrane permeability [47]. It also activates the complement system [48] and may induce an immune response upon repeated injection [49–50]. When polylysine is incorporated into DNA complexes in amounts that result in positively charged complexes, interactions with proteins (cellular or in biological fluids) membranes (at the cell surface or of internal compartments) must be considered.

As well as polylysines, a series of other DNA-binding compounds have been used in ligand/DNA complexes: polycations such as polyethylenimine [51], polyarginine, protamine, or histones [43] and intercalative compounds like bisacridine [25] or ethidium dimers [52]. Use of the intercalators in transfections in the absence of polycations was largely unsuccessful: Essential requirements like condensation of the

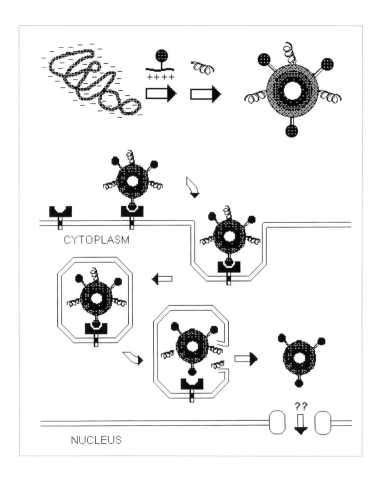

Figure 3-1. Receptor mediated DNA delivery. Negatively charged plasmid DNA is bound and condensed by a polycationic carrier molecule. The carrier molecule is linked to a receptor-binding ligand. Additional components like endosomolytic agents may be included in the DNA complex. The complex is bound to the receptor at the cell surface and internalized into endosomes. The endosomal acidification may trigger the endosomolytic agent to destabilize the lipid membrane, resulting in the release of the DNA complex into the cytoplasm. Further pathways, like the transfer of the DNA into the nucleus, remain to be characterized.

DNA and interactions with cellular phospholipids may be missing. Successful gene transfer was demonstrated using insulin as a cell binding ligand conjugated to N-acylurea albumin as a cationic carrier [30].

3.4   Delivery Across Cellular Membranes

Although complexes of DNA and polylysine-conjugated ligands can be efficiently delivered into cells, the accumulation in internal vesicles may reduce greatly the efficiency of gene transfer. For instance, transferrin or asialoglycoprotein receptor-mediated delivery into BNL Cl.2 hepatocytes results in uptake of DNA into practically all cells, but only few cells express the delivered DNA [24, 53]. The addition of the lysosomotropic agent, chloroquine, or glycerol to the transfection medium may increase transfection efficiency, probably by interfering with lysosomal degradation and enhancing the release of the DNA into the cytoplasm. In K562 cells the positive effect of chloroquine is especially strong [20]. This effect can be explained by the unusually low pH in early endosomes of K562 cells [54]. Consequently, high amounts of the weak base accumulate in the endocytic vesicles leading to the swelling and finally to the destabilization of the endosomes. However, the use of chloroquine is limited because only a limited number of cell lines respond to chloroquine, and it has toxic side effects. Incubation of cells with transferrin-polylysine/DNA or polylysine/DNA complexes in the presence of 1–1.5 M glycerol results in a strongly enhanced transfection efficiency in several cell lines and primary fibroblasts [55]. However, several cell lines, including suspension cells such as K562 and M07e, were refractory to this technique.

The inclusion of replication-defective adenoviruses has been shown to augment the levels of transferrin-mediated gene transfer up to more than 2000-fold in cell lines that express high levels of both adenovirus and transferrin receptor [56–57]. In a next step, ligand-decorated DNA complexes were provided with an endosome-destabilizing domain as a further important function of viral entry. For this purpose, (inactivated) adenovirus [32, 41, 53, 58–61] and rhinovirus [62] particles, or synthetic peptides derived from sequences of the influenza virus hemagglutinin N-terminus [24, 26, 44, 63–64] and the rhinovirus VP-1 N-terminus [62] have been attached to the DNA complex. Incorporation of these agents has been achieved by chemical or enzymatic linkage to polylysine [53, 60, 61, 65], by ionic interaction to polylysine [44], by biotinylation [59, 62] which allows subsequent binding to streptavidin-polylysine, or by an antibody bridge [58]. As a result, gene transfer to cultured cells was enhanced by up to 10 000-fold, depending on the cell type used. In T cell leukemic Jurkat cells, adenovirus-enhanced and peptide-enhanced gene transfer have similar efficiencies. In most cases the adenovirus-linked transfection complex is 5 to 100 times more efficient. Using PEI-based conjugates, high level gene expression was obtained in the absence of additional endosome-destabilizing agents [51].

3.5   Assembly of Transfection Complexes

The assembly of plasmid molecules and a large number of ligand-polycation conjugates into one supramolecular complex has to be considered as a particular critical step in receptor-mediated gene transfer. The size and modification of the polycation, the ratio of positively charged polycation to negatively charged DNA, and also the protocol of complex formation can influence the *in vitro* and *in vivo* gene transfer efficiency [3, 27, 38, 43, 66–67].

For characterization of DNA polycation complexes their electrophoretic mobility (reflects charge and size of complexes), electron microscopy (shape, size), laser light scattering (size), circular dichroism (conformation of DNA), or centrifugation technique (molecular weight and condensation) has been used. Measurements are complicated by the difficulty of generating homogenous complexes at high concentrations.

The net charge of the DNA/polycation complex affects its solubility. Complexes with either excess of polycation or DNA are stabilized in solution by the positive or negative charges. At charge ratios close to electroneutrality, hydrophobic domains of polycations like polylysine are considered responsible for low solubility in water. This may lead to aggregation and precipitation of complexes. Also disproportionation into nonstoichiometric soluble complexes and electroneutral insoluble complexes has been observed [66].

DNA/polylysine-conjugate complexes have been prepared in several ways. Wu and colleagues [13, 14] mixed the compounds at high salt concentration where electrostatic binding is greatly reduced. Reduction of the salt concentration by dialysis into physiological buffer results in a thermodynamically controlled complex formation. Charge ratios of polylysine/DNA << 1 and enhanced hydrophilicity due to the conjugated asialoglycoprotein presumably are essential for the solubility of the complex.

A different approach is described by Wagner et al. [43]: Flash mixing of rather dilute compounds in physiological phosphate-free buffer results in formation of kinetically controlled complexes. The rationale of the protocol is to prevent microheterogenicity of compound solutions and subsequently complexes. Otherwise, initial generation of unevenly (e.g. positively and negatively) charged complexes inevitably would result in the formation of aggregates. Charge ratios of polylysine/DNA of <1/2 to >2/1 have been applied. At ratios of electroneutrality or surplus of polylysine, rod-like and donut-like particles of 50 to 150 nm diameter are formed. Complexes containing transferrin-conjugated polylysine have increased solubility compared with the use of unmodified polylysine (Ogris M. and Wagner E., unpublished observation).

Interestingly, donuts of similar sizes are formed, irrespective of whether small DNA, plasmid DNA of approx. 5 kbp, or 48 kbp DNA is used in the complex formation. Using a standard expression plasmid of approx. 5 kbp, several DNA molecules are incorporated into one particle. In an attempt to generate unimolecular DNA complexes, Perales and colleagues [3, 27, 38] added polylysine conjugates slowly, in several small portions, to a vortexing solution of DNA in approx. 0.5–0.9 M salt until a charge ratio of polylysine/DNA of approx. 0.7 was reached. The protocol has been reported to generate monomeric DNA complexes with sizes of approx. 10–30 nm. These findings have been reported to be applicable for *in vivo* gene transfer applications [27, 38]. However, despite the previous encouraging efforts, major steps still have to be made to generate homogenous and stable complex formulations that mediate efficient and target-specific gene delivery.

3.6   Targeted Gene Transfer *in vitro* and *in vivo*

The concept of utilizing the specificity of a ligand-receptor interaction for targeted gene delivery is very attractive, most challenging, but contains numerous pitfalls. For instance, particular care has to be taken when polylysine is part of the DNA complexes. We observed in some cell cultures that minor changes in the DNA/polylysine-conjugate ratio of the complex (resulting in positively charged complexes) may convert a ligand-specific gene transfer into a completely unspecific process (Wagner E, unpublished data). Furthermore, positively charged DNA/polylysine complexes also strongly interact with serum proteins, such as complement factor C3, which results in inactivation of the complexes. Another important parameter for *in vivo* targeted gene transfer is the size of the complex. Phagocytotic clearance of large particles by macrophages hampers the targeting specificity of gene transfer. Large DNA particles might be applicable in administrations where the target cells are directly accessible (e.g. intravenous targetting of blood cells; local injection into soft target tissue or tumors; topical application and inhalation). Upon systemic application only very small complexes might have a chance to leave the circulation and migrate into a specific target tissue.

In vivo gene transfer using ligand-polylysine condensed DNA was reported for the first time by Wu and colleagues [14–17]. They showed that marker gene expression after i.v. injection of asialo-orosomucoid-polylysine/DNA complexes in rats was highest after 24 h. Activity then declined and after 96 h, none was detectable. However, when partial hepatectomy was performed shortly after injection of the complexes, gene expression was shown to persist for at least 11 weeks postinjection. There was no evidence of integration of the foreign plasmid DNA which appeared to persist extra-chromosomally. Using a different protocol for complex formation, Perales and colleagues [27] have recently reported prolonged expression (140 days) of a delivered gene in hepatocytes without liver surgery. In agreement with Wu and coworkers, they could not detect plasmid integration into the genomic DNA. Systemic delivery of anti-pIgR Fc-polylysine coated DNA resulted in *in vivo* expression in the lung [38].

On the other hand, intravenous application of DNA/transferrin-polylysine complexes into the tail veins of mice or rats were largely unsuccessful (Zatloukal K. and Wagner E., unpublished observations), presumably also because of physical restrictions and inactivation of the complexes by serum components. The majority of complexes were found to accumulate and be degraded by the reticuloendothelial system in liver and spleen. Only local injection of DNA or adenovirus-linked DNA/transferrin-polylysine directly into tissues or tumors produced significant gene expression (Wagner E., Kircheis R. et al., our unpublished results). Obviously, further optimization on the transfection system has to be performed in order to transfer the high *in vitro* transfection efficiency to *in vivo* applications.

3.7   Prospects of Receptor-Mediated Gene Transfer

The first versions of receptor-mediated transfection complexes have come into existence. Several functional domains involved in viral entry (see Tab. 3-2) have already been incorporated into DNA complexes and their influence on gene transfer has

Table 3-2. Examples of viral entry functions

Entry function	Virus	Viral domain, mechanism
Packaging of genome	Adenovirus, Retroviruses	Mu peptide, core particle, gag proteins
Binding to cell surface	Influenza virus	HA-1, binding to sialic acid
	Rhinoviruses	major group: ICAM, minor group: LDL-receptor
	Retroviruses	MLV: gp70-phosphate transporter HIV: gp120-CD4 of T-cells;
Internalization into the cell	Adenovirus, rhinovirus, Influenza virus, SFV	endocytosis into endosomes
	Herpes viruses, HVJ	fusion at cell surface
Release into cytoplasm	Adenovirus	endosome disruption
	Rhinovirus	formation of endosomal pore; VP-1
	Influenza virus, HIV, Sendai virus, SFV	fusion; influenza HA-2, HIV gp 41, Sendai F1, SFV E1 protein
Transfer into nucleus	Adenovirus	injection of DNA through nuclear pore
	Influenza virus	transport of RNPs into nucleus
	HIV	nuclear localization of HIV core particle
Maintenance of expression	Retroviruses	integration (integrase, LTR elements)
	Adeno-associated virus	integration (rep proteins, ITR elements)
	Herpes virus, EBV	episomal persistence (e.g. oriP, EBNA-1)

been tested. Nuclear entry and maintenance of the transfected DNA (in integrated or episomal form) in the cell are additional steps where nonviral systems may benefit from including viral elements. Specific protein domains that are responsible for active transport across the nuclear envelope have been identified, such as the SV40 large T antigen nuclear localization signal (NLS) peptide [68]. These signal peptides have been shown to carry large artificial substrates (conjugates of proteins or gold particles) from the cytoplasm to the nucleus. Insulin and some other protein ligands have been also found to be transported into the nucleus after internalization into the cell. The use of insulin for receptor-mediated delivery and nuclear transport of DNA has been reported [31]. However, at present no convincing evidence has been reported to suggest that nuclear localization domains enhance transfection efficiency. In replicating cells, the DNA may enter the nucleus during mitosis, after break-down of the nuclear envelope. Nuclear import mechanisms are considered to be more critical in nondividing cells. Retroviruses such as murine Moloney leukemia virus MLV lack specific nuclear transport domains and can infect only replicating cells, whereas the HIV core particle is actively transported into the nucleus; as a consequence, HIV can infect nondividing cells. The size of the transported nucleic acid particle may be critical. In the case of influenza virus, the viral genome is distributed over eight small RNPs that each can be translocated into nucleus. In the case of adenovirus, the capsid is too large to be translocated and accumulates at the nuclear pore: Subsequently, the DNA is released from the capsid and injected through the nuclear pore.

The expression cassette for the transgene itself must be carefully designed in terms of its control and expression. Several options are explored to achieve long-term gene maintenance. For example, retroviruses and adenovirus-associated virus stably insert their genome into the host genome. The integration mechanisms have been characterized (retrovirus: LTR sequences, integrase protein; AAV: ITR sequences, rep proteins) and may be exploited by incorporation of the corresponding nucleic acid and protein elements into a synthetic transfection complex. EBV or Herpes viruses can persist in infected cells without integrating their genome into the host. The copy number of the viral genome is controlled and maintained by a viral origin of replication and a viral nuclear retention sequence. This mechanism is utilized in episomal vector plasmids that contain the appropriate elements in the DNA sequence (e.g. EBV oriP, EBNA-1). Another option is the generation of human autonomous replicons [69] and artificial chromosomes for maintenance of large genomic sequences.

A great challenge is the assembly of the gene transfer particles in a way that the individual entry functions synergize but do not interfere with each other. Also the appropriate timing during entry into the cell is important. For instance, endosome-disruption domains should be active only after internalization of the complex and not at the cell surface. In addition, incorporation of this function should not result in large complexes that would limit specificity and *in vivo* applications. Some of the receptor-mediated transfection systems have already been found highly efficient in *ex vivo* gene transfer. Others have been applied for *in vivo* gene transfer, with first encouraging results. Further optimization will be required, but it can be expected that beside viral vectors and completely artificial methods, optimized versions of synthetic target-specific complexes will play a major role in gene therapy.

References

1. Cotten, M. and Wagner, E, (1993) Non-viral approaches to gene therapy. *Curr. Opin. Biotechnol.* 4: 705–10.
2. Ledley, F. (1994) Non-viral gene therapy. *Curr. Opin. Biotechnol.* 5: 626–36.
3. Perales, J. C. , Ferkol, T., Molas, M. et al. (1994) An evaluation of receptor-mediated gene transfer using synthetic DNA-ligand complexes. *Eur. J. Biochem.* 226: 255–66.
4. Felgner, J. H. , Kumar, R., Sridhar, C. N. et al. (1994) Enhanced gene delivery and mechanism studies with a novel series of cationic lipid formulations. *J. Biol. Chem.* 269: 2550–61.
5. Loeffler, J., Behr, J. P. (1993) Gene transfer into primary and established mammalian cell lines with lipopolyamine-coated DNA. *Methods Enzymol.* 217: 599–618.
6. Boussif, O., Lezoualc'h, F., Zanta, M. A. et al. (1995) A novel, versatile vector for gene and oligonucleotide transfer into cells in culture and *in vivo*: polyethylenimine. *Proc. Natl. Acad. Sci. USA* 92: 7297–301.
7. Haensler, J., Szoka, F. C. (1993) Polyamidoamine cascade polymers mediate efficient transfection of cells in culture. *Bioconjugate Chem.* 4: 372–9.
8. Cheng, S. Y., Merlino, G. T., Pastan IH (1983) A method for coupling of proteins to DNA: synthesis of α_2-macroglobulin-DNA conjugates. *Nucl. Acid. Res.* 11: 659–69.
9. Stavridis, J. C., Deliconstantinos, G., Psallidopoulos, M. C. et al. (1986) Construction of transferrin-coated liposomes for *in vivo* transport of exogenous DNA to bone marrow erythroblasts in rabbits. *Exp. Cell. Res.* 164: 568–72.
10. Stavridis, J. C., Psallidopoulos, M. C., Armenakas, N. A. et al. (1984) Expression of the human β-δ globin gene in rabbits. *Cell. Mol. Biol.* 30: 209–16.
11. Wang, C. Y., Huang, L. (1987) pH-sensitive immunoliposomes mediate target-cell specific delivery and controlled expression of a foreign gene in mouse. *Proc. Natl. Acad. Sci. USA* 84: 7851–55.
12. Kaneda, Y., Iwai, K., Uchida, T. (1989) Increased expression of DNA cointroduced with nuclear protein in adult rat liver. *Science* 243: 375–78.
13. Wu, G. Y., Wu, C. H. (1987) Receptor-mediated *in vitro* gene transformation by a soluble DNA carrier system. *J. Biol. Chem.* 262: 4429–32.
14. Wu, G. Y., Wu, C. H. (1988) Receptor-mediated gene delivery and expression *in vivo*. *J. Biol. Chem.* 263: 14621–24.
15. Wu, C., Wilson, J., Wu, G, (1989) Targeting genes: delivery and persistent expression of a foreign gene driven by mammalian regulatory elements *in vivo*. *J. Biol. Chem.* 264: 16985–16987.
16. Wu, G. Y., Wilson, J. M., Shalaby, F. et al. (1991) Receptor-mediated gene delivery *in vivo*. Partial correction of genetic analbuminema in Nagase rats. *J. Biol. Chem.* 266: 14338–14342.

17. Wilson, J. M., Grossman, M., Cabrerea, J. A. et al. (1992) A novel mechanism for achieving transgene persistence *in vivo* after somatic gene transfer into hepatocytes. *J. Biol. Chem.* 267: 11483–89.

18. Wilson, J. M., Grossman, M., Wu, C. H. et al. (1992) Hepatocyte-directed gene transfer *in vivo* leads to transient improvement of hypercholesterolemia in low density lipoprotein receptor-deficient rabbit. *J. Biol. Chem.* 267: 963–7.

19. Chowdhury, N. R., Wu, C. H., Wu, G. Y. et al. (1993) Fate of DNA targeted to the liver by asialoglycoprotein receptor mediated endocytosis *in vivo*: prolonged persistence in cytoplasmic vesicles after partial hepatectomy. *J. Biol. Chem.* 268: 11265–71.

20. Cotten, M., Laengle-Rouault, F., Kirlappos, H. et al. (1990) Transferrin-polycation-mediated introduction of DNA into human leukemic cells: stimulation by agents that affect the survival of transfected DNA or modulate transferrin receptor levels. *Proc. Natl. Acad. Sci. USA* 87: 4033–37.

21. Wagner, E., Zenke, M., Cotten, M. et al. (1990) Transferrin-polycation conjugates as carriers for DNA uptake into cells. *Proc. Natl. Acad. Sci. USA* 87: 3410–14.

22. Cotten, M., Wagner, E., Birnstiel, M. L. (1993) Receptor mediated transport of DNA into eukariotic cells. *Methods Enzymol.* 217: 618–44.

23. Wagner, E., Curiel, D., Cotten, M. (1994) Delivery of drugs, proteins and genes into cells using transferrin as a ligand for receptor-mediated endocytosis. *Adv. Drug Del. Rev.* 14: 113–36.

24. Plank, C., Zatloukal, K., Cotten, M. et al. (1992) Gene transfer into hepatocytes using asialoglycoprotein receptor mediated endocytosis of DNA complexed with an artificial tetra-antennary galactose ligand. *Bioconjugate Chem.* 3: 533–39.

25. Haensler, J., Szoka, F. C. (1993) Synthesis and characterization of a trigalactosylated bisacridine compound to target DNA to hepatocytes. *Bioconjugate Chem.* 4: 85–93.

26. Midoux, P., Mendes, C., Legrand, A. et al. (1993) Specific gene transfer mediated by lactosylated poly-L-lysine into hepatoma cells. *Nucl. Acid. Res.* 21: 871–8.

27. Perales, J. C., Ferkol, T., Beegen, H. et al. (1994) Gene transfer *in vivo*: sustained expression and regulation of genes introduced into the liver by receptor-targeted uptake. *Proc. Natl. Acad. Sci. USA* 91: 4086–90.

28. Merwin, J. R., Noell, G. S., Thomas, W. L. et al. (1994) Targeted delivery of DNA using YEE(GalNAcAH)3, a synthetic glycopeptide ligand for the asialoglycoprotein receptor. *Bioconjugate Chem.* 5: 612–20.

29. Wadhwa, M. S., Knoell, D. L., Young, A. P. et al. (1995) Targeted gene delivery with a low molecular weight glycopeptide carrier. *Bioconjugate Chem.* 6: 283–91.

30. Huckett, B., Ariatti, M., Hawtrey, A. O. (1990) Evidence for targeted gene transfer by receptor-mediated endocytosis: Stable expression following insulin-directed entry of *neo* into HepG2 cells. *Biochem. Pharmacol.* 40: 253–263.

31. Rosenkranz, A. A., Yachmenev, S. V., Jans, D. A. et al. (1992) Receptor-mediated endocytosis and nuclear transport of a transfecting DNA construct. *Exp. Cell. Res.* 199: 323–9.

32. Cotten, M., Wagner, E., Zatloukal, K. et al. (1993) Chicken adenovirus (CELO virus) particles augment receptor-mediated DNA delivery to mammalian cells and yield exceptional levels of stable transformants. *J. Virol.* 67: 3777–85.

33. Batra, R. K., Wang-Johanning, F., Wagner, E. et al. (1994) Receptor-mediated gene delivery employing lectin-binding specificity. *Gene Ther.* 1: 255–60.

34. Chen, J., Gamou, S., Takayanagi, A. et al. (1994) A novel gene delivery system using EGF

receptor-mediated endocytosis. *FEBS Lett.* 338: 167–9.

35. Trubetskoy, V. S., Torchilin, V. P., Kennel, S. et al. (1992) Use of N-terminal modified poly(L-lysine)-antibody conjugates as a carrier for targeted gene delivery in mouse lung endothelial cells. *Bioconjugate Chem.* 3: 323–7.

36. Trubetskoy, V. S., Torchilin, V. P., Kennel, S. et al. (1992) Cationic liposomes enhance targeted delivery and expression of exogenous DNA mediated by N-terminal modified poly(L-lysine)-antibody conjugate in mouse lung endothelial cells. *Biochim. Biophys. Acta* 1131: 311–3.

37. Ferkol, T., Kaetzel, C. S., Davis, P. B. (1993) Gene transfer into respiratory epithelial cells by targeting the polymeric immunoglobulin receptor. *J. Clin. Invest.* 92: 2394–400.

38. Ferkol, T., Perales, J. C., Eckman, E. et al. (1995) Gene transfer into the airway epithelium of animals by targeting the polymeric immunoglobulin receptor. *J. Clin. Invest.* 95: 493–502.

39. Buschle, M., Cotten, M., Kirlappos, H. et al. (1995) Receptor-mediated gene transfer into T-lymphocytes via binding of DNA/CD3 antibody particles to the CD3 protein complex. *Hum. Gene Ther.* 6: 753–61.

40. Rojanasakul, Y., Wang, L. Y., Malanga, C. J. et al. (1994) Targeted gene delivery to alveolar macrophages via Fc receptor-mediated endocytosis. *Pharmaceut. Res.* 11: 1731–6.

41. Baatz, J. E., Bruno, M. D., Ciraolo, P. J. et al. (1994) Utilization of modified surfactant-associated protein B for delivery of DNA to airway cells in culture. *Proc. Natl. Acad. Sci. USA* 91: 2547–2551.

42. Ross, G. F., Morris, R. E., Ciraolo, G. et al. (1995) Surfactant protein A-polylysine conjugates for delivery of DNA to airway cells in culture. *Hum. Gene Ther.* 6: 31–40.

43. Wagner, E., Cotten, M., Foisner, R. et al. (1991) Transferrin-polycation-DNA complexes: The effect of polycations on the structure of the complex and DNA delivery to cells. *Proc. Natl. Acad. Sci. USA* 88: 4255–59.

44. Plank, C., Oberhauser, B., Mechtler, K. et al. (1994) The influence of endosome-disruptive peptides on gene transfer using synthetic virus-like gene transfer systems. *J. Biol. Chem.* 269: 12918–24.

45. Shen, W. C., Ryser, H. J. P. (1978) Conjugation of poly-L-lysine to albumin and horseradish peroxidase: a novel method of enhancing the cellular uptake of proteins. *Proc. Natl. Acad. Sci. USA* 75: 1872–6.

46. Pruzanski, W., Saito, S. (1978) The influence of natural and synthetic cationic substances on phagocytic activity of human polymorphonuclear cells. *Exp. Cell. Res.* 117: 1–13.

47. Arnold, L. J., Dagan, A., Gutheil, J. et al. (1979) In *Proc. Natl. Acad. Sci. USA* 76: 3246–50.

48. Plank, C., Mechtler, K., Szoka, F., Wagner E. (1996) Activation of the complement system by synthetic DNA complexes: A potential barrier for intravenous gene delivery. *Hum. Gene Ther.* 7: 1437–1446.

49. Vermeersch, H., Remon, J. P. (1994) Immunogenicity of poly-D-lysine, a potential polymeric drug carrier. *J. Control. Release* 32: 225–9.

50. Findeis, M. A., Merwin, J. R., Spitalny, G. L. et al. (1993) Targeted delivery of DNA for gene therapy via receptors. *Trends Biotechnol.* 11: 202–5.

51. Kircheis, R., Kichler, A., Wallner, G. et al. (1997) Coupling of cell-binding ligands to polyethylenimine for targeted delivery. *Gene Ther.* 4: 409–418.

52. Wagner, E., Cotten, M., Mechtler, K. et al. (1991) DNA-binding transferrin conjugates as functional gene-delivery agents: synthesis by linkage of polylysine or ethidium homod-

imer to the transferrin carbohydrate moiety. *Bioconjug.Chem.* 2: 226–31.

53. Zatloukal, K., Wagner, E., Cotten, M. et al. (1992) Transferrinfection: A highly efficient way to express gene constructs in eukariotic cells. *Ann. N. Y. Acad. Sci.* 660: 136–53.

54. Sipe, D. M., Jesurum, A., Murphy, R. F. (1991) Absence of Na$^+$,K$^+$-ATPase regulation of endosomal acidification in K562 erythroleukemia cells. *J. Biol. Chem.* 266: 3469–74.

55. Zauner, W., Kichler, A., Schmidt, W. et al. (1996) Glycerol enhancement of ligand-polylysine/DNA transfection. *Biotechniques* 20: 905–913.

56. Curiel, D. T., Agarwal, S., Wagner, E. et al. (1991) Adenovirus enhancement of transferrin-polylysine-mediated gene delivery. *Proc. Natl. Acad. Sci. USA* 88: 8850–54.

57. Cotten, M., Wagner, E., Zatloukal, K. et al. (1992) High-efficiency receptor-mediated delivery of small and large (48 kb) gene constructs using the endosome disruption activity of defective or chemically inactivated adenovirus particles. *Proc. Natl. Acad. Sci. USA* 89: 6094–98.

58. Curiel, D. T., Wagner, E., Cotten, M. et al. (1992) High-efficiency gene transfer by adenovirus coupled to DNA-polylysine complexes. *Hum. Gene Ther.* 3: 147–54.

59. Wagner, E., Zatloukal, K., Cotten, M. et al. (1992) Coupling of adenovirus to transferrin-polylysine/DNA complexes greatly enhances receptor-mediated gene delivery and expression of transfected genes. *Proc. Natl. Acad. Sci. USA* 89: 6099–103.

60. Cristiano, R. J., Smith, L. J., Kay, M. A. et al. (1993) Hepatic gene therapy: Efficient gene delivery and expression in primary hepatocytes utilizing a conjugated adenovirus-DNA complex. *Proc. Natl. Acad. Sci. USA* 90: 11548–52.

61. Wu GY, Zhan, P., Sze LL et al. (1994) Incorporation of adenovirus into a ligand based DNA carrier system results in retention of original receptor specificity and enhances targeted gene expression. *J. Biol. Chem.* 269: 11542–6.

62. Zauner, W., Blaas, D., Küchler, E. et al. (1995) Rhinovirus mediated endosomal release of transfection complexes. *J. Virol.* 69: 1085–92.

63. Oberhauser, B., Plank, C., Wagner, E. (1995) Enhancing endosomal exit of nucleic acids using pH-sensitive viral fusion peptides. *In*: S. Akhtar (ed.), *Delivery Strategies for Antisense Oligonucleotide Therapeutics*. Florida: CRC Press Inc., pp. 247–268.

64. Wagner, E., Plank, C., Zatloukal, K. et al. (1992) Influenza virus hemagglutinin HA-2 N-terminal fusogenic peptides augment gene transfer by transferrin-polylysine/DNA complexes: Towards a synthetic virus-like gene transfer vehicle. *Proc. Natl. Acad. Sci. USA* 89: 7934–8.

65. Fisher, K. J., Wilson, J. M. (1994) Biochemical and functional analysis of an adenovirus-based ligand complex for gene transfer. *Biochem. J.* 299: 49–58.

66. Kabanov, A. V., Kabanov, V. A. (1995) DNA Complexes with polycations for the delivery of genetic material into cells. *Bioconjugate Chem.* 6: 7–20.

67. McKee, T. D., DeRome, M. E., Wu GY et al. (1994) Preparation of asialoorosomucoid-polylysine conjugates. *Bioconjugate Chem.* 5: 306–11.

68. Kalderon, D., Richardson, W. D., Markham, A. F. et al. (1984) Sequence requirements for nuclear location of simian virus 40 large-T antigen. *Nature* 311: 33–38.

69. Haase, S. B., Heinzel, S. S., Calos, M. P. (1994) Transcription inhibits the replication of autonomously replicating plasmids in human cells. *Mol. Cell. Biol.* 14: 2516–24.

4 Liposomes in Gene Therapy

O. Bagasra, M. Amjad., M. Mukhtar

4.1   Introduction

Recent advances in molecular and cellular biology have upgraded the idea of clinical human gene therapy from mere speculation to a clear possibility. The concept of gene therapy is about attacking human disease at the level of underlying genetic mechanisms and defects. As Aposhian stated, in 1970, "If one considers the purpose of a drug to restore the normal function of some particular process in the body, then DNA would be considered the ultimate drug" [1]. Implementation of the concept of DNA as an ultimate drug has involved, among other important aspects, development of gene delivery vehicles, optimization of gene delivery methods and studies to determine stability and expression of the desired gene. In this regard, a number of gene delivery systems, as described in earlier chapters, have been tried. Use of liposomes as a gene transfer method is the focus of this chapter but representation of their role deserves a detailed introduction. Therefore this article describes their preparation and role as a vehicle to deliver DNA molecules and the possible delivery mechanism of encapsulated materials.

Liposomes can be defined as closed vesicles, enclosing an internal aqueous compartment which is separated from the external medium by a lipid bilayer membrane. Phospholipids are components of lipid bilayer membrane, however, advances in liposomology have made it possible to employ a wide variety of lipid incorporation in the lipid bilayer. The choice of lipid incorporation in bilayer is based on their capability to enhance the properties of liposomes in terms of membrane permeability, prevention of aggregation, fusion, stability etc. Different types of liposomes have been characterized based on their size, morphology, method of preparation, composition and function [2].

4.2   Preparation of liposomes

A variety of methods have been reported to prepare liposomes. Of these, we describe a few briefly. For detailed description the reader is referred to articles by New [2, 3] and references therein. Conventional, methods of liposome preparation have

T. Blankenstein (ed.) Gene Therapy
©1999, Birkhäuser Verlag Basel

includes a) mechanical dispersion, b) solvent dispersion and C) detergent solubilization. Basically, these methods are based on the nature of components used for liposome membrane preparation, solubility of the compound to be entrapped (in terms of water or lipid soluble) and the size of liposomes generated.

4.2.1   Mechanical Dispersion Method

These methods include drying lipid onto the side of a glass container, hydrating by addition of an aqueous medium and then dispersing the swelled lipid by shaking by one of several means. This process leads to the formation of multilamellar vesicles. It is possible to control the size of liposomes by applying a specific mode of shaking. The aqueous volume enclosed within the lipid membrane is usually only a small proportion of the total volume used for swelling. In this regard, the mechanical dispersion method leads to waste of water soluble compounds to be entrapped, although the yield of captured material per ml per gram lipid may be satisfactory.

4.2.2   Solvent Dispersion Method

This method of liposome preparation, first reported by Batzri and Korn (1973), relys on solubilization of lipids comprising liposome membrane in an organic solvent and later, on mixing with aqueous phase containing material that get entrapped within the liposome [4]. This method is further sub-categorized based on the use of water-miscible or immisible organic solvent. This procedure gives rise to a high proportion of a particular size of unilamellar or multilamellar vesicles whose yield is dependent on proper mixing of lipid with solvent. The degree of encapsulation of aqueous medium depends on the type of organic solvent used. A major advantage of this method is the simplicity and low risk of degradative changes in sensitive lipids. However, the degree of solubility of lipids in organic solvent, the volume of organic solvent that can be introduced into the medium and the subsequent removal of the solvent are drawbacks of this method. Altogether, these factors result in limitation of the quantity of lipid dispersed which may lead to a low percentage of encapsulation.

4.2.3   Detergent Solubilization Method

This method is based on solubilization of lipids in detergent which is then brought into contact with the aqueous phase containing the material to be entrapped. This results in formation of micelles, and is a critical step in preparation of liposomes using this method. Size, shape and synthesis of a particular form of micelle (bilayer, lamellar or spherical) is dependent on the nature (ionic, non-ionic or amphoteric), and the concentration of detergent and lipid involved. Detergent solubilization methods are preferably used to prepare liposomes with lipophilic proteins inserted into the membrane. However, in general, detergent methods are inefficient in terms of percent entrapment.

4.3   Liposomes Can Deliver Nucleic Acids

As described above versatility of liposome preparation methods have made feasible tailoring of these structures in a variety of ways to incorporate virtually any compound for delivery to mammalian cells [5]. A number of attributes like size, surface charge, composition and bilayer fluidity of liposomes contribute toward their versatility as therapeutic agents carriers to cell. Mode of entrapment of DNA by liposomes have led to their categorization as positively charged or cationic and negatively charged or anionic liposomes. Both of these kinds been used for DNA transfer. Positively charged liposomes hold the DNA by the peripheral positive charge, whereas in negatively charged liposomes, DNA is entrapped in the aqueous interior. Negatively charged liposomes, in some cases, being able to destabilize liposomal membrane at low-pH, are also known as pH-sensitive liposomes.

4.3.1   Cationic Liposomes

Cationic liposomes usually consist of a positively charged lipid, a co-lipid and a cationic amphiphile, e.g. dioleyl phosphatidyalethanolamine (DOPE) or N-[1-(2, 3-dioleyloxy)propyl]-N,N,N,-trimethylammoniumchloride (DOTMA) first reported by Felgner et al. (1987), to transfect cells in culture [6]. Cationic amphiphiles have often been confused with cationic liposomes. The latter denomination suggests the occurrence of spherical lipid particles with an aqueous interior neither of which is needed when cationic amphiphiles are used for transfection [7]. These amphiphiles are known for efficient transfection. Because of this attribute, DOTMA has been commercialized as a transfectant (Life Technologies Gaithersburg, MD, USA). It has been reported to form lipid-DNA complexes with high loading efficiency. Use of specific amphiphiles as DNA transfer molecules have shown an efficient transfection than Ca^+ phosphate or DEAE-dextran methods [6]. These findings are based on the expression of a specific reporter genelike, chloramphenicol acetyltransferase (CAT), Luciferase (Luc) and beta-galactosidase. An additional feature which makes cationic amphiphiles and their analogues/derivatives more attractive is that they undergo degradation after transfection, suggesting reduced toxicities of these compounds. Examples of such compounds include fluorescent coumarin labeled analogues of cationic derivatives of cholesterol and diacyl glycerol such as cholesteryl (4-tri-

methylammonio) butanoate (ChoTB) and 1,2-dioleoyl-3-(4-trimethyl ammonio) bu-
tano-sn-glycerol (DOTB) [8]. A variety of other such cationic molecules have been
used for transfection and an account about usage of these and other molecules have
recently been reported by Remy et al. (1995) [7].

4.3.2   Anionic Liposomes

Anionic liposomes were developed as a means of cytoplasmic delivery of DNA and
have been classified as intrinsically pH-sensitive liposomes or external non-lipid
trigger liposomes.The pH-sensitivity of these liposomes is dependent on the combination of phosphatidylethanolamine (PE), either with one of a number of acidic
amhiphiles and titrable polymer [9, 10] or titrable synthetic peptide [11]. External
non-lipid trigger liposomes are potentially useful but successful delivery using this
type of acid-sensitivity has yet to be demonstrated.

4.4   Mechanism of DNA Delivery

As far as delivery of nucleic acids into the cells is concerned, cationic liposomes have
been employed usually for *in vitro* as well as for *in vivo* gene transfer [12]. The mechanism involves electrostatic interactions between negatively charged nucleic acids and
positively charged phospholipids in such a way that the liposome, carrying entrapped
DNA, on the whole, has a net positive charge facilitating binding to the negatively
charged cell surface [13]. Delivery of DNA to cells is therefore, dependent upon characteristics of the DNA carrying vehicle. For example, cationic amphiphiles, such as
lipopolyamines, are positively charged and have a self aggregating hydrocarbon tail
linked to a polycationic head which is able to condense DNA into discrete particles
[14]. It has been suggested that cationic lipid-coated particles may take either of two
routes reminiscent of viral infection; a) membrane fusion at the most curved edge of
the cell surface or b) spontaneous endocytosis [15]. The fate of these DNA coated
particles in the cells is unknown. Based on transcription of the gene being transfected, it appears that some of the nucleic acid must reach the nucleus [13, 16].

pH-sensitive liposomes have been suggested to be internalized by endocytosis, a
main pathway for cellular uptake. Internalization of liposomes into endosomes
takes place following their binding to the cell surface. Once formed, the endosomes
proceed to the lysosomes for degradation. Based on the time course involved endosomes have been categorized as early or late endosomes. The pH-range in early endosomes is more acidic than the external milieu and gradually becomes even more
acidic in late endosomes [17, 18].

pH-insensitive liposomes can pass through endosomes intact and hence are degraded in lysosomes [19]. Liposome-encapsulated antigens are processed in lyso-

somes, recycled and presented to T cells, whereas pH-sensitive liposomes collapse and give rise to cytoplasmic delivery of encapsulated material. Three possible ways have been suggested for cytoplasmic delivery; a) destabilization or rupture of endosome membranes, b) leakage into endosome lumen and c) fusion with endosome membrane.

Endocytosis pathways have also been used for targeted delivery, making use of specific ligands which bind to membrane-based receptors present on target cells. One such example is the delivery of asialofetuin-labeled liposomes (AF-liposomes) encapsulating plasmid pCMV beta DNA to rat hepatocytes. This delivery process was shown to become more efficient in the presence of the amphiphile DOPE [20]. Cristiano and Curiel (1996) reported usage of synthetic vectors to deliver a gene via receptor mediated endocytic pathway [21]. These vectors consist of two linked functional domains: a ligand domain and a DNA-binding domain. These two domains achieve specific binding to receptors on target cells and binding and incorporation of the transgene containing the plasmid into the vector complex, respectively. Polylysine has been one of the several candidates used to bind DNA, whereas ligands used for targeting include transferrin (TF), epidermal growth factor (EGF) and surfactant-associated protein B etc. A detailed account of receptor-mediated gene delivery is described in this volume in the 3rd chapter by Wagner et al.

4.5   Applications of pH-Sensitive Liposomes

Use of pH-sensitive liposomes to transfect mouse L-tk cells (lacking thymidine kinase) with an exogenous thymidine kinase (TK) gene have been reported (51). The plasmid, pPCTK-64, delivered by the liposomes, contains the TK gene which is under the control of a cAMP-dependent promoter, allowing controlled expression of the enzyme. This system has also been exploited for *in vivo* targeted delivery of a bacterial toxin gene to ascites [51].

There are several other applications of pH-sensitive liposomes like delivery of fluorescent dye to murine L-929 fibroblast cells [52] and delivery of the anti-tumor drug, ara-C [53–56]. Despite their potential as carriers, the poor *in vivo* stability of the pH-sensitive liposomes has been a limiting factor in their widespread use and further development.

4.6   Liposome Mediated Nucleic Acid Transfer

Liposomes have been used often for gene transfer, both *in vitro* and *in vivo*. For gene therapy purposes, cationic liposomes have shown efficient gene delivery in *in vitro* situations and encouraging gene expression *in vivo* [22–26]. Cationic liposome-DNA complexes are being used to deliver a gene for cystic fibrosis. The gene responsible for this ailment is the cystic fibrosis transmembrane conductance regulator gene CFTR [27]. A number of studies support the potential for cationic lipid-mediated gene transfer in gene therapy of cystic fibrosis in humans' [22, 28, 29]. For gene transfer into lungs, a recombinant prostaglandin G/H (PGH) gene-cationic liposome complex demonstrated protection of rabbit lungs against endotoxin injury and endotoxin-induced pulmonary hypertension [30].

Direct injection of liposome-DNA formulations into tumor masses demonstrated the feasibility of gene delivery and expression as well as generation of a cellular immune response [31]. Delivery of the human leukocyte antigen gene, HLA B7, aided by liposomes, has been reported to correct melanoma [32, 24]. In cancer gene therapy, the degree of growth inhibition of malignant cells appeared to be dependent both on the lipid to DNA ratio and the total DNA complex administered to the cells [33].

Liposome-mediated gene transfer is also targeted to various tissues such as endothelial tissue and the gastrointestinal tract. In the case of endothelial cells, catheter-mediated cationic lipid-DNA complexes showed *in vivo* transduction and expression into arterial walls [34, 35]. The possibility of gene delivery into various tissues of the gastrointestinal tract, with the help of a liposome-lacZ marker gene, reveal an opportunity to study gastrointestinal physiology and the possibility of gene therapy for these disorders [36]. It has also been suggested that liposomes might be used for delivery of genes to the liver [37]. Cationic liposomes showed an efficient transfection into primary cultured murine hepatocytes compared with other conventional methods [38].

One of the important uses of liposomes is the transfer of antisense oligonucleotides and ribozymes, inside the cell, for controlling viral or cellular gene expression. Antisense technology faces a number of problems such as degradation of oligonucleotides by nucleases and poor membrane permeability. There are a number of factors which affect the intracellular availability of antisense nucleotides to achieve antisense activity. A number of studies have demonstrated that the liposomal carrier can transfer effectively, antisense oligonucleotides inside the cell without any degradation [39, 40]. Encapsulation of oligonucleotides in antibody-targeted liposomes (immunoliposomes) have been very helpful in targeted delivery and control of HIV-1 replication in chronically infected CEM cells [41, 42]. Similarly, in ribozyme technology, potential application of liposome-ribozymes complex is another way to circumvent the problem of intracellular degradation. Delivery of anti HIV-1 drugs into one of the major viral producing tissues, i.e. lymphoid tissues, was also accomplished by utilizing liposome-encapsulated, anti-reverse transcriptase enzyme inhibitor 2', 3'-dideoxyinosine [43].

Due to potential safety hazards associated with systemic localization of delivered DNA, use of processed mRNA in gene therapeutic approaches is also being envisioned. Liposomes are one of the potential candidates for transfer of processed mRNA [44].

Certain modifications in liposomal structures have shown their enhanced transfer and targeting capacity [45, 46]. An *in vivo* direct gene transfer mediated by Haemagglutinating virus of Japan (HVJ), also called Sendai virus, and liposomes combines the characteristics of liposomes as DNA carrier and the viral (HVJ virus) fusion properties [47–49]. This modification enhances the delivery of the therapeutic gene to the host cell nucleus, avoiding the shuttling of the liposome-DNA complex through endosomes. An additional modification involves mixing of DNA with a non-histone, chromosomal high mobility group-1 protein (HMG-1). That increases the nuclear translocation and expression of the gene [50].

4.7   Conclusion

We have summarized various factors involved in the role of liposomes in gene therapy, including liposome preparation and their role as a delivery vehicle for DNA and a variety of macromolecules. We have also described the use of various categories of liposomes including cationic, anionic, their efficiency of transfection and receptor-mediated DNA delivery via endocytosis.

Conventional methods of liposome preparation suffer mainly from lower encapsulation efficiency, however, due to better transfection efficiency, cationic liposomes and cationic DNA molecule carriers, such as lipopolyamines, have been reported to be efficient compared with other categories of liposome. Although new and potentially more useful cationic DNA carrier molecules are being developed, several parameters need to be better elucidated and improved before taking advantage of the role of these molecules as a carrier for genetic material. Receptor-mediated targeted delivery of DNA molecules holds promises for the future use of liposomes. For conventional liposome preparations, several considerations such as *in vivo* targetibility, transfection efficiency and DNA carrying capacity need to be further explored.

Accomplishment of gene therapy would demand delivery of DNA molecules to different types of cells. In terms of transfection efficiency and toxicity, various studies have shown that different cell types behave differently toward a specific type of liposome. Therefore, the molecular basis of this variability needs to be studied.

There are certain unanswered questions regarding delivery of liposome mediated genetic material. Liposomes-mediated delivery of DNA into the cytoplasm is well documented but the genetic material is only active inside the nucleus. A number of recent studies have shown expression of genetic material transported by liposomes, however the complete mechanism for transfer of liposome-mediated genetic material into the nucleus needs to be unraveled so that improvements can be made to make it more efficient.

References

1. Aposhian, H. V. (1970) The use of DNA for gene therapy-the need, experimental approach, and implications. *Perspec. Biol. Med.* 14: 987–1008.
2. New, R. R. C. (1995) Influence of liposome characteristcs on their properties and fate. *In*: The liposomes. CRC Press Inc. Boca Raton, Florida, pp 3–20.
3. New, R. R. C. (1990) Preparation of liposomes. *In*: R. R. C. New (ed.), *Liposomes A Practical Approach*. IRL Press. Oxford university Press, New York, pp 3–20.
4. Batzri, S. and Korn, E. D. (1973) Single bilayer liposomes prepared without sonication. *Biochim. Biophys. Acta* 298: 1015–1019.
5. Gregoriadis, G., Florence, A. T. (1993) Liposome technology, 2nd edition, RCR press Inc. Boca Raton.
6. Felgner, P. L., Gadek, T. R., Holm, M., Roman, R., Chan, H. W., Wenz, M., Northop, J. P., Ringold, G. M. and Danielson, M. (1987) Lipofectin: A highly efficient, lipid-mediated DNA-transfection procedure. *Proc. Natl. Acad. Sci. USA* 84: 7413–7417.
7. Remy, J. S., Sirlin, C. and Behr, J. P. (1995) Gene transfer with cationic amphiphiles. *In*: The liposomes. CRC Press Inc. Boca Raton, Florida, pp. 159–170.
8. Leventis, R. and Silvius, J. R. (1990) Interaction of mammalian cells with lipid dispersions containing novel metabolizable cationic amphiphiles. *Biochim. Biophys. Acta* 1023: 124–132.
9. Tirrell, D., A.,Takigawa, D., Y. and Seki, K. (1985) pH-sensitization of phospholipid vesicles via complexation with synthetic poly(carboxylic acids). *Ann. N.Y. Acad. Sci.* 446: 237.
10. Uster, P. S. and Deamer, D. (1985) pH-dependent fusion of liposomes using titrable polycations *Biochemistry* 24: 1–8.
11. Parente, R., A., Nir, S. and Szoka, F. C. (1988) pH-dependent fusion of phosphatidylcholine small vesicles: induction by a synthetic amhiphilic peptide J.*Biol. Chem.* 263: 4724.
12. Legendre, J. -Y. and Szoka, F. C., Jr. (1992) Delivery of plasmid DNA into mammalian cell lines using pH-sensitive liposomes: Comparison with cationic liposomes. *Pharm. Res.* 9: 1235–1242.
13. Felgner, P. L. and Ringold, G. M. (1989) Cationic liposome-mediated transfection. *Nature* 337: 387–388.
14. Behr, J. P. (1986) DNA strongly binds to micelles and vesicles containing lipopolyamines or lipointercalatants. *Tet. Lett.* 27: 5861–64.
15. Haywood, A. M. (1975) 'Phagocytosis' of Sendai virus by model membranes. *J. Gen. Virol.* 29: 63–68.
16. Friend, D. S., Debs, R. J. and Duzgunes, N. (1990) Interactions between DOTMA liposomes, CV-1 and U937 cells and their isolated nuclei. 30th annual meeting of the American Society for Cell Biology, San Diego. *J. Cell Biol.* 111: 663.
17. Mellman, I., Fuchs, R. and Helenius, A. (1986) Acidification of the endocytic and exocytic pathways. *Ann.Rev. Biochem.* 55: 663–700.
18. Schmid, S. L., Fuchs, R., Male, P. and Mellman, I. (1988) Two distinct subpopulations of endosomes ionvolved in membrane recycling and transport to lysosomes. *Cell* 52: 73–83.
19. Harding, C. V., Collins, D., Slot, J. W., Geuze, H. J. and Unanue, E. R. (1991) Liposome-encapsualted antigens are processed in lysosomes, recycled and presented to T cells. *Cell* 64: 393.

20. Hara, T., Kuwasawa, H., Aramak, Y., Takada, S., Koike, K., Ishidate, K., Kato, H. and Tsuchiya, S. (1996) Effects of fusogenic and DNA-binding amphilphilic compounds on the receptor mediated gene transfer into hepatic cells by asialofetuin-labeled liposomes. *Biochem. Biophys. Acta-Biomemb.* 1278N1: 51–58.
21. Cristiano R. J., and Curiel, D. T. (1996) Strategies to accomplish gene delivery via the receptor-mediated endocytosis pathway. *Cancer Gene Ther.* 3: 49–57.
22. Canonico, A. E., Conary, J. T., Meyrick, B., Brigham, K. L. (1994) Aerosol and intravenous transfection of human alpha-1 antitrypsin gene to lungs of rabbits. *Amer. J. Respir. Cell Mol. Biol.* 10: 24–29.
23. Canonico, A. E., Plitman, J. D., Conary, J. T., Meyrick, B. O. and Brigham, K. L. (1994) No lung toxicity after repeated aerosol or intravenous delivery of plasmid-cationic liposome complexes. *J. Appl. Physiol.* 77: 415–419.
24. Nabel, E. G., Yang, Z. Y., Muller, D., Chang, A. E., Gao, X., Huang, L., Cho, K., Nabel, G. J. (1994) Safety and toxicity of catheter gene delivery to the pulmonary vasculature in a patient with metastatic melanoma. *Hum. Gene Ther.* 5: 1089–1094.
25. Yoshimura, K., Rosenfeld, M. A., Nakamura, H., Scherer, E. M., Pavirani, A., Lecocq, J. P., Crystal, R. G. (1992) Expression of the human cystic fibrosis transmembrane conductance regulator gene in the mouse lung after *in vivo* intratracheal plasmid mediated gene transfer. *Nucl. Acid. Res.* 20: 3233–3240.
26. Zhu, N., Liggitt, D., Liu, Y., Debs, R. (1993) Systemic gene expression after intravenous DNA delivery into adult mice. *Science* 261: 209–211.
27. Hyde, S. C., Gill, D. R., Higgins, C. F., Trezise, A. E. O., Macvinish, L. J., Cuthbert, A. W., Ratcliff, R., Evans, M. J., Colledge, W. H. (1993) Correction of the ion-transport defect in cystic fibrosis transgenic mice by gene therapy. *Nature* 362: 250–255.
28. Logan, J. J., Bebok, Z., Walker, L. C., Peng, S. Y., Felgner, P. L., Siegal, G. P., Frizzell, R. A., Dong, J. Y., Howard, M., Matalon, S., Lindsey, J. R., Duvall, M. and Sorscher, E. J. (1995) Cationic lipids for reporter gene and CFTR transfer to pulmonary epithelium. *Gene Ther.* 2: 38–49.
29. Stribling, R., Brunette, E., Liggitt, D., Gaensler, K., Debs, R. (1992) Aerosol gene delivery *in vivo*. Proc Natl. Acad. Sci. USA 89: 11277–11281.
30. Conary, J. T., Parker, R. E., Christman, B. W., Faulks, R. D., King, G. A., Meyrick, B. O., Brigham, K. L. (1994) Protection of rabbit lungs from endotoxin injury by *in vivo* hyperexpression of the prostaglandin G/H synthase gene. *J. Clin. Invest.* 93: 1834–1840.
31. Plautz, G. E., Yang, Z. Y., Wu BY, Gao, X., Huang, L. and Nabel GJ (1993) Immunotherapy of malignancy by *in vivo* gene transfer into tumors. *Proc Natl Acad Sci. USA* 90: 4645–4649.
32. Nabel, G. J., Nabel, E. G., Yang, Z. Y., Fox, B. A., Plautz, G. E., Gao, X., Huang, L., Shu, S., Gordon, D., Chang, A. E. (1993) Direct gene transfer with DNA liposome complexes in melanoma: expression, biologic activity and lack of toxicity in humans. *Proc. Natl. Acad. Sci. USA* 90: 11307–11311.
33. Hofland, H. and Huang, L. (1995) Inhibition of human ovarian-carcinoma cell proliferation by liposome-plasmid DNA complex. *Biochem. Biophys. Res. Commun.* 207: 492–496.
34. Nabel, E. G., Yang, Z. Y., Plautz, G., Forough, R., Zhan, X., Haudenschild, C. C., Maciag, T., Nabel, G. J. (1993) Recombinant fibroblast growth factor 1 promotes intimal hyperplasia and angiogenesis in arteries *in vivo*. *Nature* 362: 844–846.
35. Nabel, E. G., Plautz, G. and Nabel, G. J. (1992) Transduction of a foreign histocompat-

ibility gene into the arterial wall induces vasculitis. *Proc. Natl. Acad. Sci. USA* 89: 5157–5161.

36. Schmid, R. M., Weidenbach, H., Draenert, G. F., Lerch, M. M., Liptay, S., Schorr, J., Beckh, K. H. and Adler, G. (1994) Liposome-mediated *in vivo* gene-transfer into different tissues of the gastrointestinal tract. *Z. Gastroenterol.* 32: 665–670.

37. Leibiger, B., Leibiger, B., Sarrach, D., Zuhlke, H. (1991) Expression of exogenous DNA in rat liver cells after liposome mediated transfection *in vivo*. *Biochem. Biophys. Res. Commun.* 174: 1223–1231.

38. Watanabe, Y., Nomoto, H., Takezawa, R., Miyoshi, N., Akaike, T. (1994) Higly efficient transfection into primary cultured mouse hepatocytes by use of cation-liposomes: An application for immunization. *J. Biochem.* 116: 1220–1226.

39. Lappalainen, K., Urtti, A., Jaaskelainen, I., Syrjanen, K. (1994) Cationic liposomes mediated delivery of antisense oligonucleotides targeted to HPV 16 E7 messenger RNA in Caski cells. *Antivir. Res.* 23: 119–130.

40. Thierry, A. R., Dritschilo, A. (1992) Intracellular availability of unmodified, phosphorothioated and liposomally encapsulated oligodeoxynucleotides for antisense activity. *Nucl. Acid. Res.* 20: 5691–5698.

41. Zelphati, O., Imbach, J. L., Signoret, N., Zon, G., Rayner, B., Leserman, L. (1994) Antisense oligonucleotides in solution or encapsulated in immunoliposomes inhibit replication of HIV-1 by several different mechanisms. *Nucl. Acid. Res.* 22: 4307–4314.

42. Zelphati, O., Wagner, E. and Leserman, L. (1994) Synthesis and anti-HIV activity of thiocholesteryl-coupled phosphodiester antisense oligonucleotides incorporated into immunoliposomes. *Antivir. Res.* 25: 13–25.

43. Harvie, P., Desormeaux, A., Gagne, N., Tremblay, M., Poulin, L., Beauchamp, D. and Bergeron, M. G. (1995) Lymphoid tissues targeting of liposome-encapsulated 2',3'-dideoxyinosine. *AIDS* 9: 701–707.

44. Lu, D., Benjamin, R., Kim, M., Conry RM, and Curiel DT (1994) Optimization of methods to achieve mRNA-mediated transfection of tumor cells *in vitro* and *in vivo* employing cationic liposome vectors. Cancer *Gene Ther.* 1: 245–252.

45. Puyal, C., Milhaud, P., Bienvenue, A., Philippot, J. R. (1995) A new cationic liposome encapsulating genetic material: A potential delivery system for polynucleotides. *Eur. J. Biochem.* 228: 697–703.

46. Zhou, X. H., Huang, L. (1994) DNA transfection mediated by cationic liposomes containing lipopolylysine-characterization and mechanism of action. *Biochim. Biophys. Acta-Biomembranes* 1189: 195–203.

47. Kaneda, Y., Iwai, K., Uchida, T. (1989) Increased expression of DNA co-introduced with nuclear protein in adult rat liver. *Science* 243: 375–378.

48. Morishita, R., Gibbons, G. H., Ellison, K. E., Nakajima, M., Zhang, L., Kaneda, Y., Ogiara, T., Dzau, V. J. (1993) Single intraluminal delivery of antisense CDC2 kinase and proliferating cell nuclear antigen oligonucleotides results in chronic inhibition of neointimal hyperplasia. *Proc. Natl. Acad. Sci. USA* 90: 8474–8478.

49. Tomita, N., Higaki, J., Ogihara, T., Kondo, T., Kaneda, Y. (1994) A novel gene-transfer technique mediated by HVJ (Sendai virus), nuclear protein and liposomes. *Cancer Detect. Prevent.* 18: 485–491.

50. Kato, K., Nakanishi, M., Kaneda, Y., Uchida, T., Okada, Y. (1991) Expression of hepatitis B virus surface antigen in adult rat liver: co-introduction of DNA and nuclear protein by a simplified liposome method. *J. Biol. Chem.* 266: 3361–3364.

51. Wang, C. Y. and Huang, L. (1987) pH-sensitive immunoliposomes mediate target-cell-specific delivery and controlled expression of a foreign gene in mouse. *Proc. Natl. Acad. Sci. USA* 84: 7851–5.
52. Connor, J. and Huang, L. (1985) Efficient cytoplasmic delivery of a fluorescent dye by pH-sensitive immunoliposomes. *J. Cell Biol.* 101: 582.
53. Brown, P. M. and Silvius, J. R. (1990) Mechanism of delivery of liposome-encapsulated cytosine arainoside to CV-1 cells *in vitro*. Fluorescence-microscopic and cytotoxicity studies. *Biochim. Biophys. Acta* 1023: 341–351.
54. Collins, D. and Huang L. (1987) Cytotoxicity of diphteria toxin A fragment to toxin-resistant murine cells delivered by pH-sensitive immunoliposomes. *Cancer Res.* 47: 735–739.
55. Collins, D., Litzinger, D. and Huang, L. (1990) Structural and functional comparisons of pH-sensitive liposomes composed of phosphatidylethanolamine and three different diacylsuccinylglycerols. *Biochim. Biophys. Acta* 1025: 234–242.
56. Collins, D., Maxfield, F. R. and Huang, L. (1989) Immunoliposomes with different acid sensitivities as probes for the cellular endocytic pathway. *Biochim. Biophys. Acta* 987: 47–55.

5 Recombinant Adeno-Associated Virus (r AAV) Vectors

M. Hallek, C.-M. Wendtner, R. Kotin, D. Michl, E.-L. Winnacker

Abstract

Adeno-associated virus (AAV) is a single-stranded DNA dependovirus of the family of *Parvoviridae* which has promising features as a vector for somatic gene therapy. Different recombinant (r) AAV vectors have been generated which seem to have some advantages in comparison with other vectors, like the ability to transduce terminally differentiated and non-dividing cells, the lack of any apparent pathogenicity, a low immunogenicity, a relatively high stability of transgene expression and the potential of targeted integration. Recent improvements of rAAV packaging now allow the generation of sufficient quantities of rAAV for clinical trials. Pre-clinical studies with rAAV are currently performed not only for the treatment of a variety of inherited, monogenic defects such as β-thalassemia, sickle cell anemia, Fanconi anemia, chronic granulomatous disease, Gaucher disease, metachromatic leukodystrophy, or cystic fibrosis, but also for acquired diseases like infection with HIV and non-Hodgkin's lymphoma. The diversity of these studies indicates that rAAV might have a broad range of clinical applications. A first clinical trial with rAAV vectors has been started for cystic fibrosis. While several important issues including safety, tissue tropism and methods to achieve site-specific integration need further clarification, rAAV seems to have a sufficient number of advantages to be considered seriously as a future gene therapy vector.

5.1 Introduction

For the majority of the current clinical gene therapy trials, replication-incompetent viral vectors, derived from retroviruses and adenoviruses, are used [1]. Adeno-associated virus (AAV) is a newer, promising viral vector which seems able to transduce terminally differentiated and non-dividing cells including bone-marrow progenitor cells, alveolar stem cells, bronchial epithelial cells and glial cells at relatively high frequency [2, 3]. In 1994 the first clinical trial with an AAV vector containing the human CFTR gene was approved by the Recombinant DNA Advisory Committee (RAC) of the US National Institutes of Health (NIH) for the treatment of cystic fibrosis [2]. Excellent reviews on the biology and genetics of AAV have been published [2–8]. Therefore, the main focus of this article is to review latest improvements and potential clinical applications of rAAV vectors.

5.2   Biology of Wild Type (wt) AAV

AAV was discovered initially as a co-infecting agent during an adenovirus outbreak, with no apparent pathogenicity contributed by AAV [9–11]. To date, no apparent human disease caused by AAV has been detected. On the contrary, AAV seems to be protective against bovine papillomavirus and adenovirus (Ad) mediated cellular transformation [12–14]

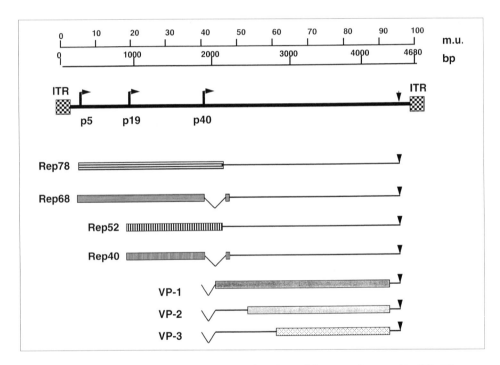

Figure 5-1. Map of the AAV genome (adapted with modifications from Kotin [3]). The AAV genome encompasses 4680 bp, divided into 100 map units (mu). Indicated are the two inverted terminal repeats (ITRs), the three viral promoters at map position 5 (p5), 19 (p19) and 40 (p40) and the common polyadenylation signal at map position 96 (vertical arrow). The open reading frames are represented by rectangles, untranslated regions by solid lines and the introns by carats. Large Rep proteins (Rep78 and Rep68) under control of the p5 promoter and small Rep proteins (Rep52 and Rep40) driven by the p19 promoter exist in spliced and unspliced isoforms. The cap genes encoding three different capsid proteins (VP-1, VP-2, VP-3) are under control of the p40 promoter.

AAV requires co-infection with a helper virus for replication: Either adenovirus or Herpesviridae (Epstein-Barr virus, herpes simplex virus 1 and 2, cytomegalovirus) can function as a helper for AAV productive infection [7, 15, 16]. The linear single-stranded AAV genome is 4680 nucleotides and contains two open reading frames encoding structural and non-structural proteins (Fig. 5-1). The symmetrical organization of the terminal 145 nucleotides of the genome permit the ends to base pair intramolecularly, into a Y- or T-shaped arrangement. These inverted terminal repeats (ITR) are the only *cis* element required for replication, encapsidation and targeted integration of the AAV genome. The non-structural protein products of the *rep* genes, *Rep* 68 and *Rep* 78, regulate AAV DNA replication, AAV gene expression and mediate targeted integration of AAV DNA into chromosome 19 (*v.i.*). In the absence of helper virus co-infection, *Rep* 68 or *Rep* 78 regulates AAV gene expression and replication negatively. Thus, AAV is not defective for replication in the sense that a complementing gene product is provided by the helper virus. *Rep* 68 and *Rep* 78, acting as part of a molecular switch, turn-off AAV gene expression and replication under non-permissive conditions. The presence of a helper virus reverses the inhibitory effects of the *Rep* proteins on AAV gene expression. *Rep* 68 and *Rep* 78 biochemical activities include sequence specific DNA binding, sequence and strand specific endonuclease and ATP dependent DNA helicase [17–20]. The ITR is a substrate for *Rep* and all three activities are involved in a process called terminal resolution which allows the linear AAV genome to replicate. The *cap* genes encode the capsid proteins which exist in three different isoforms, VP-1, VP-2, and VP-3. While the mechanisms of AAV replication are understood to a certain extent, less is known about the viral entry mechanism or the intracellular transport of virions after infection. Recently, binding studies with [^{125}I]-labelled AAV particles allowed the identification of a 150-kDa cell membrane protein believed to be the putative cellular receptor of AAV [21].

5.3 Recombinant AAV Vectors

Some of the qualities of AAV desirable for gene therapy include:
- Non-pathogenicity in humans [3, 10].
- Broad host range of wtAAV should allow delivery of foreign DNA to a wide range of mammalian cells [22].
- Stable expression of the transgene as shown in animal models [23].
- Low risk of insertional mutagenesis, due to the weak activation of downstream genomic sequences or intrinsic promoters by the ITRs [24]. Since the transcriptional elements within the ITRs are oriented towards the viral genome and not toward the cellular DNA, it has been presumed that the transcriptional activation of cellular oncogenes is unlikely to occur [3].
- Relatively easy purification and concentration of the virus due to the stable,

naked icosahedral structure [25].

- Infection of non-dividing cells including hematopoietic progenitor cells and post-mitotic neurons [26–28].
- Low immunogenicity of rAAV, therefore, long-term expression of transgenes and repeated infections of target cells are possible [23, 25].

The first rAAV vectors were constructed by cloning the AAV2 genome into bacterial plasmids, replacing part of the *cap* gene with a transgene under the control of the AAV p40 promoter. The capsid gene was provided by co-transfection of a wild-type AAV genome and constructs were packaged with full length wtAAV [29]. However, the packaging of these constructs produced high amounts of wtAAV and the cloning capacity of the vectors was limited. A major breakthrough has been the design of two packaging plasmids, psub201 and pAAV/Ad [29, 30]: psub201 contains unique restriction sites which were engineered inboard and outboard of the ITR to facilitate insertion of the transgene and thus, completely deletes both AAV ORFs. The helper plasmid pAAV/Ad contains both the *rep* and *cap* gene driven by adenoviral terminal repeats instead of AAV ITRs, thus, reducing the probability of homologous recombination and the generation of wtAAV [29, 30]. Since the AAV ITRs can easily form hairpin structures which are unstable in bacteria, Srivastava and colleagues constructed a packaging plasmid, pWN-1, in which the cloning sites for the transgene were separated from the AAV-ITRs, thus increasing the stability in bacterial cells [31].

5.3.1 Recent Improvements of rAAV Packaging

The titers of recombinant viral particles produced, using the above described packaging protocols, were too low for clinical gene therapy. Substantial progress has been made with the development of improved packaging systems to allow the production of higher titer recombinant vectors yielding titers of 3×10^{11} particles/ml (Tab. 5-1). One of these newer packaging systems is characterized by a helper plasmid, containing the SV40 origin of replication, thereby generating high copy numbers of the *rep* and *cap* genes in the presence of SV40 large T-antigen provided by the packaging cell line COS7 [25]. With this new helper plasmid pSVoriAAV, the yield of rAAV particles was increased approximately sixty fold over a non-replicating helper plasmid. In addition, the yield of rAAV particles produced per packaging cell was increased to more than one thousand particles per cell compared with approximately one rAAV particle per cell described in conventional packaging systems [25, 32]. Other new packaging systems, using adenovirus-polylysine-DNA complexes instead of calcium phosphate or electroporation mediated DNA transfer with subsequential adenovirus infection of the packaging cell line, could also slightly increase virus titers (to $\approx 3 \times 10^6$ transducing units/ml) [33].

Packaging of rAAV vectors is still cumbersome because two separate plasmids, i.e. the transgene and the helper plasmid, need to be transfected into packaging cells and subsequent infection with adenovirus is required. To overcome this major obstacle, several groups have tried to establish packaging cell lines by stable or inducible expression of *rep* and *cap* genes. However, the generation of packaging cell lines constitutively expressing *rep* and *cap* genes has been hampered by the cytotoxic or cytostatic properties of the *Rep* protein [34–36]. Some cell lines providing AAV *Rep* functions were not efficient enough, producing low rAAV titers in the range of 10^3 infectious particles per ml [37, 38]. Cell lines expressing *Rep*78, under the control of the glucocorticoid-responsive mouse mammary tumor virus promoter, showed no cytotoxic effects of the *Rep* protein, induced efficient rAAV DNA replication and initiated the assembly of rAAV particles although the particles were noninfectious. This defect could be overcome by transient, high-level expression of individual *Rep* proteins [39]. Presumably, the assembly of infectious rAAV requires higher levels of *Rep* than those provided by the inducible promoters. Recently, Clark and colleagues described a simplified method of generating rAAV with packaging cell lines stabily transfected with a tripartite plasmid containing the AAV *rep* and *cap* genes driven from their internal promoters, a neomycin selection marker as well as the ITR-flanked transgene [40]. In this cell line, rAAV is produced by infecting solely the packaging cell line with adenovirus yielding titers up to 1.2×10^{10} particles per ml starting from 4×10^8 packaging cells. Thus, the yield of this procedure is approximately 30 viral particles per packaging cell. It should be pointed out that HeLa cells were used to package rAAV: The HeLa cells used for packaging had the disadvantage to induce only low levels of *Rep* expression, because they lack the adenoviral E1A gene which regulates the AAV p5 and p19 promoters. Nevertheless, this study shows that it is feasible to generate *Rep*-expressing helper cell lines for the production of rAAV vectors. However, it remains to be shown that sufficiently high levels of *rep* expression can be achieved, with these packaging cell lines, to produce reasonable amounts of infectious rAAV particles.

Another strategy to decrease the number of 'hits' for successful rAAV packaging has been the constant expression of rAAV genomes on an EBV episome in 293 packaging cells. Therefore, infectious rAAV particles could be produced by only two 'hits': transfection of the helper plasmid coding for *rep* and *cap* and infection with adenovirus [41]. A similar technique to improve the packaging efficiency was the generation of cell populations containing stably integrated and rescuable rAAV genomes in 293 cells [42]. The packaging cell lines were subsequently transfected with the helper plasmid, wherein the *rep* gene was driven from the human immunodeficiency virus (HIV) long terminal repeat (LTR) promoter instead of the endogenous AAV p5 promoter, since the expression from the HIV promoter would not be inhibited by the *Rep* proteins themselves as in the case of the AAV p5 promoter [43]. These improvements yielded a total particle titer of near 2×10^{11}/ml, beginning with 2×10^7 packaging cells, thus, representing a yield of 10^4 particles per cell [42].

Table 5-1. Recent improvements to achieve high-yield rAAV packaging

Cell line	total cell number	genomic particle titer	infectious particle titer	genomic particle: infectious particle ratio	genomic particles per packaging cell	Reference
HeLa	4×10^8	1.2×10^{10}	1.2×10^8	100: 1	30	40
HEK 293	2×10^7	2×10^{11}	1×10^{10}	20: 1	1×10^4	75
COS-7	2×10^8	2×10^{11}	1×10^9	200: 1	1×10^3	25

Another major advance would be the generation of a helper virus free packaging system. Colosi *et al.* transiently transfected, into the packaging cells, the adenovirus genes necessary for AAV gene expression, DNA replication and packaging [44]. For this purpose, 293 cells were transfected with adenovirus E2a, E4 and VA genes. Apparently, this method allowed the packaging of rAAV at an efficiency comparable with standard procedures of helper virus infection during packaging. No wild type adenovirus contamination was seen after packaging [44].

5.3.2 Tissue Specificity of rAAV Vectors

An important characteristic and consideration for the use of any viral vector system is its natural tropism. For both wtAAV and rAAV, only limited studies are published (Tab. 5-2). Flotte and colleagues tested the *in vivo* distribution of a rAAV-p5lacZ vector in mice after intraperitoneal application. In these studies, rAAV seemed to be tropic for the upper respiratory tract and the non-lymphoid areas of the spleen, while in other organs like the pancreas, no expression was detectable. However, this tissue distribution was due to specificity of the promoter rather than to distinct DNA distribution [42]. In another study, murine bone marrow cells, transduced with rAAV containing the lacZ gene, were injected into lethally irradiated mice and the tissue distribution of rAAV was monitored. One hour after bone marrow transplant, rAAV was detected in the liver: At a later time point, rAAV was seen in other organs like bone marrow, kidney and spleen [45]. Topic (stereotactic or bronchoscopic) applications of rAAV into the brain or lungs showed high local transgene expression (Tab. 5-2). Importantly, preliminary data from studies in Rhesus monkeys did not provide evidence for rAAV transduction in the gonads [46]. Nevertheless, more comprehensive studies of the natural tropism of rAAV as well as of rAAV and tissue targeting using a different promoter are necessary.

Table 5-2. Tissue distribution of recombinant AAV vectors after in vivo application in animal models

Tissue	Transgene	Species	Expression level (absent, low, intermediate, high)	Reference
bone marrow	lacZ	mouse	+	45
brain (caudate nucleus, amygdala, striatum, hippocampus)		rat	high	28
brain (striatum)	tyrosine hydroxylase	rat	high	28
bronchial epithelial	CFTR	Rhesus macaques	high	46
bronchial epithelium	CFTR lacZ	rabbit rabbit	high high	63, 65
kidney	lacZ	mouse	intermediate	42
kidney	lacZ	mouse	+	45
liver	lacZ	mouse	low	42
liver	lacZ	mouse	high	45
lung (airways)	lacZ	mouse	high	42
lung	CAT	rat	low	76
Pancreas	lacZ	mouse	absent	42
Peritoneum	lacZ	mouse	absent	42
spleen (non-lymphoid)	lacZ	mouse	high	42
spleen (lymphoid follicles)	lacZ	mouse	absent	42
spleen	lacZ	mouse	+	45

+ = expression observed, but data on expression preliminary.

5.3.3   Clinical Applications of rAAV Vectors

1. Hematologic diseases and gene transfer into hematopoietic progenitor cells

The potential of rAAV to transduce both non-dividing quiescent cells and dividing cells is still a controversial issue. It was reported that rAAV infected non-dividing cells at almost equivalent efficiency compared with actively proliferating cells [26]. It was hypothesized that rAAV virions may traverse through nuclear pores, due to their small size, without the need of mitosis-mediated nuclear membrane dissolution for nuclear localization. However, since gene transcription requires a double-stranded DNA template, the single-stranded virion DNA has to be converted into duplex form. Accordingly, Russell and colleagues have found that AAV vector genomes can enter and persist in stationary phase cells although transduction preferentially occurs in cells that have already entered S phase, because of the requirement of polymerase for the single to double strand conversion [47].

Whatever the mechanism of transduction in non-dividing cells, rAAV seems to be an attractive vector to target stem cell-like populations. rAAV allowed transduction of murine bone marrow cells with a neomycin resistance gene (NeoR) [48] and of human CD34+ hematopoietic progenitor cells with the lacZ reporter gene [27]. Furthermore, Miller and colleagues accomplished the transduction of human progenitor-derived erythroid cells with the human γ-globin gene [49]. Therefore, rAAV vectors are potential tools for the therapy of human hemoglobinopathies like sickle cell disease and β-thalassemia (Tab. 5-3) [50]. For both diseases, the γ-globin gene transcription seems to be critical, resulting in an increased fetal hemoglobin synthesis and consequently in a more efficient erythropoiesis and decreased hemolysis [51]. β-thalassemia is characterized by inadequate β-globin chain synthesis. In heterozygous individuals, β-globin production is decreased, relative to α-globin, without lethal consequences (thalassemia minor). In homozygous individuals, β-globin synthesis is impaired to the extent that only poorly hemoglobinized, and thus defective, erythrocytes are produced. This results in hemolysis, anemia and early death of untreated patients (thalassemia major) [52]. Homozygous **sickle cell anemia** is caused

Table 5-3. Use of recombinant AAV vectors for ex vivo gene transfer

Cell type or cell lines	Species	Transgene	Promoter	Ref.
hematopoietic progenitor cells				
CD34+ progenitor cells	human	γ-globin	CMV	27
"	human	FACC	RSV	54
"	Rhesus macaques	γ-globin	LCR site II	77
"	mouse	neo^R	HSVtk	48
lymphoid cells				
EBV-transformed B cells and T cell clones	human	neomycin phosphotrans-ferase (neo^R)	?	78
B cells (CGD)	human	p47 phox	CMV	55
LP-1, RPMI 8226 (multiple myeloma)	human	B7-1, B7-2	CMV	25, 67

U937 (monocytoid)	human	neomycin phospho-trans ferase	murine sarcoma virus (MSV) LTR	79
NC37 (lymphoblastoid)	human	neo^R	MSV LTR	79
myeloid cells				
K562 (erythroleukemia)	human	γ-globin	HS-2	24
"	human	anti-α-globin	SV40 α-globin	80
"	human	neo^R	MSV LTR	79
KG1, HEL, HL60 (myeloid leukemia)	human	neo^R	MSV LTR	79
MO7e (megakaryoblastic leukemia)	human	GMCSF	SV40	81
MEL (erythroleukemia); 293	murine human	β-globin	HS2-4	53
bronchial and airway epithelial cells				
IB3 (CF bronchial epithelium)	human	CFTR	AAV p5	63
primary CF nasal polyps	human	CFTR	AAV p5	23
primary CF nasal polyps; primary neonatal foreskin	human	alkaline phosphatase, lacZ	LTR	82
cells of the urogenital tract				
HEK 293 (embryonic kidney cells)	human	anti-sense HIV	RSV-LTR	83
"	human	neo^R	MSV LTR	79
COS7 (kidney)	green monkey	GMCSF	SV40	81
R3327 (prostate); MBT-2 (bladder)	rat	IL-2	CMV	84
Other cell types				
GM0877 (primary fibroblast from a type 2 Gaucher disease patient)	human	glucocerebrosidase	SV40	57
MLD 577γ (primary cell line from patient with metachromatic leukodystrophy)	human	arylsulfatase	SV40	57
BLKCL.4 (fibroblasts)	murine	neo^R	MSV LTR	79

neo^R = neomycin phosphotransferase; MSV LTR = murine sarcoma virus long terminal repeat.

by a mutant hemoglobin (HbS, $\alpha_2\beta_2$S) and results in hemoglobin polymerization with subsequent vaso-occlusive disorders and multi-organ damage. HbF ($\alpha_2\gamma_2$), and thus γ-globin, gene transfer has an inhibiting effect on polymerization and HbS precipitation within erythroid cells. The efficient transduction and expression of the γ-globin gene, linked to a globin regulatory sequence, of which the locus control region (LCR), HS2 site, is located 5' to the ϵ-globin gene, was accomplished *in vitro* using the erythroleukemic cell line K562 [24]. It was shown that the rAAV-transduced γ-globin gene exhibited normal regulation upon hemin induction of erythroid maturation and was expressed at high levels, nearly equivalent to native γ-globin gene expression. In a recent study, the potential of rAAV vectors coding for a human β-globin gene together with DNaseI hypersensitive sites of the LCR was shown both in hematopoietic and non-hematopoietic cells [53].

5.3.3.1

Fanconi anemia is characterized by pancytopenia and a predisposition to malignancy due to chromosomal instability [54]. With rAAV-mediated expression of the FACC (Fanconi anemia complementation group C) gene in human peripheral blood progenitor cells, a normal DNA repair with consecutive phenotypic correction and survival of progenitor colonies could be achieved [54]. **Chronic granulomatous disease** (CGD) represents another inherited hematopoietic disorder and is characterized by a defect of the phagocyte-specific NADPH-oxidase. Using immortalized B cell lines of these patients, a successful gene transfer of the p47phox gene accounting for nearly a third of CGD mutations was accomplished by rAAV transduction [55].

5.3.3.2 Neurologic diseases

There are interesting applications for the use of rAAV vectors in neurology because they seem to transduce terminally differentiated neural tissue. Kaplitt et al. showed that rAAV transduced post-mitotic brain tissue for up to three months [28]. A rodent model of Parkinson´s disease was established by unilaterally denervating the striatum of rats and then challenging the animals with dopaminergic agents which resulted in asymmetrical rotation, related to the striatal dopamine deficit. Stereotactic injection of rAAV vectors, coding for tyrosine hydroxylase, into denervated striatum resulted in significant behavioural recovery. Therefore, it is anticipated that AAV vectors, expressing the tyrosine hydroxylase gene, may be useful for somatic gene therapy of Parkinson´s disease. Many other applications, including stereotactic targeting of brain tumors with AAV vectors coding for suicide genes, seem feasible in the near future.

5.3.3.3 Metabolic diseases

Lysosomal storage diseases are another target for somatic gene therapy [56]. Expression of defective genes, encoding lysosomal enzymes, results in pathogenic storage of their substrates within the lysosomes. The most common lysosomal storage disease is *Gaucher disease*, characterized by a deficiency of glucocerebrosidase. Another examined disorder is *metachromatic leukodystrophy* caused by deficiency of arylsulfatase A activity. Both lysosomal storage diseases are thought to be treated by enabling the expression of sufficient levels of normal enzyme in order to overcome the congenitally missing enzyme function. Recombinant AAV vectors were used to supplement the glucocerebrosidase and arylsulfatase A gene in murine and human primary fibroblasts, resulting in significant and long-lasting expression *in vitro* [57]. Clinical studies making use of this technology are expected in the near future.

5.3.3.4 Cystic fibrosis

Cystic fibrosis (CF) is currently the only clinical application for which a clinical trial with AAV vectors has been started. CF is a common, autosomal recessive disease caused by mutations in the gene encoding the CF transmembrane conductance regulator (CFTR) [58, 59]. The CFTR is a Cl channel that is regulated by cAMP-dependent phosphorylation and by intracellular nucleotides [60]. Mutations in the CFTR cause a loss of function in the CFTR Cl channels which results in a defective cAMP-mediated Cl transport across affected epithelia [60]. Somatic gene therapy aims to correct the defect by overexpressing the normal CFTR cDNA in the airways [61]. In a first step, successful targeting of airway epithelial cells by rAAV was demonstrated using an AAVp5neo reporter gene vector [62]. Using the IB3-1 CF-defective bronchial epithelial cell line, transduction of the CFTR gene, subcloned into a rAAV vector, was achieved. This effect was associated with the correction of the small linear chloride conductance and the outwardly rectifying chloride channel, both being critical for the CF defect [63, 64]. Furthermore, rAAV-CFTR particles were instilled into lungs of rabbits using a fiberoptic bronchoscope: This resulted in transgene expression for up to six months after the application, without detectable side effects from gene transfer [23]. In a recent study, AAVlacZ particles were applied via tracheobronchoalveolar lavage into newborn rabbit lungs and showed that gene expression occurred, preferentially, in alveoli in the alveolar epithelial progenitor cells with no apparent inflammatory side effects [65]. In another animal model, the CFTR gene was transduced by recombinant AAV particles into lungs of Rhesus macaques using a similar endobronchial administration route and long-lasting expression of the CFTR gene product was achieved without evidence of inflammation or toxicity [46]. The first clinical trial using AAV-based vector technology was started in the USA in September 1994 after RAC approval [2].

5.3.3.5 Neoplasia

Although cancer currently represents the most frequent target of gene therapy trials [1, 66] rAAV vectors have not been used frequently for this purpose. Our group has used rAAV vectors to transduce the T cell co-stimulatory molecules B7-1 and B7-2 into lymphoid tumor cell lines which were negative for these two cell surface antigens: High efficiency gene transfer was achieved (up to 80% of cells after two days). rAAV-transduced, B7-1 or B7-2 positive tumor cells elicited a severalfold higher T cell response than untransduced or NeoR transduced controls [67].

5.4   Future Perspectives

As with any vector system, rAAV will not be the ideal gene vehicle for *all* gene therapy applications. Moreover, many issues still require further clarification and improvement. At present, the predominant limitation of the rAAV vectors are size constraints of approximately 4.7 kb for the transgene which exclude the development of recombinant vectors for important diseases like Duchenne´s muscular dystrophy and make other interesting applications like F.VIII or IX more difficult. Efforts should be made to overcome these cloning constraints, although solutions to this problem are certainly not trivial. The production of sufficient titers and amounts of rAAV for large animal protocols remains technically challenging. Substantial improvements in vector production have been reported recently but all remain labor intensive and costly.

Another important topic will be the safety of rAAV compared with other viral vectors. Retroviral systems based on Moloney murine leukemia virus were shown to induce T cell leukemia and lymphomas in mice and in primates [68]. This was most probably due to wild-type retrovirus generation during the packaging procedure resulting in insertional mutagenesis as the pathogenic mechanism. So far, no pathology could be attributed to wtAAV but the potential for toxicity of rAAV and wtAAV has not been unequivocally established. Since rAAV probably does not integrate site-specifically, the possibility of insertional mutagenesis remains. However, as outlined above, the AAV ITRs have only weak intrinsic promoter activity for cellular sequences because the transcriptional elements within the ITRs are oriented towards the viral but not the cellular genome. It is possible that *rep*-deleted rAAV vectors seem to persist in an episomal state rather than integrating unspecifically into the host genome [69]. Further, yet unclarified, safety concerns about AAV-based gene therapy have been raised regarding both horizontal transmission of the virus due to the high stability of the virion and *in vivo* rescue of recombinant vectors, if the rAAV-infected recipient is infected simultaneously with wtAAV and adenovirus [2].

Another potential problem, in particular for immunocompromised hosts, would be contamination with wild-type adenovirus required for the generation of rAAV in standard packaging protocols. So far, adenovirus contaminants cannot be separated from rAAV particles without a significant loss in rAAV titer because most techniques use heat to inactivate adenovirus. Due to the different density and size of both viruses, it should be possible to achieve adenovirus-free AAV preparations with no loss in rAAV titer by refined separation techniques. There are newer packaging systems which use replication-defective adenovirus, which incorporates a rAAV vector in its E1 region, while rAAV particles can be rescued in the E1-complementary 293 cell line without adding wt adenovirus [55]. The above mentioned packaging method by Colosi and colleagues, which use 293 cells transfected with the essential adenoviral genes to package rAAV, may eventually replace all wt adenovirus helper functions and allow adenovirus-free packaging [44].

Large scale production is still cumbersome and too inefficient for clinical applications. Production of efficient packaging cell lines, stably expressing *Rep*, was hampered by the propensity of *Rep* proteins to suppress cell growth. In most packaging protocols, it is still necessary to co-transfect packaging cells with the transgene and the packaging plasmid and to co-infect it with the helpervirus, i.e. adenovirus. One interesting approach to overcome these limitations is the use of adenovirus-polylysine-DNA complexes which allow the simultaneous delivery of rAAV vector, AAV complementation vector and adenovirus [33].

Finally, rAAV might have the unique potential for targeted integration into a defined region of human chromosome 19q13.3-qter, if the mechanism of targeted integration is better understood [70, 71]. The capability of targeted integration may provide safer, prolonged expression of the transgene. Recently, characterized activities of *Rep* 68 and *Rep* 78 have established that the DNA binding specificity and endonuclease actvities are not restricted to the hairpin conformer of the AAV ITR. DNA binding is sequence specific, however, endonuclease activity is much less efficient on non-hairpinned substrates [72, 73]. *Rep* binds not only to a GCTC repeat motif in AAV ITRs but also to a specific region at the AAVS1 locus, thus mediating a complex formation between heterologous DNA substrates [20]. Further support for the involvement of Rep proteins in targeted integration come from *in vitro* results demonstrating that uni-directional DNA synthesis is initiated from a cloned chromosome 19 subfragment in the presence of *Rep* 68 or *Rep* 78 [73]. Based on the biochemical activities of *Rep* proteins and analysis of proviral structures, it is likely that AAV targeted integration occurs by a copy-choice recombination process [3]. Studies are under way which use this integration promoting ability of *Rep* while preventing the toxic and growth-suppressing effects of *Rep*, e.g. moderate exposure of cells to *Rep* by liposomes loaded with *Rep* during the infection with rAAV or similar techniques enabling simultaneous expression of *Rep* during the infection procedure [74]. This would not only diminish the risk of insertional mutagenesis but also enhance stable expression of the transgene.

In conclusion, rAAV represents a promising gene transfer system for somatic gene therapy. While several issues, including safety and efficacy of packaging of rAAV, need further clarification and refinement, this vector system has a sufficient number of advantages to be seriously considered as a future gene therapy vector. The first clinical AAV-based gene transfer studies, in patients with cystic fibrosis, will show whether or not first generation rAAV vectors are sufficiently safe and efficient to promote further efforts and research on this interesting viral vector.

Acknowledgements

We would like to thank Dr. B. Safer, Professor P.C. Scriba, Professor B. Emmerich, and Dr. P. Heinrich for their continuous support, as well as our coworkers Dr. C. Bogedain, Dr. J. Chiorini, M. Braun-Falco, Dr. R. Buhmann, A. Doenecke, C. Heberger, Dr. G. Maass, E. Mangold, Dr. A. Nolte, U. Scheer, and R.Schilling for help and stimulating discussions. Work presented in this review has been supported in part by grants of the Wilhelm Sander-Stiftung (to M.H., E. -L. W., and C. M. W.) and of the NIH (to R.K.).

References

1. Marshall, E. M. (1995) Gene therapy's growing pains. *Science* 269: 1050–1055.
2. Flotte, T. R., Carter, B. J. (1995) Adeno-associated virus vectors for gene therapy. *Gene Therapy* 2: 357–362.
3. Kotin, R. (1994) Prospects for the use of adeno-associated Virus as a vector for human gene therapy. *Human Gene Ther* 5: 793–801.
4. Berns, K. I. (1990) Parvovirus replication. *Microbiol Rev* 54: 316–329.
5. Berns, K. I. (1991) Parvoviridae and their replication. *In*: B. N. Fields and D. M. Knipe (eds), *Fundamental Virology*. Raven Press, New York, pp. 817–37.
6. Carter, B. J. (1992) Adeno-associated virus vectors. *Curr. Opin. Biotechnol.* 3: 533–539.
7. Muzyczka, N. (1992) Use of adeno-associated virus as a general transduction vector for mammalian cells. *Curr. Topic. Microbiol. Immunol.* 158: 97–129.
8. Samulski, R. J. (1993) Adeno-associated virus: integration at a specific chromosomal locus. *Curr. Opin. Genet. Develop.* 3: 74–80.
9. Blacklow, N. R. (1988) Adeno-associated viruses of humans. *In*: J. Pattison (ed.), *Parvoviruses and Human Disease*. CRC Press, Boca Raton, FL, pp. 165–174.
10. Blacklow, N. R., Hoggan, M. D., Sereno, M. S., Brandt, C. D., Kim, H. W., Parrott, R. H., Chanock, R. M. (1971) A seroepidemiologic study of adenovirus-associated virus infections in infants and children. *Am. J. Epidemiol.* 94: 359–366.
11. Blacklow, N. R., Hoggan, M. D., Kapikian, A. Z., Austin, J. B., Rowe, W. P. (1968) Epidemiology of adeno-associated virus infection in a nursery population. *Am. J. Epidemiol.* 89: 368–78.
12. Hermonat, P. L. (1989) The adeno-associated virus Rep78 gene inhibits cellular transformation by bovine papilloma virus. *Virology* 172: 253–261.
13. Mayor, H. D., Houlditch, G. S., Mumford, D. M. (1973) Influence of adeno-associated virus on adenovirus-induced tumors in hamsters. *Nature* 241: 44–46.
14. Khlief, S. N., Myers, T., Carter, B. J., Trempe, J. P. (1991) Inhibition of cellular transfor-

mation by the adeno-associated virus rep gene. *Virology* 181: 738–741.

15. Buller, R. M. L., Janik, J. E., Sebring, E. D., Rose, J. A. (1981) Herpes simplex virus type I and II completely help adeno-virus associated virus replication. *J. Virol.* 40: 241–7.

16. McPherson, R. A., Rosenthal, L. J., Rose, J. A. (1985) Human cytomegalovirus completely helps adeno-associated virus replication. *Virology* 147: 217–222.

17. Im, D. -S., Muzyczka, N. (1990) The AAV origin binding protein REP68 is an ATP-dependent site-specific endonuclease with DNA helicase activity. *Cell* 61: 447–457.

18. Chiorini, J. A., Weitzman, M. D., Owens, R. A., Urcelay, E., Safer, B., Kotin, R. M. (1994) Biologically active rep proteins of adeno-associated virus type 2 produced as fusion proteins in Escherichia coli. *J. Virol.* 68: 797–804.

19. Chiorini, J. A., Yang, L., Safer, B., Kotin, R. M. (1995) Determination of adeno-associated virus Rep68 and Rep78 binding sites by random sequence oligonucleotide analysis. *J. Virol.* 69: 7334–7338.

20. Weitzman, M. D., Kyöstiö, S. R. M., Kotin, R. M., Owens, R. A. (1994) Adeno-associated virus (AAV) Rep proteins mediate complex formation between AAV DNA and its integration site in human DNA. *Proc. Natl. Acad. Sci. USA* 91: 5808–5812.

21. Mizukami, H., Muramatsu, S., Young, N. S., Brown, K. F. (1995) Adeno-associated virus type 2 binds to a 150 kilodalton cell membrane glycoprotein (abstr.). *Blood* 86 (Suppl. 1): 240a.

22. Hermonat, P. L., Muzyczka, N. (1984) Use of adeno-associated virus as mammalian DNA cloning vector: transduction of neomycin resistance into mammalian tissue culture cells. *Proc. Natl. Acad. Sci. USA* 81: 6466–6470.

23. Flotte, T. R., Afione, S. A., Conrad, C., McGrath, S. A., Solow, R., Oka, H., Zeitlin, P. L., Guggino, W. B., Carter, B. J. (1993) Stable *in vivo* expression of the cystic fibrosis transmembrane conductance regulator with an adeno-associated virus vector. *Proc. Natl. Acad. Sci. USA* 90: 10613–10617.

24. Walsh, C. E., Liu, J. M., Xiao, X., Young, N. S., Nienhuis, A. W., Samulski, R. J. (1992) Regulated high level expression of a human γ-globin gene introduced into erythroid cells by an adeno-associated virus vector. *Proc. Natl. Acad. Sci. USA* 89: 7257–7261.

25. Chiorini, J. A., Wendtner, C. M., Urcelay, E., Safer, B., Hallek, M., Kotin, R. M. (1995) High-efficiency transfer of the T cell co-stimulatory molecule B7-2 to lymphoid cells using high-titer recombinant adeno-associated virus vectors. *Hum. Gene Ther.* 6: 1531–41.

26. Podsakoff, G., Wong, K. J., Chatterjee, S. (1994) Efficient gene transfer into non-dividing cells by adeno-associated virus-based vectors. *J. Virol.* 68: 5656–5666.

27. Goodman, S., Xiao, X., Donahue, R. E., Moulton, A., Miller, J., Walsh, C., Young, N. S., Samulski, R. J., Nienhuis, A. W. (1994) Recombinant adeno-associated virus mediated gene transfer into hematopoietic progenitor cells. *Blood* 84: 1492–1500.

28. Kaplitt, M. G., Leone, P., Samulski, R. J., Xiao, X., Pfaff, D. W., O'Malley, K. L., During, M. J. (1994) Long-term gene expression and phenotypic correction using adeno-associated virus vectors in the mammalian brain. *Nat. Genet.* 8: 148–154.

29. Samulski, R. J., Berns, K. I., Tan, M., Muzyczka, N. (1982) Cloning of adeno-associated virus into pBR322: rescue of intact virus from the recombinant plasmid in human cells. *Proc. Natl. Acad. Sci. USA* 79: 2077–2081.

30. Samulski, R. J., Chang, L. -S., Shenk, T. (1987) A recombinant plasmid from which an infectious adeno-associated virus genome can be excised *in vitro* and its use to study viral replication. *J. Virol.* 61: 3096–3101.

31. Nahreini, P., Woody, M. J., Zhou, S. Z., Srivastava, A. (1993) Versatile adeno-associated

virus 2-based vectors for constructing recombinant virions. *Gene* 124: 257–262.

32. Samulski, R. J., Chang, L. -S., Shenk, T. (1989) Helper-free stocks of recombinant ade-no-associated viruses: normal integration does not require viral gene expression. *J. Virol.* 63: 3822–3828.

33. Mamounas, M., Leavitt, M., Yu, M., Wong-Staal, F. (1995) Increased titer of recombi-nant AAV vectors by gene transfer with adenovirus coupled to DNA-polylysine com-plexes. *Gene Ther.* 2: 429–432.

34. Mendelson, E., Smith, M. G., Miller, I. L., Carter, B. J. (1988) Effect of a viral rep gene on transformation of cells by an adeno-associated virus vector. *Virology* 166: 612–615.

35. Caillet-Fauquet, P., Perros, M., Brandenburger, A., Spegelaere, P., Rommelaere, J. (1990) Programmed killing of human cells by means of an inducible clone of parvoviral genes encoding non-structural proteins. *EMBO J.* 9: 2989–2995.

36. Yang, Q., Chen, F., Trempe, J. P. (1994) Characterization of cell lines that inducibly ex-press the adeno-associated virus rep proteins. *J. Virol.* 68: 4847–4856.

37. Vincent, K. A., Moore, G. K., Haigwood, N. L. (1990) Replication and packaging of HIV envelope genes in a novel adeno-associated virus vector system. *In: Vaccines 90.* Cold Spring Harbor Laboratory Press, Cold Spring Harbor, pp. 353–359.

38. Trempe, J. P., Yang, Q. (1993) Characterization of a cell line that expresses the AAV repli-cation proteins (abstr.). Proceedings *Fifth Parvovirus Workshop* 5-3. Crystal River, FL.

39. Hölscher, C., Hörer, M., Kleinschmidt, J. A., Zentgraf, H., Bürkle, A., Heilbronn, R. (1994) Cell lines inducibly expressing the adeno-associated virus (AAV) rep gene: re-quirements for productive replication of rep-negative AAV mutants. *J. Virol.* 68: 7169–7177.

40. Clark, K. R., Voulgaropoulou, F., Fraley, D. M., Johnson, P. R. (1995) Cell lines for the production of recombinant adeno-associated virus. *Hum. Gene Ther.* 6: 1329–1341.

41. Lebkowski, J. S., McNally, M. A., Okarma, T. B. (1992) Production of recombinant ade-no-associated virus vectors. *United States Patent* 5: 414.

42. Flotte, R. T., Barraza-Ortiz, X., Solow, R., Afione, S. A., Carter, B. J., Guggino, W. B. (1995) An improved system for packaging recombinant adeno-associated virus vectors capable of *in vivo* transduction. *Gene Ther.* 2: 29–37.

43. Beaton, A., Palumbo, P., Berns, K. I. (1989) Expression from the adeno-associated virus p5 and p19 promoters is negatively regulated *in trans* by the *rep* protein. *J. Virol.* 63: 4450–4454.

44. Colosi, P., Elliger, S., Elliger, C., Kurtzman, G. (1995) AAV vectors can be efficiently pro-duced without helper virus (abstr.). *Blood* 86 (Suppl.1): 627a.

45. Ponnazhagan, S., Wang, X. -S., Srivastava, A., Yoder, M. C. (1995) Adeno-associated virus 2-mediated gene transfer and expression in murine hematopoietic progenitor cells *in vivo* (abstr.). *Blood* 86 (Suppl 1): 240a.

46. Flotte, T. R., Conrad, C., Reynolds, T., Afione, S., Adams, R., Allen, S., Guggino, W. B., Carter, B. J. (1995) Preclinical evaluation of AAV vectors expressing the human CFTR cDNA (abstr.). *J. Cell. Biochem.* 21A: 364.

47. Russell, D. W., Miller, A. D., Alexander, I. E. (1995) Adeno-associated virus vectors pref-erentially transduce cells in S phase. *Proc. Natl. Acad. Sci. USA* 91: 8915–8919.

48. Zhou, S. Z., Broxmeyer, H. E., Cooper, S., Harrington, M. A., Srivastava, A. (1993) Adeno-associated virus 2-mediated gene transfer in murine hematopoietic progenitor cells. *Exp. Hematol.* 21: 928–933.

49. Miller, J. L., Donahue, R. E., Sellers, S. E., Samulski, R. J., Young, N. S., Nienhuis, A. W.

(1994) Recombinant adeno-associated virus (rAAV)-mediated expression of a human gamma-globin gene in human progenitor-derived erythroid cells. *Proc. Natl. Acad. Sci. USA* 91: 10183–7.

50. Walsh, C. E., Liu, J. M., Miller, J. L., Nienhuis, A. W., Samulski, R. J. (1993) Gene therapy for human hemoglobinopathies. *Proc. Soc. Exp. Biol. Med.* 204: 289–300.

51. Perrine, S. P., Ginder, G., Faller, G. V., Dover, G. H., Ikuta, T., Witkowska, H. E., Cai, S. -P., Vichinsky, E. P., Oliveri, N. F. (1993) A short-term trial of butyrate to stimulate fetal-globin-gene expression in the β-globin disorders. *N. Engl. J. Med.* 328: 81–86.

52. Weatherall, D. J., Clegg, J. B. (1981) *The thalassemia syndromes*. Blackwell, London.

53. Einerhand, M. P. W., Antoniou, M., Zolotukhin, S., Muzyczka, N., Berns, K. I., Grosveld, F., Valerio, D. (1995) Regulated high-level human β-globin gene expression in erythroid cells following recombinant adeno-associated virus-mediated gene transfer. *Gene Ther.* 2: 336–343.

54. Walsh, C. E., Nienhuis, A. W., Samulski, R. J., Brown, M. G., Miller, J. L., Young, N. S., Liu, J. M. (1994) Phenotypic correction of Fanconi anemia in human hematopoietic cells with a recombinant adeno-associated virus vector. *J. Clin. Invest.* 94: 1440–1448.

55. Thrasher, A. J., Alwis Md, Casimir CM, Kinnon, C., Page, K., Lebkowski, J., Segal AW, Levinsky RJ (1995) Generation of recombinant adeno-associated virus (rAAV) from an adenoviral vector and functional reconstitution of the NADPH-oxidase. *Gene Ther.* 2: 481–485.

56. Kay, M. A., Woo SLC (1994) Gene therapy for metabolic disorders. *Trends Genet.* 10: 253–7.

57. Wei, J. -F., Wei, F. -S., Samulski, R. J., Barranger, J. A. (1994) Expression of the human glucocerebrosidase and arylsulfatase A genes in murine and patient primary fibroblasts transduced by an adeno-associated virus vector. *Gene Ther.* 1: 261–268.

58. Boat, T. F., Welsh, M. J., Beaudet al. (1989) Cystic fibrosis. *In*: C. R. Scriver, A. L. Beaudet, W. S. Sly, D. Valle (eds), *The Metabolic Basis of Inherited Disease*. McGraw-Hill, Inc., New York, pp. 2649–2680.

59. Riordan, J. R., Rommens, J. M., Kerem, B. -S., Alon, N., Rozmahel, R., Grzelczak, Z., Zielenski, J., Lok, S., Plavsic, N., Chou, J. -L., Drumm, M. L., Iannuzzi, M. C., Collins, R. S., Tsui, L. -C. (1989) Identification of the cystic fibrosis gene: cloning and characterization of complementary DNA. *Science* 245: 1066–73.

60. Riordan, J. R. (1993) The cystic fibrosis transmembrane conductance regulator. *Ann. Rev. Physiol.* 55: 609–630.

61. Alton EWFW, Geddes, D. M. (1995) Gene therapy for cystic fibrosis: a clinical perspective. *Gene Therapy* 2: 88–95.

62. Flotte, T. R., Solow, R., Owens, R. A., Afione, S., Zeitlin, P. L., Carter, B. J. (1992) Gene expression from adeno-associated virus vectors in airway epithelial cells. *Am. J. Respir. Cell. Mol. Biol.* 7: 349–356.

63. Flotte, T. R., Afione, S. A., Solow, R., Drumm, M. L., Markakis, D., Guggino, W. B., Zeitlin, P. L., Carter, B. J. (1993) Expression of the cystic fibrosis transmembrane conductance regulator from a novel adeno-associated virus promoter. *J. Biol. Chem.* 268: 3781–3790.

64. Egan, M., Flotte, T., Afione, S., Solow, R., Zeitlin, P. L., Carter, B. J., Guggino, W. B. (1992) Defective regulation of outwardly rectifying Cl channels by protein kinase A corrected by insertion of CFTR. *Nature* 358: 581–584.

65. Zeitlin, P. L., Chu, S., Conrad, C., McVeigh, U., Ferguson, K., Flotte, T. R., Guggino,

W. B. (1995) Alveolar stem cell transduction by an adeno-associated viral vector. *Gene Ther.* 2: 623–31.

66. Culver, K. W., Blaese, R. M., Gene therapy for cancer. *Trends Genet.* 1994) 10: 174–178.
67. Wendtner, C. M., Mangold, E., Braun-Falco, M., Chiorini, J. A., Krause, A., Emmerich, B., Scriba, P. C., Winnacker, E. -L., Kotin, R. M., Hallek, M. (1995) Efficient, stable and functional expression of costimulatory molecules B7-1 (CD80) and B7-2 (CD86) in human lymphoma cells by an improved adeno-associated virus (AAV) vector (abstr.). *J. Molec. Med.* 73:B13–14.
68. Donahue, R. E., Kessler, S. W., Bodine, D., McDonagh, K., Dunbar, C., Goodman, S., Agricola, B., Byrne, E., Raffeld, M., Moen, R., Bacher, J., Zsebo, K. M., Nienhuis, A. W. (1992) Helper virus induced T cell lymphoma in nonhuman primates after retroviral mediated gene transfer. *J. Exp. Med.* 176: 1125–1135.
69. Flotte, T. R., Afione, S. A., Zeitlin, P. L. (1994) Adeno-associated virus vector gene expression occurs in nondividing cells in the absence of vector DNA integration. *Am. J. Respir. Cell. Mol. Biol.* 11: 517–21.
70. Kotin, R. M., Siniscalco, M., Samulski, R. J., Zhu, X., Hunter, L., Laughlin, C. A., McLaughlin, S., Muzyczka, N., Rocchi, M., Berns, K. I. (1990) Site-specific integration by adeno-associated virus. *Proc. Natl. Acad. Sci. USA* 87: 2211–2215.
71. Kotin, R. M., Menninger, J. C., Ward, D. C., Berns, K. I. (1991) Mapping and direct visualisation of region-specific viral DNA integration site on chromosome 19q13-qter. *Genomics* 10: 831–834.
72. Chiorini, J. A., Wiener, S. M., Owens, R. A., Kyostio, S. R. M., Kotin, R. M., Safer, B. (1994) Sequence requirements for stable binding and function of adeno-associated virus 2 Rep68 protein on the ITR. *J. Virol.* 68: 7448–7457.
73. Urcelay, E., Ward, P., Wiener, S. M., Safer, B., Kotin, R. M. (1995) Asymmetric replication *in vitro* from a human sequence element is dependent on adeno-associated virus Rep protein. *J. Virol.* 69: 2038–2046.
74. Natsoulis, G., Surosky, R., Godwin, S., McQuiston, S., Kurtzman, G. (1995) A system for specifically targeting genes to chromosome 19 (abstr.). *Blood* 86 (Suppl 1): 236a.
75. Flotte, R. T., Barraza-Ortiz, X., Solow, R., Afione, S. A., Carter, B. J., Guggino, W. B. (1995) An improved system for packaging recombinant adeno-associated virus vectors capable of *in vivo* transduction. *Gene Ther.* 2: 29–37.
76. Flotte, T. R. (1993) Prospects for virus-based gene therapy for cystic fibrosis. *J. Bioenerg. Biomembr.* 25: 37–42.
77. Miller, J. L., Walsh, C. E., Samulski, R. J., Young, N. S., Nienhuis, A. W. (1993) Transfer and expression of the human γ-globin gene in purified hematopoietic progenitor cells from Rhesus bone marrow using a recombinant adeno-associated virus (rAAV) vector (abstr.). Proceedings *Fifth Parvovirus Workshop* Crystal River, FL.
78. Muro-Cacho, C. A., Samulski, R. J., Kaplan, D. (1992) Gene Transfer in human lymphocytes using a vector based on adeno-associated virus. *J. Immunother.* 11: 231–37.
79. Lebkowski, J. S., McNally, M. M., Okarma, T. B., Lerch, L. B. (1988) Adeno-associated virus: a vector system for efficient introduction and integration of DNA into a variety of mammalian cell types. *Mol. Cell. Biol.* 8: 3988–96.
80. Ponnazhagan, S., Nallari, M. L., Srivastava, A. (1994) Suppression of human α-globin gene expression mediated by the recombinant adeno-associated virus 2-based vectors. *J. Exp. Med.* 179: 733–738.
81. Luo, F., Zhou, S. Z., Cooper, S., Munski, N. C., Boswell, H. S., Broxmeyer, H. E.,

Srivastava, A. (1993) Adeno-associated virus 2-mediated transfer and functional expression of a gene encoding the human granulocyte-macrophage colony-stimulating (abstr.). *Blood* 82 (Suppl. 1): 303a.

82. Halbert, C. L., Alexander, I. E., Wolgamot, G. M., Miller, A. D. (1995) Adeno-associated virus vectors transduce primary cells much less efficiently than immortalized cells. *J. Virol.* 69: 1473–1479.

83. Chatterjee, S., Johnson, P. R., Wong, J. K. K. (1992) Dual-target inhibition of HIV-1 *in vitro* by means of an adeno-associated virus antisense vector. *Science* 258: 1485–1488.

84. Philip, R., Brunette, E., Kilinski, L., Murugesh, Deepa, McNally, M. A., Ucar, K., Rosenblatt, J., Okarma, T. B., Lebkowski, J. S. (1994) Efficient and sustained gene expression in primary T lymphocytes and primary and cultured tumor cells mediated by adeno-associated virus plasmid DNA complexed to cationic liposomes. *Mol. Cell. Biol.* 14: 2411–2418.

6 Gene Transfection Using Particle Bombardment

H. Tahara, T. Kitagawa, T. Iwazawa, M. T. Lotze

Abstract

A number of non-viral methods have been applied to transfer foreign genes into mammalian somatic tissues or organs using non-viral vectors. Among these techniques, particle-mediated gene transfer (PMT) or "gene gun" has been developed and promising results have been reported in the last few years. Other approaches include use of naked DNA injection (particlularly into muscle), protein coupled DNA to target cellular receptors, calcium phosphate transfection, electroporation and liposomal gene delivery. Depending in part on the cell type, PMT appears to be the most efficient of these techniques with transfection efficiencies of 1–15%. In this article, we summarize the characteristics of this approach, present our preliminary results and discuss possible future application of this gene transfer technology.

6.1   Introduction

Viral vectors (retrovirus, adenovirus, and adeno-associate virus) have been the predominant means of gene delivery in clinical protocols to date. In these studies, safety has been confirmed in many settings including the direct delivery of adenoviral vectors and the application of retroviral producer cell lines directly into the human brain to achieve *in vivo* transduction. However, to ensure safety in the conduct of these protocols, extensive testings of the viral reagents to be used and close monitoring of the patients treated has been required by regulatory agencies. These risks include the possibility of insertional mutagenesis and developing replication competent retroviral (RCR) after recombination. Beside the safety issues, retroviral vectors have some shortcomings including their incapability of transducing non-proliferating cells and low efficiency of transduction when used *in vivo*.

To overcome these problems, a number of methods have been applied to transfer foreign genes into mammalian somatic tissues or organs using non-viral vectors. These include the direct injection of naked plasmid DNA into muscles [1], delivery of DNA complexes with specific protein carriers [2], co-precipitation of DNA with calcium phosphate [3], and conjugation of DNA with various form of liposome formulations [4, 5]. Along with these techniques, particle-mediated gene transfer (PMT) or "gene gun" has been developed and some promising results have been reported in recent years, particularly in vaccination protocols. In this article, we will summarize the characteristics and discuss future applications of this gene transfer strategy.

6.2   Design of gene gun – How does it work?

This technique of PMT was developed initially to accomplish genetic transformation of various crop plants [6–10]. Soon after this technology was demonstrably successful in plant gene transfection, it was applied to mammalian somatic tissues [11]. The investigators used a high-voltage electric discharge device to be used for the acceleration of DNA-coated gold particles into living cells [7,8]. With this device, a motive force was generated to accelerate the DNA-coated gold particles to high velocity, enabling efficient penetration of target cells, tissues, or organs in culture as well as the tissue or organs *in vivo*.

6.2.1   Generation of a Motive Force

To generate a motive force, two approaches have been utilized, 1) the electric shock wave; and 2) later with a helium-pulse.

6.2.1.1 Electric gene gun

The DNA-coated particles were suspended in ethanol and placed evenly onto a thin Mylar film. The Mylar sheet was then placed adjacent to two closely spaced electrodes, where an electric arc, generated by a high-voltage discharge, provided the motive force. The force used to accelerate the DNA-coated particles to high velocity could be partially controlled by varying the discharge voltage (3–25 kV)

6.2.1.2 Helium-pulse gene gun

The DNA-coated particles were suspended in ethanol and placed evenly onto the inner surface of the plastic tube, and were delivered into the tissue by the helium gas pulse (150–400 psi). Although the "electric" gene gun enables fine control of the motive force, the helium gas pulse gene gun has been the prevailing device of choice lately because of the ease of handling when compared with the electric gene gun. Thus, the results described in this manuscript were obtained largely using the helium gas pulsed gene gun unless noted otherwise. This device is now available from BioRad.

6.2.2 Coating of Gold Beads with DNA

For both types of gene guns, plasmid DNA was precipitated onto 1.5 to 3.0 µm gold particles (beads) which are chemically inert in most, if not all, biological systems. The size of the bead determines the mass weight. Therefore, the larger bead, which has a larger mass weight, delivers a greater motive force when the same acceleration is applied. Thus a higher penetration force for the gold beads used in gene gun is related to bead size. We are currently using the gold beads with about 2.0 µm in size which have been shown to be useful for getting high expression of the transfected gene (un-published data).

6.2.3 Conversion of Motive Force into Particle Projection (Electric or Helium-Driven)

When the electric gene gun was used, the DNA-coated particles were then re-suspended in ethanol and deposited evenly onto a thin Mylar film. This Mylar sheet was placed over the exhaust port of an explosion chamber and a shock wave was generated by a high voltage electric discharge. The projected Mylar sheet was stopped by a stainless steel screen, while the associated DNA-gold particles continued to be projected through the meshwork toward the target cells. The DNA-coated gold particles then penetrated the cell or tissues effectively, due to this force.

When the helium-driven gene gun was used, particles were suspended in a solution of 0.1 mg of polyvinyl pyrrolidone per ml in absolute ethanol. This DNA/gold particle preparation was coated onto the inner surface of a Tefzel tubing using a tube loader, and the tubing was cut into 0.5-inch segments. Then, each segment contained 1.25 µg plasmid DNA coated onto 0.5 mg gold beads. These gold beads, adhering to the inner surface of the tube, were driven by the helium pressure released through the tube. Metal mesh was placed between the bullet-containing tube and the target to distribute the gold beads over at the target cells or tissue.

6.2.4 Adjustment of Velocity and Resulting Distribution of Beads

These adjustments could be performed by varying; 1) the discharge voltage of the electric device or the pressure of the helium gas pulse, 2) the bead density, and 3) the bead size. We usually adjust bead density to 2×10^6 beads (0.1 mg) per cm^2 after performing optimization experiments and have used this for various applications without major modification with consistent results. Bead size has been modified to optimize the depth of the penetration balancing with the damage to the cells. Gold beads ranging from 1.5 to 3.0 μm are currently used for these purposes.

6.3   Expression and Efficiency of Gene Gun Transfection

Similar to other non-viral gene transfer techniques, PMT provides transient transfection of expressed genes. The frequency of stable gene transfer was estimated in CHO cells and MCF-7 cells to be 1.7×10^{-3} and 6×10^{-4}, respectively. However, these numbers can be compared favorably with transfection frequencies commonly obtained with other DNA-mediated gene transfer methods including electroporation, calcium phosphate, and liposomal delivery [11]. Optimal conditions for gene gun transfection has been examined by our group and others [11, 12]. Although the level of expression and efficacy of gene gun transfection are not significantly different in different types of target cells, optimization of conditions, especially the motive force, of transfection is recommended for each cell type.

6.3.1   *In vitro* transfection

Cells in culture conditions, adherent or non-adherent cell lines, or cells freshly isolated from tissue, can be efficiently transfected using this PMT. Transfection efficiency and expression after transfection have been reported previously [11], and also examined by our group (unpublished observation).

6.3.1.1 Efficiency

Efficiency of this transfection strategy has been reported previously in detail [11]. In this study, MCF-7 cells, from a human mammary carcinoma cell line, were transfected with PMT using expression plasmids containing an *Escherichia coli* β–galactosidase (β–Gal) gene sequence driven by cytomegalovirus (CMV) immediate early promoter or mouse mammary tumor virus (MMTV) promoter. Under their conditions, only 3–5% of MCF-7 cells exhibited β–Gal activity following transfection with MMTV-β–Gal, but 25% of MCF cells exhibited β–Gal activity when transfected with CMV-β–Gal. Thus, it was shown that the efficiency of PMT is higher than the level of expression of the transgene which depends on the promoter driving the gene. In freshly isolated cells derived from the mammary gland of rats, β–Gal activity was observed in 3–4% of the cells after transfection.

6.3.1.2 Expression

Optimization of transfection for higher expression was performed using various murine cell lines including NIH3T3 (mouse fibroblast cell line), MC38 (mouse colon cancer cell line), and MCA205 (mouse sarcoma cell line). Figure 6-1A shows the IL-12 expression from MC38 after PMT with IL-12 expression plasmid under different conditions using the helium driven gene gun. IL-12 expression after PMT with 200 psi was about 20 ng/10^6 cells/48 h and it was higher with 300 psi or more. When the cells were transfected with a pressure of 500 psi or more, the expression levels remained comparable with those at 300 psi but damage to the cells were more apparent. Thus, we concluded that the maximal transgene expression from MC38 can be obtained when the helium gas pressure is 300 psi. Optimization of the PMT conditions were performed using other cell lines including MCA205 (Fig. 6-1B) and

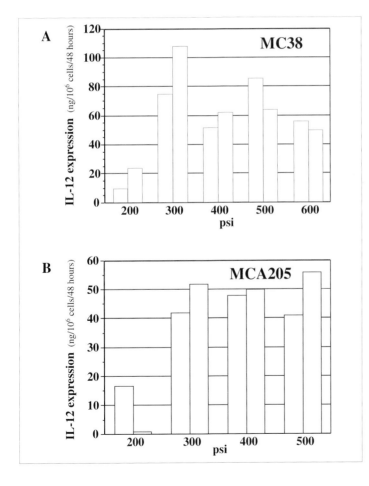

Figure 6-1. Effective cytokine gene expression after particle mediated gene transfer (PMT or gene gun). A: MC38, mouse colon cancer cell line, as a target. Expression of mIL-12 was measured with ELISA specific for heterodimeric mouse IL-12. Levels of mIL-12 expression were calculated and the total amount of mIL-12 expressed from 10^6 target cells in 48 h after PMT of mIL-12 expression plasmid. B: MCA205, mouse sarcoma cell line, as a target.

NIH3T3. These results showed that the pressure needed for the optimal expression in these cells was similar to that of the MC38 tumor cells.

Most of the clinical protocols using gene transfected tumor cells, used radiated tumor cells, mainly due to the concern for unlimited proliferation of gene modified cells. To investigate whether such radiation can affect the expression after PMT, target cells were irradiated before or after PMT and the transgene expression was examined (Fig. 6-2). Expression of mIL-10 transgene was not reduced significantly, in any of the conditions, in the first 48 h, but rather increased in some cases following MC38 transfection. The duration of expression was also examined and appears to be somewhat longer in radiated target cells (data not shown). Thus, PMT is suitable even in protocols which use irradiation of target cells.

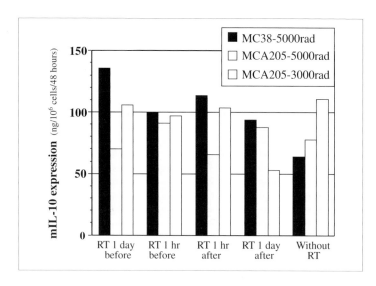

Figure 6-2. Irradiation increases cytokine gene expression after PMT (gene gun) in some cell types. The murine tumors MC38 or MCA205 were irradiated at the given time points with PMT of an mIL-10 expression plasmid. Expression of the cytokine (mIL-10) was measured with mIL-10 ELISA.

6.3.2   *In vivo* transfection

PMT can be performed successfully to transfect target tissue *in vivo*. Yang *et al*. have shown the feasibility of this approach [11]. They bombarded rat liver, mouse skin and muscle *in vivo* using the expression plasmids containing chloramphenicol acetyltransferase (CAT) and demonstrated that significant levels of CAT activity could be detected in target tissues. Our group confirmed that a significant transgene production could be achieved after PMT *in vivo* in mouse skin using an expression plasmid of mIL-10. Mice were shaved on their flank, marked, and bombarded by the helium-driven gene gun to the shaved area with varied pressures. The skin of the bombarded mice was harvested at various time periods after bombardment, mechanically minced, chemically lyzed and homogenized. These homogenized samples were centrifuged and the resultant supernatant was measured for mIL-10 by ELISA to determine *in vivo* expression of the transgene (Fig. 6-3A, B). When the expression was measured 48 h after PMT, significantly higher expression was detected when a pressure of 200 psi or more was used, but caused burn-like damage to the skin when a pressure of 600 psi or more was employed. Transfection using 300 psi consistently showed effective gene expression without obvious tissue damage. The expression of transgenes was examined over time after *in vivo* PMT (Fig. 6-3B). In this experiment, mIL-10 expression reached the maximum value 20 h after bombardment, decreased rapidly, and became undetectable within 48 h. These results confirmed that the gene transfected with the gene gun was expressed in a transient manner. This characteristic is advantageous in many cases, but not in others. Repeated PMT or some modification of the transfection system is required for more

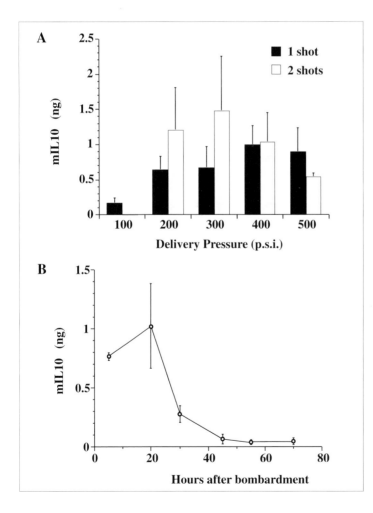

Figure 6-3 The levels and duration of cytokine gene expression after PMT (gene gun). Mouse skin was transfected with PMT in vivo with a mIL-10 expression plasmid, harvested 48 h after PMT, and examined for in situ mIL-10 expression. The cytokine was extracted from the mouse skin with the method previously described.

sustained gene expression, but stable expression appears to be difficult to achieve in *in vivo* PMT of the skin. The depth of the PMT transfection is also of interest. Rakhmilevich *et al.* showed that expression of PMT transfection was limited to the upper layer of the skin and deeper penetration was not associated with better expression in mouse skin [12].

6.4 Application of Gene Gun Transfection to Human Clinical Trials

Accumulating evidence, described in this article, strongly suggests that this gene gun mediated gene transfer can be applied in gene therapy clinical trials which do not require stable gene expression from the target cells/organs. In this manuscript, only examples of gene gun application were described to avoid redundancy.

6.4.1 *Ex vivo* strategy

The very first attempt at clinical application was to use the *ex vivo* immuno-gene therapy approach to treat cancer patients. This clinical protocol employed gene gun transfection instead of retroviral transduction which has been used in most of the existing clinical trials of this type as described by Tahara and Lotze previously [13]. It was developed based on the promising results from preclinical studies, initially using retroviral vector [14] and subsequent utilizing the gene gun [15]. The therapy involved delivery of a genetically engineered "tumor vaccine" to cancer patients. This vaccine was created by removing tumor or other target cells from a patient and irradiating them to prevent further growth. Then, a gene that encodes granulocyte macrophage colony stimulating factor (GM-CSF), a potent immunostimulatory cytokine, was transfected into the cells using the gene gun to create a vaccine preparation. Within 4 h, the vaccine preparation was injected intradermally to promote an immune response to tumor. Thus, gene gun application for an *ex vivo* cancer gene therapy strategy appears attractive, since some of the obstacles of cancer immuno-gene therapy protocols using retroviral vectors can be avoided. Furthermore, the cell types, which are previously believed to be difficult to be transduced (i.e. bone marrow-derived dendritic cells), are now can be transfected efficiently [16].

6.4.2 *In vivo* approach

In vivo gene gun approach can be applied in systems which require transient gene expression at the surface layer of the target organ/tissue. One example was to express cytokine locally adjacent to a tumor located near in the skin surface. Rakhmilevich *et al.* reported promising studies of this approach using IL-12 [12].

IL-12 is a heterodimeric cytokine which exerts many important regulatory functions on the immune reaction. These include strong induction of IFN-γ production and promotion of Th1 type responses. Our preclinical studies in mice indicated that injection of fibroblasts expressing IL-12 into a tumor can eradicate an established tumor and elicit specific systemic anti-tumor responses at the distant sites [17–19]. Phase I clinical trials of IL-12 gene therapy have been proposed using direct injection of tumors with genetically engineered autologous fibroblasts based on these promising preclinical results [20]. The study of Rakhmilevich *et al.* was performed using *in vivo* gene gun transfection instead of retrovirally transduced fibroblasts. They demonstrated a significant expression of IL-12 *in situ* at the surface of the skin after the gene gun shot. The expression of IL-12 was confined to the upper layer of the skin and not at the deeper layer adjacent to or inside the tumor site, but potent antitumor effects were shown in multiple tumor types in mice with no obvious systemic side effects. This approach might be brought into the clinic very soon after conducting additional studies to ensure the safety of this approach.

6.5 Conclusion

As described in this manuscript, PMT can achieve transient gene expression in target cells or tissue with predictable and limited damage to the cells. Although PMT has some limitations, this simple and non-viral approach can be a viable option for some types of clinical gene therapy. Improvement in level and duration of the gene expression after PMT may broaden the application of this approach in the future.

References

1. Wolff, J. A., Malone, R. W., Williams, P., Chong, W., Acsardi, G., Jani, A. and Felgmer, P. L. (1980) Direct gene transfer into mouse muscle *in vivo*. *Science* 247: 1465–1468
2. Wu., C. H., Wilson, J. M. and Wu, G. Y. (1989) Targeting genes: Delivery and persistent expression of a foreign gene driven by mammalian regulatory elements *in vivo*. *J. Biol. Chem.* 264: 16985–16987
3. Benevnisty, N. and Reshef. L. (1986) Direct injection of genes into rats and expression. *Proc. Natl. Acad. Sci. USA* 83, 9551–9555
4. Kaneda, Y., Iwai, K. and Uchida, T. (1989) Increased expression of DNA cointroduced with nuclear protein in adult liver. *Science* 243, 375–378
5. Felgner, P. L. and Ringold, G. M., (1989) Cationic liposome-mediated transfection. *Nature* 337, 387–388
6. McCabe, D. E., Swain, W. F., Martinell, B. J. and Christou, P. (1988) Stable transformation of soybean (*glycine max*) by particle acceleration. *Bio/Technology* 6: 923–926.
7. Christou, P., McCabe, D. E., Martinell, B. J. and Swain, W. F. (1990) Soybean genetic enginieering – commercial production of transgenic plants. *Trends Biotechnol.* 8: 145–151.
8. Christou, P., McCabe, D. E. and Swain, W. F. (1988) Stable transformation of soybean

callus by DNA-coated gold particles. *Plant Physiol.* 87: 671–674.

9. Klein, T. M., Harper, E. C., Svab. Z., Sanford, J. C., Fromm. M.E. and Maliga. P. (1988) Stable transformation of intact *Nicotian* cells by the particle bombardment trocess. *Proc. Natl. Acad. Sci. USA* 85, 8502–8505

10. Christou, P. Swain, W., Yang. N. -S. and McCabe, D. E. (1989) Inheritance and expression of foreign genes in transgenic soybean plants. *Proc. Natl. Acad. Sci. USA* 86, 7500–7504

11. Yang, N. -S., Burkholder, J., Roberts, B., Martinell, B. and McCabe, D. (1990) *In vivo* and *in vitro* gene transfer to mammalian somatic cells by particle bombardment. *Proc. Natl. Acad. Sci. USA* 87, 9568–9572

12. Rakhmilevich, A. L., Turner, J., Ford, M. J., McCabe, D., Sun, W. H., Sondel, P. M., Grota, K. and Yang, N. -S. (1996) Gene gun-mediated skin transfection with interleukin 12 gene results in regression of established primary and metastatic murine tumors. *Proc. Natl. Acad. Sci. USA* 93, 6291–6296

13. Tahara, H., Lotze, M. T. (1995) *In*: P. L. Chang (eds), *Cancer gene therapy; Somatic Gene Therapy*. CRC Press, London, U.K. , pp. 263–285

14. Dranoff, G., Jaffee E., Lazenby A., Golumbek, P., Levitsky, H., Brose, K., Jackson, V., Hamada H., Pardoll D., Mulligan R. (1993) Vaccination with irradiated tumor cells engineered to secrete murine granulocyte-macrophage colony-stimulating factor stimulates potent, specific, and long lasting anti-tumor immunity. *Proc. Natl. Acad. Sci. USA* 90: 3539–3543

15. Mahvi, D. M., Burkholder, J. K., Turner, J., Culp, J., Malter, M. S., Sondel, P. M., Yang, N. -S. (1996) Particle-mediated gene transfer of Granulocyte-Macrophage-Colony-Stimulating factor cDNA to tumor cells: Implications for a clinically relevant tumor vaccine. *Hum. Gene Ther.* 7, 1535–1543

16. Tuting, T., Ma, D. I., DeLeo, A. B., Lotze, M. T. and Storkus, W. J. (1998) Bone marrow-derived dendritic cells genetically modified to express tumor-associated antigens induce antitumor immunity *in vivo. J. Immunol.* 160: 1139–1147.

17. Tahara, H., Zeh, H. J. III, Storkus, W. J., Pappo, I., Watkins, S. C., Gubler, U., Wolf, S. F., Robbins, P. D., Lotze, M. T. (1994) Fibroblasts genetically engineered to secrete interleukin-12 can suppress tumor growth and induce anti-tumor immunity to a murine melanoma *in vivo. Cancer Res.* 54: 182–189

18. Tahara, H., Zitvogel, L., Storkus, W. J., Zeh, H. J.III, McKinney, T. G., Schreiber, R. D., Gubler, U., Robbins, P. D., Lotze, M. T. (1995) Effective eradication of established murine tumors with interleukin 12 (IL-12) gene therapy using a polycistronic retroviral vector. *J. Immunol.* 154: 6466–6474

19. Zitvogel, L., Tahara, H., Robbins, P. D., Storkus, W. J., Clarke, M. R., Nalesnik, M. A. and Lotze, M. T. (1995) Cancer Immunotherapy of established tumors with interleukin-12: Effective delivery by genetically engineered fibroblasts. *J. Immunol.* 155: 1393–1403

20. Tahara, H. and Lotze, M. T. (1995) IL-12 Gene therapy using direct injection of tumors with genetically engineered autologous fibroblasts. *Hum. Gene Ther.* 6: 1607–1624

II. Gene Therapy of Single Gene Defects

1 Gene Therapy for Severe Combined Immunodeficiencies

A. Aiuti, C. Bordignon

1.1   Introduction

Severe combined immunodeficiency diseases (SCIDs) are a group of primary immunodeficiencies characterized by profoundly impaired cell-mediated and humoral responses [1, 2]. Affected children typically fail to thrive and become ill with recurrent infections caused by bacteria, viruses, and opportunistic pathogens. The molecular defects have now been identified for the majority of SCID phenotypes and are summarized in Table 1-1 [3–27]. The defects affect lymphocyte receptors [28], signal transduction molecules [29], transcription factors [30], and enzymes of purine metabolism such as adenosine deaminase (ADA) [31] and purine nucleoside phosphorilase (PNP) [32]. The identification, cloning, and expression of the genes responsible for the different forms of SCID renders them potentially curable with somatic cell gene therapy. The ideal approach to gene therapy would require efficient gene transfer into the stem cell of the hematopoietic system, thus allowing the appropriate expression of the normal gene into the affected cells of hematopoietic or lymphoid lineage [33–35]. However, initial difficulties in obtaining efficient transduction of stem cells have led several investigators to approach gene transfer directly into differentiated lymphocytes [36–38]. Gene transfer into cells of the lympho-hematopoietic system for SCID is currently based on the use of retroviral vectors [33]. ADA-deficiency was the first genetic disorder to be treated with retroviral-mediated gene transfer and, to date, the only SCID treated. Here, we will review the most recent advances in gene therapy of different SCIDs, with particular regard to ADA-deficient SCID (ADA⁻ SCID), the prototype of immunodeficiencies amenable to gene therapy.

1.2   Retroviral-Vector Mediated Gene Transfer for SCID

Replication-deficient, recombinant retroviruses are in general derived from the backbone of Moloney murine leukemia virus [39–41]. These recombinant retroviruses might contain up to 9 Kb of exogenous information [35], and therefore can accommodate the cDNA encoding for the defective genes that cause inherited SCID.

T. Blankenstein (ed.) Gene Therapy
©1999, Birkhäuser Verlag Basel

Table 1-1. Congenital SCIDs potentially responsive to gene therapy

Disease	Inheritance	Clinical phenotype	Affected cells	Molecular defect	Presumed pathogenesis
X-linked SCID	XL	Combined Immunodeficiency	T cells, and NK cells (reduced); B cells (defective)	Mutations in common γ-chain of IL-2, IL-4, IL-7, IL-9, IL-15 receptors [3, 4]	Multiple defects in signaling mediated by common gamma-chain
Jak3-deficiency	AR	Combined Immunodeficiency	T cells, and NK cells (reduced); B cells (defective)	Mutations in Jak3 gene [5, 6]	Defects in intracellular signal transduction mediated by Jak3 protein tyrosine kinase
ADA deficiency	AR	Combined Immunodeficiency, chartilage defects	T cells, B cells	Defects in ADA gene [7–16]	T cell defect from toxic metabolites due to enzyme deficiency
PNP deficiency	AR	Combined Immunodeficiency, hemolytic anemia, neurologic symptoms	Primarily T cells	Defects in PNP gene [17–19]	T cell defect from toxic metabolites due to enzyme deficiency
CD8-deficiency	AR	Combined Immunodeficiency	CD8+ T cells (absent); CD4+ T cells (defective)	Mutations in Zap-70 kinase gene [20–22]	Defective signal transduction in T cells
MHC class II deficiency	AR	Combined Immunodeficiency, failure to thrive, protracted diarrhea	Lymphocytes, MHC class II expressing cells	Mutations in transcription factors for MHC class II molecules [23–27]	Defective MHC class II expression

Retroviral vectors offer several advantages including: the high efficiency of gene transfer into replicating cells; the stable integration of the transduced gene into the genome of the target cell; and the lack of further spread after gene transduction. A major disadvantage is the inability to infect non dividing-cells, which is a critical issue for gene transfer into a population of hematopoietic stem cells mostly residing in G_0/G_1. Another important issue is the potential danger of replication-competent virus production, especially in SCID patients that lack the normal immune system barrier. However, the generation of safer second and third generations of packaging cell lines [39–41], and the systematic testing of viral preparation have rendered this risk very unlikely. There has always been concern about the potential risk of insertional mutagenesis which might cause activation of cellular oncogenes or inactivation of tumor suppressor genes. Yet, there has never been evidence of a malignancy or any pathology in individuals treated with replication-defective retroviruses.

1.3   Gene Therapy of ADA-Deficient SCID

Inherited ADA-deficiency [42] represents the cause of about 25% of SCIDs and is fatal without treatment [43, 44]. The molecular defect resides, in most cases, in single base pair mutations that generate amino acid substitutions [9–12, 15], although splicing mutations [13, 16] and deletions [14] have also been reported. ADA is an intracellular enzyme that catalyzes the conversion of adenosine to inosine in the purine nucleotide cycle. In ADA⁻ SCID, purine metabolites such as deoxyadenosine, adenosine, or dATP accumulate in the tissues. The intracellular accumulation of purine metabolites is the cause of the inhibition of proliferation, differentiation, and maturation of immunocompetent cells as well as direct cell death (reviewed by Hirschhorn [31]). A form of the ADA enzyme is also localized on the lymphocyte cell surface in association with the T cell activation molecule, dipeptidyl peptidase IV (DPPIV or CD26) [45], and with the adenosine receptor 1 [46]. The extra cellular ADA enzyme, complexed to CD26, seems to be involved in T cell activation events [47] and is capable of reducing the local external concentration of adenosine, thus rendering T cells more resistant to the inhibitory effects of adenosine [48].

Allogeneic bone marrow transplantation (BMT) is the treatment of choice for ADA-deficient SCID if an HLA-identical sibling bone marrow donor is available [31]. At present, 90–100% of children receiving such a transplant can be expected to achieve engraftment with full immunological reconstitution and disease-free survival. Unfortunately, HLA-matched donors are available only for a minority of patients. Alternative BMT strategies, such as T cell depleted marrow transplant from haplotype-mismatched parental donor or unrelated histocompatible marrow have resulted in cure rates ranging from 40–70%, depending on the disease state at the time of bone marrow transplantation [44, 49]. Other sources of stem cells employed in transplantation of SCID patients include fetal tissues [50] and placental/umbilical cord blood [51]. Cord blood seems particularly promising because of its low incidence of graft versus host disease, even in HLA-mismatched situations, due to the functional immaturity of T cells [52]. *In utero* transplantation of progenitor cells is another experimental therapeutical option that may overcome many of the risks associated with BMT. Two different groups reported recently the *in utero* transplantation of T cell-depleted, paternal bone marrow progenitor cells into fetus affected by X-SCID, resulting in their immune reconstitution after birth [53, 54].

It should also be mentioned that some ADA⁻ SCID patients might not be candidates for cytoablation due to previous infectious damage to the lung or liver, or might present a milder phenotype that does not justify the risks associated with transplantation. As an alternative therapy, injection with bovine ADA conjugate to polyethylene glycol (PEG-ADA) has resulted in improvement in growth, correction of metabolic abnormalities, a variable degree of recovery of immune functions, and a decrease in the incidence of severe infections (reviewed by Hershfield [55]).

Somatic cell gene therapy has become a therapeutical option since 1990, when the first clinical trial began at the NIH on two-young girls affected by ADA-deficiency [37]. Here, we will review first the initial *in vitro* and *in vivo* pre-clinical approaches to gene therapy for ADA⁻ SCID and then focus on the results of the clinical trials.

1.3.1 Pre-Clinical Models for Gene Therapy of ADA⁻ SCID

The first step towards gene therapy of ADA-deficient SCID was the generation of retroviral vectors, encoding the human ADA gene, that allowed the transduction of human ADA-deficient T cell lines. Genetically-modified ADA⁻ T cells expressed normal levels of the ADA enzyme, thus becoming resistant to the inhibitory and toxic effects of deoxyadenosine [56, 57]. A pre-clinical model was subsequently developed in which peripheral blood lymphocytes from ADA⁻ SCID patients were transduced with a retroviral vector, encoding human ADA, and injected into immunodeficient mice [38]. Transduced human cells survived long-term in the recipient animal, expressed the normal ADA gene, and acquired a survival advantage with respect to uncorrected ADA-deficient cells. The restoration of enzyme activity in human ADA-deficient peripheral blood T cells by gene transfer induced *in vivo* the production of human immunoglobulins as well as the generation of human alloreactive and antigen-specific T lymphocytes [58]. These pre-clinical results produced important information for the approval of the first gene therapy protocols at the NIH [37] and at the San Raffaele Institute in Milan, Italy [34].

At the same time, several groups approached gene transfer into human hematopoietic stem cells, with the goal to obtain a long-lasting, definitive source of genetically-corrected lymphocytes *in vivo*. Transduction with ADA-encoding sequence was achieved *in vitro* both in human bone marrow [59, 60] and cord blood-derived clonogenic progenitor cells [61], with levels of transduction ranging from 20–80%. Efficient gene transfer was also demonstrated for the more primitive "long-term culture-initiating cells" (LTC-IC) [62] that sustain long-term bone marrow culture for at least 5 weeks. Transduction of LTC-IC with the ADA gene resulted in adequate expression of the normal enzyme in their progeny of differentiated cells [59–61]. Nevertheless, the *in vivo* relevance of these assays for gene transfer is still unclear because the relationship between *in vitro* progenitors and the *in vivo* repopulating human stem cells is still not fully proved.

Because of its relative simplicity, the mouse model was chosen initially to study gene transfer of the ADA gene into hematopoietic stem cells capable of long-term reconstitution *in vivo*. Several investigators have reported high efficient transfer and long-term expression of the human ADA gene in mice transplanted with retroviral vector-infected bone marrow cells [63–67]. Non-human primates represent a more adequate pre-clinical large-animal model to test the efficacy of gene therapy of hematopoietic stem cells [68]. The report of the first successful long-term gene transfer into monkeys [69] demonstrated the feasibility of introducing a functional human ADA gene into hematopoietic stem cells and obtaining expression in multiple hematopoietic lineages after transplantation. These data prompted the approval of a clinical gene therapy protocol using transduced bone marrow cells at the Institute for Applied Radiobiology and Immunology, The Netherlands. Since the initial report of van Beusechem and colleagues [69], considerable improvements in culture conditions and retroviral producer cell lines have been achieved, but the levels of peripheral blood cells carrying the transduced ADA gene in transplanted monkeys remain relatively low, ranging from 0.1–2% [70–73], being at least one order of magnitude less efficient with respect to the mouse model [69, 71]. The explanation for this partial success in the primate model is still unclear, but might depend on the cell cycle status of the hematopoietic stem cell, the source of stem cells (bone marrow, peripheral blood), the loss of multipotent stem cell during the transduction procedure, the differential expression of retroviral receptor on the stem cell or the requirement of a conditioning for the transplant recipient. More recently, it has been shown that bone marrow cells or peripheral blood progenitor cells, collected after treatment with granulocyte-colony stimulating factor (G-CSF) and stem cell factor (SCF), are superior targets for retroviral gene transduction in primates, suggesting that it is possible to modulate *in vivo* the functional status of the target stem cell and its susceptibility to gene transfer [74].

1.3.2   Clinical Trials for Gene Therapy of ADA⁻ SCID

The demonstration that allogeneic bone marrow transplantation can restore the cellular defects of ADA-deficiency, the cloning of the ADA gene, and the encouraging results obtained with the pre-clinical models led to the approval of four gene-therapy trials designed to test the safety and efficacy of gene therapy of ADA⁻ SCID [34, 37, 73, 75]. All the trials were based on retroviral-mediated gene transduction but differed considerably with regard to the target cells, the retroviral vector constructs, and the transduction procedure [73, 75–78] (see Tab. 1-2). Patients eligible initially for gene therapy were considered to be the ones lacking a BMT-donor and that had not gained full immunological reconstitution upon PEG-ADA treatment.

The first human gene therapy trial was initiated in 1990 at the NIH on two patients that had been previously treated with PEG-ADA [76]. The protocol was based on periodical infusions of autologous T lymphocytes expanded *ex vivo* and corrected genetically by retroviral-mediated gene transfer. After the initiation of treatment, the number of blood T cells increased to normal levels, along with the correction of cytotoxic T cell activity, cytokine production, isohemagglutinin production, and skin test responses to common antigens. ADA enzyme activity in the peripheral blood lymphocytes (PBLs) from the first treated patient increased considerably, while in the second patient ADA activity remained low, probably because of a much less efficient gene transfer in the lymphocytes of this patient.

A second protocol was initiated in 1992 at the Scientific Institute San Raffaele in Milan, Italy [77]. This protocol was designed specifically to define the relative contribution of PBLs and bone marrow progenitor cells in the long-term reconstitution of immune functions following gene transduction. To this aim, two retroviral vectors (DCA*l*, DCA*m*) were generated, encoding the human ADA minigene, differing only at a restriction site in a non-functional region of the viral LTR. The two vectors were used to transduce peripheral blood lymphocytes (DCA*l*) and bone marrow cells (DCA*m*), respectively. During the follow-up, the origin of the gene-modified cells was determined by PCR analysis and restriction enzyme digestion of genomic DNA. After 2 years of treatment, long-term survival of transduced cells was demonstrated by the presence of vector derived sequences in T and B lymphocytes, bone marrow cells, and granulocytes carrying the ADA gene. Vector-transduced cells, detected in the peripheral blood of the patients, were derived initially from transduced PBL (carrying DCA*l*). In the subsequent year, PBL and T cell clones derived from BM-specific DCA*m* vector replaced progressively by T cells containing DCA*l* vector. The frequency of gene modified T cells, clonable in the presence of G418, ranged from 0.8–8.5%. The presence of gene modified T cells restored immune functions, producing over time the normalization of total lymphocyte counts, the development of a normal T-cell receptor repertoire, and the reconstitution of cellular and humoral responses, including mitogen and antigen-specific proliferation as well as Ig production.

Placental/umbilical cord blood is another important source of haematopoietic stem cells suitable for gene transfer [61]. Three neonates have been treated by D. Kohn and colleagues [75] by transduction of the CD34⁺ cells from their umbilical-cord blood. The CD34⁺ cells were collected at birth, infected with cell-free supernatant containing a retroviral vector, and re-infused into their respective donors on

Table 1-2. Clinical trials for gene therapy of ADA-deficient SCID

Investigators	Patients	Transduced cells	Retroviral vectors	Transduction protocol
Blaese et al. [76]	2	Peripheral blood lymphocytes (PBLs)	A vector encoding for ADA under the MoMuLV LTR and neoR under the SV40 promoter (LASN)	Infection of PBLs stimulated with OKT3 (anti-CD3) antibody and IL-2 with viral supernatant
Bordignon et al. [77]	2	PBLs and bone marrow (BM) progenitors	Two vectors encoding for two copies of the ADA gene under their own promoter, plus neoR under the LTR. The two vectors differ at a restriction site to distinguish transduced cells derived from PBLs (DCA/) or BM cells (DCAm).	PBLs: cocultivation on irradiated producer after stimulation with PHA+IL-2 BM: infection of progenitor-enriched cells with viral supernatant, no cytokine added.
Kohn et al. [75]	3	Cord blood (CB) progenitor cells	Same as Blaese et al. [76]	Infection of CB CD34+ cells with viral supernatant in the presence of IL-3, SCF, IL-6
Hoogerbrugge et al. [73]	3	CD34+ BM cells	A vector expressing ADA, with no selection marker (LgAL)	Coculture of BM CD34+ cells on irradiated producer with IL-3
Onodera et al. [78]	1	PBLs	Same as Blaese et al. [76]	Same as Blaese et al. [76]

their fourth day of life. During the transduction protocol, the CD34$^+$ cells were maintained in the presence of recombinant cytokines to stimulate progenitor cell proliferation (IL-3, IL-6, and stem cell factor). In all three patients, the continued presence and expression of the ADA gene was demonstrated in leukocytes from bone marrow and peripheral blood for at least four years. Vector-containing cells were initially detected in the peripheral blood of these patients at a relatively low frequency (0.03–0.001%). In contrast, a high frequency of clonogenic progenitors containing the transduced vector (1–6%) was found in their bone marrow. Three to four years after the infusion of autologous transduced CD34$^+$ cells, the frequency of genetically modified T lymphocytes increased up to 1–10%, whereas the proportion of hematopoietic and non-T lymphoid cells containing the gene remained at lower levels (0.01–0.1%) [75a]. This selective accumulation of gene-corrected T cells might have occurred at the production level during T lymphopoiesis, and/or at the survival level, as suggested from studies in SCID-mice in which genetically corrected PBLs showed prolonged survival with respect to non-corrected cells [38]. While all patients had received lower doses of PEG-ADA in the last two years, in one of them the treatment was stopped completely for two months. During the time without PEWG-ADA, the frequency of gene corrected T lymphocytes further increased to 30% and T cell counts and PHA responses were maintained. However, responses to tetanus toxoid were lost, B lymphocyte and NK cell numbers were drastically reduced and deoxyadenosine nucleotide levels in erythrocytes increased 100-fold.

A fourth clinical trial, based solely on bone marrow gene transfer, was established in The Netherlands [73]. The experimental protocol varied from the others mainly because it used co-cultivation of bone marrow CD34$^+$ cells with the producer cell line and the addition of recombinant human IL-3 to the culture. Gene transfer efficiency ranged from 5–12%, assayed in clonogenic progenitors. Following infusion of the transduced CD34$^+$ cells, the ADA gene was detected in granulocytes and mononuclear cells for 3 months. However, the ADA gene did not persist in the peripheral blood or the bone marrow longer than 6 months. These data suggest that the gene transfer procedure did not target a sufficient number of long-lasting, multipotent progenitors. The *ex vivo* culture conditions, the low number of progenitor cells infused, and the competition with endogenous stem cells without a selective advantage, might all have contributed to the disappearance of transduced cells in these patients.

A Japanese group recently reported a T lymphocyte-directed gene therapy trial in a single patient following the identical protocol used in the NIH trial [78]. The initial results showed a frequency of circulating genetically modified cells in a range between 10 and 20% and clinical improvements similar to the ones obtained with the first patient in the NIH trial.

In addition to the differences in the protocol and target cell used for gene therapy, the 11 patients treated so far differed significantly with respect to the severity of the underlying ADA gene defect, the age, the clinical presentation of the disease and the dosage of PEG-ADA administered. All these factors may have affected gene transfer efficiency and influenced the *in vivo* survival and function of transduced cells and their role should be investigated more extensively.

In conclusion, while the biological responses to gene therapy appear evident in the majority of the patients treated, the effect on the overall clinical well-being remains difficult to evaluate because the patients have been maintained on ADA enzyme supplementation. Future studies involving progressive reduction and, eventually, discontinuation of PEG-ADA treatement, are required to establish the relative efficacy of the different gene therapy approaches for ADA⁻ SCID.

1.4 Gene Therapy for other SCIDs

The encouraging results obtained for ADA-deficient SCID represented a strong rationale for the transfer of the experience and expertise to the development of gene therapy pre-clinical models for other combined immunodeficiencies. Here, we will discuss the most recent advances in the application of somatic cell gene therapy to the correction of various molecular defects responsible for these SCIDs.

1.4.1 PNP-Deficient SCID

PNP-deficiency is a rare form of SCID that affects T cells predominantly (reviewed by Markert [32]). Like ADA, PNP is an enzyme of purine metabolism that acts on both ribonucleotides and deoxyribonucleotides. In the absence of PNP, deoxyguanosine accumulates inside the cell and is phosphorylated to deoxyGTP, causing the inhibition of cell proliferation. B cells are less sensitive than T cells to the accumulation of deoxyguanosine, probably because they use a different metabolic pathway. With the aim of establishing a pre-clinical model of gene transfer, retroviral vectors were engineered to express the murine or human PNP cDNA under transcriptional regulation of the Moloney murine leukemia virus, LTR [79, 80]. Retroviral mediated gene transfer of mouse PNP genes into human lymphocytes from PNP-deficient patients corrected their metabolic defect and partially restored their immune functions [81]. A clinical protocol for gene therapy of PNP-immunodeficiency was recently approved at our institution.

1.4.2 X-Linked SCID (X-SCID)

X-linked SCID is an inherited disorder characterized by a defect in T cell and NK differentiation; B cell maturation is preserved but B cells' function is abnormal

(Tab. 1). The molecular defect responsible for X-SCID has been identified recently as a mutation in the γ-chain of the IL-2 receptor. The finding that the abnormalities in X-SCID are more severe than those found in humans or mice, deficient in IL-2, led to the discovery that the γ-chain is also a component of the receptors for IL-4, IL-7, IL-9, and IL-15, which is now denominated as the common cytokine receptor γ-chain (reviewed by Leonard, [28]). The γ-chain interacts downstream with members of Janus family kinase, in particular Jak3, to transduce extra cellular signals leading eventually to gene activation. The treatment of choice for the minority of patients who have the availability of an HLA-compatible donor is bone marrow transplantation. In the others, HLA-mismatched transplantation results in a survival rate of 50–60%, often with residual immunodeficiency in the B cell compartment. For these reasons, several investigators have approached retroviral-mediated gene transfer of γ-chain with the aim to offer an alternative therapeutic option for X-SCID. The γ-chain has been successfully introduced into EBV-transformed B-cell lines derived from X-SCID patients, restoring surface expression of γ-chain and function to the IL-2 and IL-4 receptors as shown by reconstitution of signal transduction and cellular proliferation responses [82–84]. In addition, *in vitro* gene transfer of γ-chain into hematopoietic progenitor cells, in the presence of SCF, IL-2, and IL-7, was able to restore NK cell differentiation in X-SCID patients [85].

1.4.3   Jak3-Deficient SCID

Jak3-deficient SCID is a disease characterized by a clinical phenotype virtually identical to X-linked SCID. The molecular defects were shown recently to be in the gene encoding Jak3, an intracellular kinase bound to the lymphocyte-common γ-chain [5, 6]. This finding further substantiated the crucial role of γ-chain/Jak3 interaction in lymphocyte development and function. A vector encoding Jak3 cDNA was engineered and used to transduce B cell lines obtained from patients with Jak3-deficient SCID. Retroviral-mediated gene transfer was able to restore Jak3 expression and function, allowing normalization of cell growth of B cell lines [86]. A mouse model of Jak3-deficient SCID was recently established and used to study retroviral mediated gene transfer into hematopoietic stem cells [87]. Retroviral-mediated gene transfer into mouse hematopoietic stem cells resulted in increased numbers of lymphocytes, reversal of hypogammaglobulinemia, reconstitution of T-cell activation upon stimulation with mitogens and development of an antigen-specific immune response after immunization. Importantly, Jak3 gene-corrected lymphoid cells displayed a significant *in vivo* selective advantage, suggesting that this approach may be effective even when gene transfer efficiency is relatively low.

1.4.4 CD8-Deficiency

CD8-deficiency is an inherited autosomal recessive SCID characterized by the absence of CD8+ T cells and defective CD4+ cells in the periphery. The molecular defect results from mutations in the gene encoding Zap-70, a protein tyrosine kinase involved in T cell receptor signaling [20–22]. This rare disease is another candidate for gene therapy of lympho-hematopoietic cells and the preliminary results obtained *in vitro* are encouraging. When T-cell lines established from Zap-70-deficient patients were transduced with retroviral-vector encoding normal Zap-70, it was possible to restore Zap-70 expression and function, allowing normal T cell receptor signaling [88].

1.5 Conclusions

The *in vitro* and *in vivo* models, as well as the clinical trials for ADA⁻ SCID have demonstrated the feasibility of introducing a functional gene into peripheral blood lymphocytes and hematopoietic stem cells to restore normal immune functions in individuals affected by SCID. In addition, all these studies have clearly shown that there are no adverse effects from the administration of gene-modified cells into these patients. A critical issue to be solved remains the choice of the target cells for gene therapy. When gene modified lymphocytes were transplanted into ADA⁻ SCID patients, they produced a progressive biological improvement within a few weeks from infusion, while persisting *in vivo* for at least one to three years. Thus, relatively rapid, short-term reconstitution could be deputed to gene-modified lymphocytes. On the other hand, transduced hematopoietic stem cells seem to be responsible for long-term reconstitution of the immune system, suggesting that this approach should be considered essential for gene therapy of all SCIDs. However, an important limitation resides in the low efficiency of gene transfer into long-lasting, definitive stem cells. This is documented by the limited success of the study by Hoggerbrugge and colleagues [73], and the low frequency in the peripheral blood of transduced cells derived from hematopoietic stem cells in gene therapy or gene marking studies [34, 75, 89]. This might by partly overcome *in vivo* by a selective advantage of transduced stem cells carrying the normal gene, although the selective advantage could vary considerably for each specific disorder. Current approaches to improve *in vitro* retroviral-mediated gene transfer into hematopoietic stem cells include the use of stromal cells to create a more physiological microenvironment *in vitro* [90–92], of recombinant stromal-cell products, such as fibronectin [93], or of novel recombinant cytokines, such as the flt3-ligand [94]. Another important issue of stem cell gene therapy for SCID is whether the expression of transduced genes encoding cytokine receptors or signal transduction molecules that, unlike ADA, are strictly regulated in a tissue-specific manner, could be harmful to cells of non-lymphoid hematopoietic lineages. If this is the case, further efforts are required to ensure the expression of transduced genes in a controlled, tissue-specific manner [95] before stem cell gene transfer becomes applicable in clinical gene therapy for other SCIDs.

References

1. Cooper, M., Butler J. (1989) Primary immunodeficiency diseases. *In*: W. E. Paul (ed.), *Fundamental Immunology*. Raven Press, New York, pp. 1034.
2. Rosen, F. S., Wedgwood, R., Eibl, M. et al. (1992) Primary immunodeficiency diseases. Report of a WHO scientific group. *Immunodef. Rev* 3: 195–215.
3. Noguchi, M., Yi, H., Rosenblatt HM et al. (1993) Interleukin-2 receptor gamma chain mutation results in X-linked severe combined immunodeficiency in humans. *Cell* 73: 147–57.
4. Puck, J. M., Deschenes, S. M., Porter, J. C. et al. (1993) The interleukin-2 receptor gamma chain maps to Xq13.1 and is mutated in X-linked severe combined immunodeficiency, SCIDX1. *Hum. Mol. Genet.* 2: 1099–104.
5. Russell, S., Tayebi, N., Nakajima, H. et al. (1995) Mutation of Jak3 in a patient with SCID: essential role of Jak3 in lymphoid development. *Science* 270: 797–800.
6. Macchi, P., Villa, A., Giliani, S. et al. (1995) Mutations of Jak-3 gene in patients with autosomal severe combined immunodeficiency. *Nature* 377: 65–68.
7. Wiginton, D. A., Adrian, G. S., Friedman, R. L. et al. (1983) Cloning of cDNA sequences of human adenosine deaminase. *Proc. Natl. Acad. Sci. USA* 80: 7481–7485.
8. Valerio, D., Duyvesteyn, M. G., Meera, K. P. et al. (1983) Isolation of cDNA clones for human adenosine deaminase. *Gene* 25: 231–240.
9. Bonthron, D. T., Markham, A. F., Ginsburg, D. et al. (1985) Identification of a point mutation in the adenosine deaminase gene responsible for immunodeficiency. *J. Clin. Invest.* 76: 894–897.
10. Valerio, D., Dekker, B. M., Duyvesteyn, M. G. et al. (1986) One adenosine deaminase allele in a patient with severe combined immunodeficiency contains a point mutation abolishing enzyme activity. *EMBO J.* 5: 113–119.
11. Akeson, A. L., Wiginton, D. A., Hutton, J. J. (1989) Normal and mutant human adenosine deaminase genes. *J. Cell. Biochem.* 39: 217–228.
12. Hirschhorn, R., Tzall, S., Ellenbogen, A. (1990) Hot spot mutations in adenosine deaminase deficiency. *Proc. Natl. Acad. Sci. USA* 87: 6171–6175.
13. Akeson, A. L., Wiginton, D. A., States, J. C. et al. (1987) Mutations in the human adenosine deaminase gene that affect protein structure and RNA splicing. *Proc. Natl. Acad. Sci. USA* 84: 5947–5951.
14. Markert, M. L., Hutton, J. J., Wiginton, D. A. et al. (1988) Adenosine deaminase (ADA) deficiency due to deletion of the ADA gene promoter and first exon by homologous recombination between two Alu elements. *J. Clin. Invest.* 81: 1323–1327.
15. Hirschhorn, R., Nicknam, M. N., Eng, F. et al. (1992) Novel deletion and a new missense mutation (Glu 217 Lys) at the catalytic site in two adenosine deaminase alleles of a patient with neonatal onset adenosine deaminase-severe combined immunodeficiency. *J.*

Immunol. 149: 3107–12.

16. Santisteban, I., Arredondo, V. F., Kelly, S. et al. (1995) Three new adenosine deaminase mutations that define a splicing enhancer and cause severe and partial phenotypes: implications for evolution of a CpG hotspot and expression of a transduced ADA cDNA. *Hum. Mol. Genet.* 4: 2081–2087.

17. Goddard, J. M., Caput, D., Williams, S. R. et al. (1983) Cloning of human purine-nucleoside phosphorylase cDNA sequences by complementation in Escherichia coli. *Proc. Natl. Acad. Sci. USA* 80: 4281–4285.

18. Williams, S. R., Gekeler, V., McIvor, R. S. et al. (1987) A human purine nucleoside phosphorylase deficiency caused by a single base change. *J. Biol. Chem.* 262: 2332–2338.

19. Aust, M. R., Andrews, L. G., Barrett, M. J. et al. (1992) Molecular analysis of mutations in a patient with purine nucleoside phosphorylase deficiency. *Am. J. Hum. Genet.* 51: 763–772.

20. Arpaia, E., Shahar, M., Dadi, H. et al. (1994) Defective T cell receptor signaling and CD8+ thymic selection in humans lacking zap-70 kinase. *Cell* 76: 947–958.

21. Chan, A. C., Kadlecek, T. A., Elder, M. E. et al. (1994) ZAP-70 deficiency in an autosomal recessive form of severe combined immunodeficiency. *Science* 264: 1599–1601.

22. Elder, M. E., Lin, D., Clever, J. et al. (1994) Human severe combined immunodeficiency due to a defect in ZAP-70, a T cell tyrosine kinase. *Science* 264: 1596–1599.

23. Reith, W., Satola, S., Sanchez, C. H. et al. (1988) Congenital immunodeficiency with a regulatory defect in MHC class II gene expression lacks a specific HLA-DR promoter binding protein, RF-X. *Cell* 53: 897–906.

24. Reith, W., Barras, E., Satola, S. et al. (1989) Cloning of the major histocompatibility complex class II promoter binding protein affected in a hereditary defect in class II gene regulation. *Proc. Natl. Acad. Sci. USA* 86: 4200–4204.

25. Steimle, V., Otten, L. A., Zufferey, M. et al. (1993) Complementation cloning of an MHC class II transactivator mutated in hereditary MHC class II deficiency (or bare lymphocyte syndrome). *Cell* 75: 135–146.

26. Steimle, V., Durand, B., Barras, E. et al. (1995) A novel DNA-binding regulatory factor is mutated in primary MHC class II deficiency (bare lymphocyte syndrome). *Gene. Dev.* 9: 1021–1032.

27. Douhan J3, Hauber, I., Eibl, M. M. et al. (1996) Genetic evidence for a new type of major histocompatibility complex class II combined immunodeficiency characterized by a dyscoordinate regulation of HLA-D alpha and beta chains. *J. Exp. Med.* 183: 1063–1069.

28. Leonard, W. J. (1996) The molecular basis of X-linked severe combined immunodeficiency: defective cytokine receptor signaling. *Annu. Rev. Med.* 47: 229–239.

29. Notarangelo, L. D. (1996) Immunodeficiencies caused by genetic defects in protein kinases. *Curr. Opin. Immunol.* 8: 448–453.

30. Reith, W., Steimle, V., Mach, B. (1995) Molecular defects in the bare lymphocyte syndrome and regulation of MHC class II genes. *Immunol. Today* 16: 539–546.

31. Hirschhorn, R. (1990) Adenosine deaminase deficiency. *Immunodefic. Rev.* 2: 175–98.

32. Markert, M. L. (1991) Purine nucleoside phosphorylase deficiency. *Immunodefic. Rev.* 3: 45–81.

33. Anderson, F. (1992) Human gene therapy. *Science* 256: 808–813.

34. Bordignon, C., Mavilio, F., Ferrari, G. et al. (1993) Transfer of the ADA gene into bone marrow cells and peripheral blood lymphocytes for the treatment of patients affected by

ADA-deficient SCID. *Hum. Gene Ther.* 4: 513–20.

35. Crystal, R. (1995) Transfer of genes to humans: early lessons and obstacle to success. *Science* 270: 404–410.

36. Kantoff, P. W., Kohn, D. B., Mitsuya, H. et al. (1986) Correction of adenosine deaminase deficiency in cultured human T and B cells by retrovirus-mediated gene transfer. *Proc. Natl. Acad. Sci. USA* 83: 6563–6567.

37. Blease, M., Anderson, F., Culver, K. et al. (1990) The ADA human gene therapy clinical protocol. *Hum. Gene Ther.* 1: 327–62.

38. Ferrari, G., Rossini, S., Giavazzi, R. et al. (1991) An *in vivo* model of somatic cell gene therapy for human severe combined immunodeficiency. *Science* 251: 1363–6.

39. Danos, O., Mulligan, R. (1988) Redesign of retrovirus packaging cell lines to avoid recombination leading to helper virus production. *Mol. Cell. Biol.* 6: 2895–2902.

40. Markowitz, D., Goff, S., Bank, A. (1988) A safe packaging line for gene-transfer-separating viral genes on two different plasmids. *J. Virol.* 62: 1120–1124.

41. Miller, A. (1990) Retrovirus packaging cell lines. *Hum Gene Ther.* 1: 5–14.

42. Giblett, E., Anderson, J., Cohen, F. et al. (1972) Adenosine Deaminase deficiency in two patients with severely impaired cellular immunodeficiency. *Lancet ii* 1067–1069.

43. Hirschorn, R., Vawter, G., Kirkpatrick, J. et al. (1979) Adenosine deaminase deficiency: frequency and comparative pathology in autosomal recessive severe combined immunodeficiency. *Clin. Immunol. Immunopathol.* 14: 107–20.

44. Fischer, A. (1992) Severe combined immunodeficiencies. *Immunodefic. Rev.* 3: 83–100.

45. Kameoka, J., Tanaka, T., Nojima, Y. et al. (1993) Direct association of adenosine deaminase with a T cell activation antigen, CD26. *Science* 261: 466–469.

46. Ciruela, F., Saura, C., Canela, E. I. et al. (1996) Adenosine deaminase affects ligand-induced signalling by interacting with cell surface adenosine receptors. *FEBS Lett.* 380: 219–223.

47. Martin, M., Huguet, J., Centelles, J. J. et al. (1995) Expression of ecto-adenosine deaminase and CD26 in human T cells triggered by the TCR-CD3 complex. Possible role of adenosine deaminase as costimulatory molecule. *J. Immunol.* 155: 4630–4643.

48. Dong, R. P., Kameoka, J., Hegen, M. et al. (1996) Characterization of adenosine deaminase binding to human CD26 on T cells and its biologic role in immune response. *J. Immunol.* 156: 1349–1355.

49. O'Reilly, R., Keever, C., Small, T. et al. (1990) The use of HLA-non identical T-cell depleted marrow transplants for correction of severe combined immunodeficiency disease. *Immunodef. Rev.* 1: 273–309.

50. Touraine, J. L. (1983) European experience with fetal tissue transplantation in severe combined immunodeficiency (SCID). *Birth Defects* 19: 139–142.

51. Kurtzberg, J., Laughlin, M., Graham, M. L. et al. (1996) Placental blood as a source of hematopoietic stem cells for transplantation into unrelated recipients. *N. Engl. J. Med.* 335: 157–166.

52. Gluckman, E. (1996) Umbilical cord blood transplant in human. *Bone Marrow Transplant.* 18: 166–170.

53. Flake, A. W., Roncarolo, M. G., Puck, J. M. et al. (1996) Treatment of X-linked severe combined immunodeficiency by in utero transplantation of paternal bone marrow. *N. Engl. J. Med.* 335: 1806–1810.

54. Wengler, G. S., Lanfranchi, A., Frusca, T. et al. (1996) In-utero transplantation of parental CD34 haematopoietic progenitor cells in a patient with X-linked severe com-

bined immunodeficiency. *Lancet* 348: 1484–1487.

55. Hershfield, M. S. (1995) PEG-ADA: an alternative to haploidentical bone marrow transplantation and an adjunct to gene therapy for adenosine deaminase deficiency. *Hum. Mutat.* 5: 107–112.

56. Kantoff, P., Kohn, D., Mitsuya, H. et al. (1986) Correction of adenosine deaminase deficiency in cultured human T and B cells by retrovirus-mediated gene transfer. *Proc. Natl. Acad. Sci. USA* 83: 6563–6567.

57. Kohn, D. B., Mitsuya, H., Ballow, M. et al. (1989) Establishment and characterization of ADA deficient human T cell lines. *J. Immunol.* 142: 3971–3977.

58. Ferrari, G., Rossini, S., Nobili, N. et al. (1992) Transfer of the ADA gene into human ADA-deficient T lymphocytes reconstitutes specific immune functions. *Blood* 80: 1120–4.

59. Bordignon, C., Yu, S., Smith, C. et al. (1989) Retroviral vector-mediated high efficiency expression of adenosine deaminase in long-term cultures of ADA deficient marrow cells. *Proc. Natl. Acad. Sci. USA* 86: 6748–6752.

60. Cournoyer, D., Scarpa, M., Mitani, K. et al. (1991) Gene transfer of adenosine deaminase into primitive human hematopoietic progenitor cells. *Hum. Gene Ther.* 2: 203–13.

61. Moritz, T., Keller, D. C., Williams, D. A. (1993) Human cord blood cells as targets for gene transfer: potential use in genetic therapies of severe combined immunodeficiency disease. *J. Exp. Med.* 178: 529–36.

62. Sutherland, H. J., Eaves, C. J., Lansdorp, P. M. et al. (1991) Differential regulation of primitive hematopoietic cells in long-term cultures maintained on genetically engineered murine stromal cells. *Blood* 78: 666–672.

63. Lim, B., Williams, D., Orkin, S. (1987) Retrovirus-mediated gene transfer of human adenosine deaminase: expression of functional enzyme in murine hematopoietic stem cells *in vivo*. *Mol. Cell. Biol.* 7: 3459–3465.

64. Belmont, J., MacGregor, G., Wager-Smith, K. et al. (1988) Expression of human adenosine deaminase in murine hematopoietic cells. *Mol. Cell. Biol.* 8: 5116–5125.

65. Wilson, J. M., Danos, O., Grossman, M. et al. (1990) Expression of human adenosine deaminase in mice reconstituted with retrovirus-transduced hematopoietic stem cells. *Proc. Natl. Acad. Sci. USA* 87: 439–43.

66. van Beusechem, V., Kukler, A., Einerhand, M. P. et al. (1990) Expression of human adenosine deaminase in mice transplanted with hemopoietic stem cells infected with amphotropic retroviruses. *J. Exp. Med.* 172: 729–736.

67. Einerhand, M. P., Bakx, T. A., Kukler, A. et al. (1993) Factors affecting the transduction of pluripotent hematopoietic stem cells: long-term expression of a human adenosine deaminase gene in mice. *Blood* 81: 254–63.

68. van Beusechem, V., Valerio, D. (1996) Gene transfer into hematopoietic stem cells of non-human primates. *Hum. Gene Ther.* 7: 1649–1668.

69. van Beusechem, V., Kukler, A., Heidt PJ et al. (1992) Long-term expression of human adenosine deaminase in rhesus monkeys transplanted with retrovirus-infected bone-marrow cells. *Proc. Natl. Acad. Sci. U.s.a.* 89: 7640–7644.

70. van Beusechem, V., Bakx, T. A., Kaptein, L. C. et al. (1993) Retrovirus-mediated gene transfer into rhesus monkey hematopoietic stem cells: the effect of viral titers on transduction efficiency. *Hum. Gene Ther.* 4: 239–247.

71. Bodine, D. M., Moritz, T., Donahue, R. E. et al. (1993) Long-term *in vivo* expression of a murine adenosine deaminase gene in rhesus monkey hematopoietic cells of multiple lin-

eages after retroviral mediated gene transfer into CD34+ bone marrow cells. *Blood* 82: 1975–80.

72. van Beusechem, V., Bart Baumeister, J., Bakx, T. A. et al. (1994) Gene transfer into non-human primate CD34+CD11b-bone marrow progenitor cells capable of repopulating lymphoid and myeloid lineages. *Hum. Gene Ther.* 5: 295–305.

73. Hoogerbrugge, P. M., van Beusechem, V., Fischer, A. et al. (1996) Bone marrow gene transfer in three patients with adenosine deaminase deficiency. *Gene Ther.* 3: 179–183.

74. Dunbar, C. E., Seidel, N. E., Doren, S. et al. (1996) Improved retroviral gene transfer into murine and Rhesus peripheral blood or bone marrow repopulating cells primed *in vivo* with stem cell factor and granulocyte colony-stimulating factor. *Proc. Natl. Acad. Sci. USA* 93: 11871–11876.

75a. Kohn, D. B., Hershfield, M S., Carbonaro, D. et al. (1998) T lymphocytes with a normal ADA gene accumulate after transplantation of transduced autologous umbilical cord blood CD34+ cells in ADA-deficient SCID neonates. *Nat. Med.* 4: 775–780.

75. Kohn, D. B., Weinberg, K. I., Nolta, J. A. et al. (1995) Engraftment of gene-modified umbilical cord blood cells in neonates with adenosine deaminase deficiency. *Nat. Med.* 1: 1017–1023.

76. Blaese, R. M., Culver, K. W., Miller, A. D. et al. (1995) T lymphocyte-directed gene therapy for ADA⁻ SCID: initial trial results after 4 years. *Science* 270: 475–480.

77. Bordignon, C., Notarangelo, L. D., Nobili, N. et al. (1995) Gene therapy in peripheral blood lymphocytes and bone marrow for ADA-immunodeficient patients. *Science* 270: 470–475.

78. Onodera, M., Ariga, T., Kawamura, N. et al. (1998) Successful peripheral T-lymphocyte-directed gene transfer for a patient with severe combined immune deficiency caused by adenosine deaminase deficiency. *Blood* 91: 30–36.

79. Osborne, W. R., Miller, A. D. (1988) Design of vectors for efficient expression of human purine nucleoside phosphorylase in skin fibroblasts from enzyme-deficient humans. *Proc. Natl. Acad. Sci. USA* 85: 6851–6855.

80. Foresman, M. D., Nelson, D. M., McIvor, R. S. (1992) Correction of purine nucleoside phosphorylase deficiency by retroviral-mediated gene transfer in mouse S49 T cell lymphoma: a model for gene therapy of T cell immunodeficiency. *Hum. Gene Ther.* 3: 625–31.

81. Nelson, D. M., Butters, K. A., Markert, M. L. et al. (1995) Correction of proliferative responses in purine nucleoside phosphorylase (PNP)-deficient T lymphocytes by retroviral-mediated PNP gene transfer and expression. *J. Immunol.* 154: 3006–3014.

82. Candotti, F., Johnston, J. A., Puck, J. M. et al. (1996) Retroviral-mediated gene correction for X-linked severe combined immunodeficiency. *Blood* 87: 3097–3102.

83. Taylor, N., Uribe, L., Smith, S. et al. (1996) Correction of interleukin-2 receptor function in X-SCID lymphoblastoid cells by retrovirally mediated transfer of the gamma-c gene. *Blood* 87: 3103–3107.

84. Hacein-Bey, H., Cavazzana, C. M., Le DF et al. (1996) gamma-c gene transfer into SCID X1 patients' B-cell lines restores normal high-affinity interleukin-2 receptor expression and function. *Blood* 87: 3108–3116.

85. Cavazzana Calvo, M., S., H. -B., de Saint Basile, G. et al. (1996) Role of interleukin-2 (IL-2), IL-7, and IL-15 in natural killer cell differentiation from cord blood hematopoietic progenitor cells and from gamma chain transduced severe combined immunodeficiency X1 bone marrow cells. *Blood* 88: 3901–3909.

86. Candotti, F., Oakes, S. A., Johnston, J. A. et al. (1996) *In vitro* correction of JAK3-deficient severe combined immunodeficiency by retroviral-mediated gene transduction. *J. Exp. Med.* 183: 2687–2692.
87. Bunting, K. D., Sangster, M. Y., Ihle, J. N. et al. (1998) Restoration of lymphocyte function in Janus kinase 3-deficient mice by retroviral-mediated gene transfer. *Nat. Med.* 4: 58–63.
88. Taylor, N., Bacon, K. B., Smith, S. et al. (1996) Reconstitution of T cell receptor signaling in ZAP-70-deficient cells by retroviral transduction of the ZAP-70 gene. *J. Exp. Med.* 184: 2031–2036.
89. Dunbar, C. E., Cottler, F. M., O'Shaughnessy, J. A. et al. (1995) Retrovirally marked CD34-enriched peripheral blood and bone marrow cells contribute to long-term engraftment after autologous transplantation. *Blood* 8535: 3048–3057.
90. Moore, K. A., Deisseroth, A. B., Reading, C. L. et al. (1992) Stromal support enhances cell-free retroviral vector transduction of human bone marrow long-term culture-initiating cells. *Blood* 79: 1393–9.
91. Nolta, J. A., Smogorzewska, E. M., Kohn, D. B. (1995) Analysis of optimal conditions forr retroviral-mediated transduction of primitive hematopoietic cells. *Blood* 1: 101–110.
92. Xu LC, Karlsson, S., Byrne ER et al. (1995) Long-term *in vivo* expression of the human glucocerebrosidase gene in nonhuman primates after CD34+ hematopoietic cell transduction with cell-free retroviral vector preparations. *Proc. Natl. Acad. Sci. USA* 92: 4372–4376.
93. Hanenberg, H., Xiao, L., Dilloo, D. et al. (1996) Colocalization of retrovirus and target cells on specific fibronectin fragments increases genetic transduction of mammalian cells. *Nat. Med.* 2: 876–882.
94. Lyman, S. D., James, L., van den Bos, T. et al. (1993) Molecular cloning of a ligand for the flt3/flk-2 tyrosine kinase receptor: A proliferative factor for primitive hematopoietic cells. *Cell* 75: 1157–1167.
95. Ferrari, G., Salvatori, G., Rossi, C. et al. (1995) A retroviral vector containing a muscle-specific enhancer drives gene expression only in differentiated muscle fibers. *Hum. Gene Ther.* 6: 733–742.

2 Models of Lysosomal Storage Disorders Useful for the Study of Gene Therapy

J. Barranger, M. J. Vallor

2.1   Introduction

Lysosomal storage disorders are a group of monogenic inherited disorders that are caused by the deficiency of a catabolic enzyme. There is one exception, known as I-cell disease, where the defect is in an enzyme involved in the biosynthesis of the mannose-6-phosphate structure, required for the routing of most of the enzymes to the lysosome [1]. Localized within the organelle by a highly specialized transport system, the enzymes reside within the lysosome and are responsible for the degradation of macromolecular polymers or complex molecules that must be disposed of. Failure to degrade these molecules results in storage of the compounds and disease. The principle compounds involved in the lysosomal storage disorders are glycosaminoglycans (mucopolysaccharide), glycolipids and glycoproteins [2, 3]. Substrate accumulation is usually followed by cell dysfunction and death as well as the development of severe and often ultimately lethal symptoms in the patient, particularly when the central nervous system is involved. Many of these enzymatic deficiencies are found also in animals, (Tab. 2-1) [4–22] providing defined and controlled models for study. Bone marrow transplantation has been studied in some of these models and correlations have been drawn to the human diseases (Tab. 2-2). These results are useful in considering strategies designed to transfer genes to bone marrow stem cells [23, 24]. The majority of these animal models are found naturally, in domestic species, where selective inbreeding promotes the production of homozygotes for recessive mutant alleles. Many examples were discovered in pet and farm animals [25]. For diseases not yet discovered in animals, the advent of gene targeting in pluripotent mouse embryonic stem (ES) cells [26] has allowed the experimental development of at least three murine examples of human lysosomal storage conditions: glucocerebroside storage (Gaucher disease), cerebroside sulfatide storage (metachromatic leukodystrophy) and hexosaminidase A deficiency (Tay Sachs disease) [7–9] Additionally, gene targeted murine models of sphingomyelinase deficiency (Niemann-Pick disease), mucopolysaccharidosis VI and ceramide trihexosidase deficiency (Fabry disease) have been reported [10–12] but are not yet fully described in the published literature. Two animal models of lysosomal storage provide excellent examples of the types of studies that may be carried out with animal

Table 2-1 Some animal models of lysosomal storage diseases

Species	Enzyme deficiency	Human disease counterpart [ref.]
Rodent-Mouse	β glucuronidase	Mucopolysaccharoidosis VII (Sly Disease) [4]
	Galactosylceramidase	Globoid Cell Leukodystrophy [5]
	Unknown, cholesterol transport	Niemann Pick Disease Type C [6]
	Arylsulfatase A *	Metachromatic Leukodystrophy [7]
	Glucocerebrosidase*	Gaucher Disease [8]
	Hexosaminidase A*	Tay Sachs Disease [9]
	Sphingomyelinase*	Niemann-Pick Disease [10]
	Arylsulfatase B*	Mucopolysaccharoidosis VI (Marotoux Lamy Disease) [11]
	Ceramide Trihexosidase*	Fabry Disease [12]
Rodent-Rat	Lysosomal Acid Lipase	Wolman's Disease [13]
Cat	α Mannosidase	α Mannosidosis [14]
	β Hexosaminidase A	Type II G_{M2} Gangliosidosis [15]
	Arylsulfatase B	Mucopolysaccharoidosis VI [16]
Dog	β Galactosidase	G_{M1} Gangliosidosis [17]
	Fucosidase,	Fucosidosis [18]
	Iduronidase	Hurler Disease [19]
Sheep	β Galactosidase and αNeuraminidase	G_{M1} Gangliosidosis [20]
Goat	β Mannosidase	β Mannosidosis [21]
Bovine	α Mannosidase	α Mannosidosis [22]

models of lysosomal storage diseases, as well as some of the problems associated with their use.

2.2 The Gaucher Mouse – Interpretation of Pathology

One of the most well studied lysosomal storage diseases, Gaucher disease, is caused by the deficiency of the glucocerebrosidase (GC) enzyme [27]. In most patients, presentation is usually not associated with neurological symptoms (type 1 disease) but there is typically liver, spleen and skeletal involvement. Storage of the GC substrate, glucocerebroside, is largely confined to the cells of the reticuloendothelial system, particularly tissue macrophages. The disease is variable in the severity of its symptoms and age of onset, with patients sometimes being diagnosed in late adulthood. Effective treatments are available which can partly or wholly ameliorate the disease

Table 2-2 Lysosomal storage disorders: results of bone marrow transplantation

Enzyme deficiency	Disorder	Results
Iduronidase	Human MPS 1	Improved intellectual function; decreased storage in liver, spleen; improved skeletal deformity and mobility
Iduronidase	Canine MPS 1	Improved longevity, mobility; decreased storage and excretion of GAG
Arylsulfatase A	Human MLD	Infantile, late infantile forms – not improved Juvenile form – improved longevity and intellectual function
Fucosidase	Canine fucosidosis	Decreased storage and fucosyl compounds; improvement in CNS and PNS
α-Mannosidase	Human mannosidosis	Improved life span; stable neurologic signs for about 7 mths
α-Mannosidase	Feline mannosidosis	Improved storage in viscera and brain; improved neurologic function
Arylsulfatase B	Human MPS VI	Decreased visceral storage; improvement in cardiovascular system and skeleton
Galactocerebrosidase	Human globoid cell leukodystrophy (Krabbe)	Some improvement in longevity, decreased storage in brain
Galactocerebrosidase	Murine globoid cell leukodystrophy	Increased longevity; decreased storage in brain
Cerebroside β-galactosidase	GM 1 gangliosidosis	Decreased storage; increased longevity; improved CNS
Glucocerebrosidase	Glucosylceramide lipidosis (Gaucher)	Decreased visceromegaly; improved longevity in Type 1 disease
Sphingomyelase	Sphingomyelin lipidosis (Niemann-Pick)	Decreased storage in viscera; no change in CNS
Glucuronidase	Canine MPS VII	Improved longevity; increased mobility
Glucuronidase	Murine MPS VII	Increased longevity; decreased visceral storage; no change in CNS, PNS or eye

symptoms. These include both enzyme replacement therapy [28, 29] and bone marrow transplantation (BMT) [30]. However, despite our understanding of the causes of Gaucher disease, many uncertainties remain, concerning the fundamental pathobiology of the disease process and the cellular mechanisms of its treatment. For these reasons, an animal model is desirable. However, since no naturally occuring example of glucocerebroside storage in animals has been described, a glucocerebrosidase-deficient mouse was produced using gene targeting techniques. Gene targeting utilizes the capacity of mammalian cells to undergo homologous recombination, replacing regions of the genome with corresponding sequences of exogenous DNA. (Fig. 2-1) Gene targeting vectors are constructed using fragments of the targeted gene [26, 31], as well as two selectable markers. One marker serves both to disrupt the coding sequence of the gene and to allow "positive" selection for vector integration. The second marker provides "negative" selection in that, successful targeting removes this element and random integrants which have retained it are selected against [26]. Successful homologous recombination, thus, introduces the positive selectable marker into the endogenous coding sequence, disrupting the function

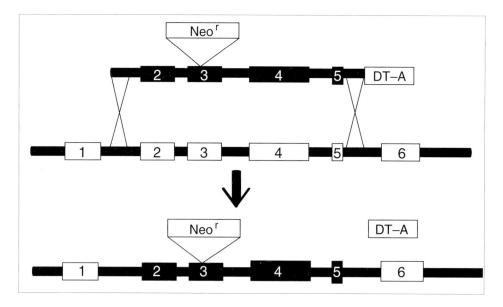

Figure 2-1. Gene Targeting of Lysosomal Enzymes Through Homologous Recombination
The gene targeting construct is composed of a piece of genomic DNA corresponding to the targeted enzyme gene. A positive selectable marker is inserted into the coding region of the gene, in this case neomycin phosphotransferase (neo^r) into exon 3. A negatively selectable marker, in this case diphtheria toxin A-chain (DT-A)is placed at either end of the region of homology. After this construct is introduced into mouse embryo stem (ES) cells, homologous recombination inside the region of homology will cause the neo^r gene to be retained and introduced into the mouse genome. The DT-A gene will be excluded in homologous recombinants. The neo^r gene will then both disrupt the function of the targeted gene, and allow selection of clones which have integrated the construct. Non-homologous integration of the gene targeting vector will result in the DT-A gene being retained, and those cells expressing it will be selected against.

of the gene. These vectors are introduced into ES cells, which are pluripotent cell lines, derived from blastocysts, usually of the 129 strain of mice [32]. (Fig. 2-2) Following selection and molecular identification of correctly gene targeted ES cell clones, the recombinant ES cells are introduced into a recipient blastocyst which they colonize [33]. These blastocysts are re-implanted into a foster mother and carried to term. Mice developing from embryos which have been successfully colonized by the ES cells are "chimeric", in that their tissues are partly ES cell derived and partly derived from the recipient blastocyst. Some of these mice will be capable of transmitting the ES cells and their targeted gene mutations, through the germline and into heterozygotes of the next generation. Heterozygotes are further mated to produce mice which are homozygous for the disrupted gene locus and essentially, exhibit no expression of the targeted gene [8, 34].

The first animal model of lysosomal storage produced, using gene targeting, was the glucocerebrosidase deficient mouse [8]. A naturally occuring canine example of Gaucher disease was described in the early 1970s, but was not able to be propagated 35). This animal exhibited neurovisceral storage of glucocerebroside but no organomegaly or skeletal involvement. The gene-targeted glucocerebrosidase deficient mouse was described in 1992. The neomycin phosphotransferase (neo^r) gene was introduced into exons 9 and 10 of the mouse GC gene. Homozygotes with this disruption exhibited cyanosis, poor feeding and movement, thin skin with rugations, and death in the first hours of life [8]. Glucocerebroside storage was seen in macrophages of the liver, bone marrow, spleen and brain but there was not appreciable organomegaly. The short life span of these mice probably did not allow time for the development of more profound organomegaly and bone lesions. However, adult mice which are treated with the glucocerebrosidase inhibitor, β conduritol epoxide, develop visceral glucocerebroside storage cells but never develop organomegaly [36]. Therefore, it is therefore possible, that glucocerebroside metabolism and the physiologic effects of its storage may be different in mice and humans. Indeed, the first animal model developed using gene targeting was for Lesch Nyhan disease (hypoxanthine phosphoribosyl transferase deficiency), in which it was found that homozygotes for the gene disruption exhibited no phenotype at all [34]. Inhibition of a related metabolic pathway was required for the Lesch Nyhan mouse model to display the overt pathology associated with the human disease [37].

The transgenic glucocerebrosidase deficient mouse model is representative of a rare subset of very severe Gaucher disease [37]. The skin rugations in particular are interesting, as they have been observed only in these cases, which are not typical of Type 2 (neurologic) disease. It has been hypothesized that GC activity may be important in maintaining skin permeability barrier integrity [38]. Thus, while this model may not exhibit all of the commonly observed symptoms of Gaucher disease, it may provide some insight into glucocerebroside metabolism that was previously unknown.

Figure 2-2. Production of Chimeric Mice and Gene Targeted Mice with ES Cells
Murine ES cells are transfected with the gene targeting construct and subjected to selection for homologous recombination. After targeted clones are picked and characterized, homologous recombinant ES cells are microinjected into blastocyst stage mouse embryos of a different strain than the 129 of the ES cells. These embryos are reimplanted into the uterus of a pseudopregnant foster mother mouse. Embryos which have been colonized by the ES cells, and come to term can be recognized by coat color chimerism; a "blotchy" coat. These chimeras are bred, and those which have incorporated ES cells into their germline can pass these cells, along with the gene disruption onto the heterozygous F1 generation. Mating of heterozygotes will produce mice homozygous for the gene disruption, and it is these mice which are deficient for the lysosomal enzyme which has been gene targeted. The homozygous mouse pictured is shown exhibiting organomegaly for clarity, although no gene targeting model of lysosomal storage disease has exhibited this phenotype.

The lethal phenotype of the GC gene disruption precludes the use of this model in most studies. In an attempt to abrogate this lethal phenotype, PCR mutagenesis was used to install the most common human Gaucher disease mutation (N370S) directly onto the mouse gene [39]. The selectable *neo^r* gene was located to the region just outside the 3' end of the mouse GC gene so that the gene function would not be destroyed. Homozygotes for this mutant GC locus die and are resorbed *in utero*, due to the disruption of a previously unknown gene, metaxin, which lies less than a kilobase downstream of the mouse GC gene [40]. The metaxin gene was not affected in the first gene targeted mouse and heterozygotes of both targeted loci were mated. These matings produced mice with one functional metaxin gene and one mutant, but otherwise intact, GC gene. An important and unanticipated finding was that the N370S mutant GC was not found to rescue the lethal phenotype of the GC gene disruption and these mice exhibited a pathology identical to mice with the homozygous gene disruption [40]. This outcome may indicate that other important genes may have been affected during homologous recombination, either in the original knockout [8], or in the disruption involving both the GC and metaxin genes [40]. The GC and metaxin genes lie in a tight group along with the thrombospondin 3 gene [40]: Other genes may also lie closely upstream of GC. Furthermore, it is also possible that the N370S mutation affects the murine enzyme more severely than the human protein (the genes are only 80% homologous between species [41]). Finally, there may be differences between the human and murine physiologies, such that the mutant enzyme is rendered nonfunctional in mice. Rescue of the lethal phenotype may be better effected with a human mutant transgene [42], rather than trying to impose this mutation directly on the murine protein.

So far, none of the animal models of Gaucher disease have exhibited the organomegaly and skeletal involvement prominent in the human disease. It may be that organomegaly and skeletal phenotypes will not be seen in Gaucher disease in non-primate systems. If these models are to be used to examine Gaucher disease

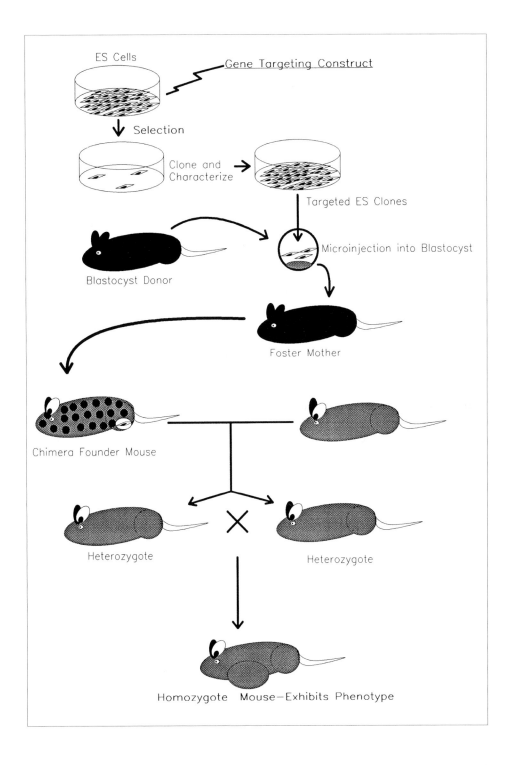

therapies, simple clearance of the stored lipid may be the only experimental parameter available to judge clinical effectiveness. It can be seen, however, that interspecies differences, in physiology, may complicate the use of animal models of lysosomal storage diseases and that definitions of specific diseases, as they present themselves in the animal, may have to be modified. Indeed, the gene targeting models of arylsulfatase A and hexosaminidase A deficiencies also exhibit lipid storage only, with no other apparent physiological change. It can be seen that the effects of enzyme deficiency in humans and the corresponding animal models may be very different, particularly in the case of gene targeting, where the complete absence of enzyme is not biochemically reflective of most disease states. Even in these models, dissimilar pathologies in human and animal systems may offer insights into aspects of lysosomal storage diseases and their metabolic and physiological effects, that are not apparent in most human cases.

2.3   The Mucopolysaccharidosis Type VII Mouse – Studies of treatment

Mouse models of human genetic diseases, produced by gene targeting, often exhibit very different pathologies than the human counterparts (see above). However, most of the naturally occuring animal models of human lysosomal storage diseases are phenotypically similar to the human examples [4–6, 12–22]. One of the most successful of these models has been the mucopolysaccharidosis VII mouse [4]. Mucopolysaccharidosis Type VII (MPS VII, Sly Syndrome) is caused by a deficiency in β glucuronidase, which results in storage of non-degraded heparan and dermatan sulfates (glycosaminoglycan) in the lysosomes [43]. Clinical manifestations include course facies, cornal opacity, shortened stature, skeletal deformity, mental retardation and developmental delay. An animal model is of particular value in studying MPS VII, as only 25 patients have been reported since the disease was first described [43]. A mouse model of β glucuronidase deficiency was described in C57Bl/6 mice at Jackson Laboratory in 1989 [4]. This model displays an exceptional resemblance to human MPS VII pathology, including shortened nose and other skeletal deformities, short length, corneal opacity and extensive cytoplasmic vacuolization in endothelial cells of the spleen and liver. The storage material was, later, identified, biochemically, as glycosaminoglycan. Additionally, the animals lack white adipose tissue, which is not a feature of the human disease.

The MPS VII mouse model has been valuable in examining novel therapeutic approaches to the treatment of the disease, particularly with bone marrow transplantation (BMT) and gene therapy. BMT has been the sole effective treatment developed for the mucopolysaccharidoses, although, MPS VII has not yet been treated in this way in humans [43]. The MPS VII mouse is an excellent system to approach BMT studies because of the ready availability of syngeneic bone marrow from the normal C57Bl/6 mouse strain. BMT experiments in MPS VII mice have resulted in the increase in liver β glucuronidase activity to 15–20% of normal [44], which would be expected if all of the endogenous liver Kupfer cells had been replaced by donor cells. Brain β glucuronidase activity was increased to 5–7% of normal. A variety of cell types exhibited dramatic or complete reduction of lysosomal storage material, including spleen, liver, cornea, perivascular and meningeal cells. The life span of treated mice was also increased three-fold and was similar to that of normal mice receiving the same treatment.

The high morbidity and mortality associated with human allogeneic BMT makes the concept of gene transfer to hematopoeitic cells an attractive alternative for treatment of the lysosomal storage diseases. In this approach, copies of the genes for deficient enzymes are introduced into the affected individual's own bone marrow stem cells, typically using viral-based vectors [45]. These stem cells are then re-introduced into the bone marrow, where they engraft and repopulate the blood with enzyme competent cells. The problems of rejection and graft vs. host disease are thus, minimized since the transplanted tissues are autologous. Several diseases are being studied with BMT/gene transfer already [45]. MPS VII mouse bone marrow has been treated with retroviral vectors [46, 47]. Use of retroviral vectors allowed expression of β glucuronidase for longer than six months and extended the life spans of treated mice between three and four-fold [46, 47]. The actual levels of β glucuronidase expression in treated animals was less than 5% of normal in all tissues examined. This low level of enzymatic activity correction was explained by histological staining of β glucuronidase in the livers and spleens of treated mice, which revealed that only the bone marrow-derived macrophages were positive. However, in spite of the low amount of enzyme activity, dramatic decreases in lysosomal storage were observed in liver and spleen [46, 47]. This observation has suggested that a small number of enzyme competent cells may be sufficient to ameliorate lysosomal storage by 'cross correction', in which enzyme activity is transferred from these cells into the storage cells [46]. This finding indicates that lysosomal storage diseases, in general, may be amenable especially to BMT/gene therapy. Lysosomal storage was not diminished in cornea or kidney, two tissues which did respond to conventional BMT [44, 46]. While BMT would seem to be a good approach to MPS VII treatment, it was found that neuron and glial cell storage of glycosaminoglycan was not affected, indicating that bone marrow-directed treatments will not be sufficient to alleviate the neurological effects of MPS VII completely. In other disorders, however, BMT may be useful in ameliorating neurologic complications [14, 48].

Direct gene transfer to brain may be required to reverse the pathology encountered in some of the neurodegenerative lysosomal storage diseases. The β glucuronidase gene has been delivered to brain cells utilizing herpes simplex virus 1 (HSV1)-based gene vectors, which are neurotropic [49]. MPS VII mice were innoculated, corneally, with β glucuronidase expressing HSV1 vectors. Following innoculation, β glucuronidase positive cells were detected in the trigeminal ganglia and brainstem, showing that the virus vector could spread from the point of innoculation into other parts of the central nervous system. β glucuronidase activity was not detectable but expression of the β glucuronidase gene was observed for four months and the number of positive cells did not diminish. An alternative approach to expression of β glucuronidase in brain relies on pluripotential, neuroglial progenitor cells [50]. These cells have been isolated from the brains of neonatal mice and when injected into MPS VII mouse CNS after birth, can contribute extensively to both neural and glial cell populations in the developing brain. These cells migrate to various regions of the brain and express β glucuronidase throughout the life of the transplanted animal. Glycosaminoglycan storage is reduced substantially in all areas of the brain colonized by the progenitor cells, even in the cerebrocortical neurons, which are derived from the MPS VII cells of the recipient. This observation suggests that the mechanism of cross correction may also be active in brain, where the normal, progenitor-derived glial cells provide β glucuronidase activity to the MPS VII neurons. These results suggest that long term correction of MPS VII in brain may be feasible, if a human line of neuroglial progenitor cells can be developed.

Another approach to the treatment of MPS VII in mice has combined gene therapy with enzyme replacement [51]. A number of lysosomal diseases are thought to be amenable to treatment with an exogenous supply of the deficient enzyme, which is taken into the storage cells by endocytosis. Patients with Gaucher disease were the first, with a lysosomal storage disorder, to be treated successfully with purified enzyme [28, 29]. The cost of production for this drug is high, however, as the enzyme, glucocerebrosidase, is not abundant in any human tissue. A way of overcoming this problem is to use gene therapy techniques to allow the patient's own tissues to express and excrete the deficient enzyme. Enzyme secreted from an artificial "organoid" would then be taken up from the blood stream on a continuous basis, similar to the method of "cross correction" described earlier. In the MPS VII mouse model, primary skin fibroblasts were transfected with a retroviral vector that expressed β glucuronidase [52]. These cells were embedded into a collagen matrix and implanted into the abdominal space of MPS VII mice. After vascularization, this organoid was shown to express β glucuronidase for up to five months. The levels of β glucuronidase activity were shown to be up to 6% of normal in most tissues. Again, this low level of enzyme activity was sufficient to correct lysosomal storage lesions completely, in spleen and liver. The skeletal and neurological effects of MPS VII were unchanged however.

MPS VII is diagnosed very infrequently in humans. Therefore, it would be difficult to evaluate the effects of different types of therapies from studies in MPS VII patients. The MPS VII mouse model has proven to be extremely valuable in studies of this nature, because of its strong resemblance to the human disease pathology. Many of the lysosomal storage disorders result from enzyme deficiencies in similar metabolic pathways [3], and the results seen in this model may be able to be extrapolated to other lysosomal storage diseases, for which animal models do not exist. For example, the phenotypic effects of delivery of β glucuronidase to the blood-stream, via the artificial organoid, closely resemble the clinical effects of enzyme therapy for Gaucher disease in humans [28, 29]. It is possible then, that the encouraging results observed in MPS VII mouse gene therapy studies would be observed in human gene therapy studies in patients with lysosomal storage diseases.

2.4   Conclusion

It can be seen that animal models of lysosomal storage disease provide a powerful investigative tool, particularly when the animal phenotype closely matches the disease symptoms in humans, as seen in the MPS VII mouse. A variety of therapeutic strategies may be assessed without resorting to human clinical experimentation. However, caution must be exercised in the use of animal models, as animal physiologies may differ from human, leading, therefore, to unexpected other symptoms. This is especially evident in gene targeted mouse models of lysosomal storage disease, considering the lethal results of glucocerebrosidase deficiency in mice and the lack of overt morphological effects in the hexosaminidase A and arylsulfatase A-deficient mouse models. However, as seen in the skin abnormalities of the GC-deficient mouse, animal models which differ from the human phenotype, can provide additional insight into the disease pathology that is not possible in humans.

2.5   Acknowledgement

This work was supported by a grant from the National Institutes of Health (DK43709). We would like to thank Margaret Jasko for assistance in preparing the manuscript.

References

1. Kornfeld, S., Sly, W. S. (1995) I-cell disease and pseudo-Hurler polydystrophy: disorders of lysosomal enzyme phosphorylation and localization: *In*: C. R. Scriver, A. L. Beaudet, W. S. Sly, D. Valle (eds), *The Metabolic Basis of Inherited Disease*. McGraw-Hill, Inc., New York, p. 2495.
2. *In*: Barranger JA, Brady, RO (eds) (1984) *Molecular Basis of Lysosomal Storage Disorders*, Orlando, Academic Press Ltd, pp xvii-xix.
3. *In*: Scriver, C. R., Beaudet A. L., Sly, W. S., Valle, D. (eds) (1995) *The Metabolic and Molecular Basis of Inherited Disease*, New York, McGraw Hill Inc., see specific chapters.
4. Birkenmeier, E. H., Davisson, M. T., Beamer, W. G., Ganschow, R. E., Vogler, C. A., Gwynn, B., Lyford, K. A., Maltais, L. M., Wawrzyniak, C. J. (1989) Murine mucopolysaccharidosis type VII. Characterization of a mouse with β glucuronidase deficiency. *J. Clin. Invest.* 83: 1258.
5. Duchen, L. W., Eicher, E. M., Jacobs, J. M., Scaravilli, F., Teixeira, F. (1980) A globoid

type of leukodystrophy in the mouse: The mutant twitcher. *In*: N. Bauman (ed.), *Neurological Mutations Affecting Myelination*. Amsterdam, Elsevier, p. 107.

6. Pentchev, P. G., Boothe, A. D., Kruth, H. S., Weintroub, H., Stivers, J., Brady, R. O. (1984) A genetic storage disorder in BALB/C mice with a metabolic block in esterification of exogenous cholesterol. *J. Biol. Chem.* 259(9): 5784.

7. Gieselmann V, personal communication.

8. Tybulewicz, V. L. J., Tremblay, M. L., LaMarca, M. E., Willemsen, R., Stubblefield, B. K., Winfield, S., Zablocka, B., Sidransky, E., Martin, B. M., Huang, S. P., Mintzer, K. A., Westphal, H., Mulligan, R. C., Ginns, E. I. (1992) Animal model of Gaucher's disease from targeted disruption of the mouse glucocerebrosidase gene. *Nature* 357: 407.

9. Yamanaka, S., Johnson, M. D., Grinberg, A., Westphal H. Crawley, J. N., Taniike, M., Suzuki, K., Proia, R. L. (1994) Targeted disruption of the hexa gene results in mice with biochemical and pathologic features of Tay-Sachs disease. *Proc. Natl. Acad. Sci. USA* 91(21): 9975.

10. personal communication.

11. personal communication.

12. personal communication.

13. Kuriyama, M., Yoshida, H., Suzuki, M., Fujiyama, J., Igata, A. (1990) Lysosomal acid lipase deficiency in rats: lipid analyses and lipase activities in liver and spleen. *J. Lipid Res.* 31: 1605.

14. Walkley, S. U., Thrall, M. A., Dobrenis, K., Huang, M., March, P. A., Siegel, D. A., Wurzelmann, S. (1994) Bone marrow transplantation corrects the enzyme defect in neurons of the central nervous system in a lysosomal storage disease. *Proc Natl Acad Sci. USA* 91: 2970.

15. Muldoon, L. L., Neuwelt, E. A., Pagel, M. A., Weiss, D. L. (1994) Characterization of the molecular defect in a feline model for Type II GM2-Gangliosidosis (Sandhoff Disease). *Am. J. Pathol.* 144(5): 1109.

16. Jezyk, P. F., Haskins, M. E., Patterson, D. F., Mellman, W. J., Greenstein, M. (1977) Mucopoly-saccharidosis in a cat with arylsulfatase B deficiency: A model of Maroteaux-Lemay syndrome. *Science* 198: 834.

17. Alroy, J., Orgad, U., De Gasperi, R., Richard, R., Warren CD, Knowles, K., Thalhammer JG, Raghaven SS (1992) Canine GM1 gangliosidosis. A clinical, morphologic, histochemical and biochemical comparison of two different models. *Am. J. Pathol.* 140: 675.

18. Healy, P. J., Farrow, B. R. H., Nicholas, F. W., Hedberg, K., Ratcliffe, R. (1984) Canine fucosidosis: a bichemical and genetic investigation. *Res. Vetrin. Sci.* 36: 354.

19. Shull, R. M., Helman, R. G., Spellacy, E., Constantopoulous, G., Munger, R. J., Neufeld, E. F. (1984) Morphologic and biochemical studies of canine mucopolysaccharidosis I. *Am. J. Pathol.* 114: 487.

20. Prieur, D. J., Ahern Rindell, A. J., Murnane, R. D. (1991) Ovine GM1-gangliosidosis. *Am J. Pathol.* 139: 1511.

21. Jones, M. Z., Cunningham, J. G., Dade, A. W., Dawson, G., Laine, R. A., Williams, C. S. F., Alessi, D. M., Mostowski, U. V., Vorrow, J. R. (1982) *In*: *Animal models of inherited metabolic diseases*. New York, Alan R. Liss Inc. p. 165.

22. Jolly, R. D., Thompson, K. G. (1978) The pathology of bovine mannosidosis. *Vet. Pathol.* 15: 141.

23. Ohashi, T., Boggs, S. S., Robbins, P. D., Bahnson, A. B., Patrene, K., Wei, F. S., Wei, J. F., Li, J., Lucht, L., Fei, Y., Clark, S., Kimak, M., He, H., Mowery-Rushton, P., Barranger

JA (1992) Efficient transfer and sustained high expression of the human glucocerebrosidase gene in mice and their functional macrophages following transplantation of bone marrow transduced by a retroviral vector. *Proc. Natl. Acad. Sci. USA* 89: 11332.

24. Nimgaonkar M, Mierski J, Beeler M, Kemp A, Lancia J, Mannion-Henderson J, Mohney T, Bahnson A, Rice E, Ball ED, Barranger JA: Cytokine mobilization of peripheral blood stem cells in patients with Gaucher disease with a view to Gene Therapy. *Exp. Hematol.; in press.*

25. Desnick, R. J., McGovern, M. M., Schuchman, E. H., Haskins, M. E. (1982) *In: Animal models of human metabolic diseases.* New York, Alan R. Liss Inc. p. 27.

26. Mansour SL, Thomas KR, Capecchi MR: Disruption of the proto-oncogene *int-2* in mouse embryo derived stem cells: a general strategy for targeting mutations to non-selectable genes. *Nature* 336: 348, 1988.

27. Barranger, J. A., Ginns, E. H. (1989) *In:* C. R. Scriver, A. L. Beaudet, W. S. Sly, D. Valle (eds), *The Metabolic Basis of Inherited Disease.* McGraw-Hill, Inc., New York, p. 2641.

28. Barranger, J. A., Ohashi, T., Hong, C. M., Tomich, J., Aerts JFGM, Tager, J. M., Nolta, J. A., Sender, L. S., Weiler, S., Kohn, D. B. (1989) Molecular pathology and therapy of Gaucher disease. *Jap. J. Inher. Met. Dis.* 51: 45.

29. Barton, N. W., Brady, R. O., Dambrosia, J. M., Bisceglie, A. M., Doppelt, S. H., Hill, S. C., Mankin, H. J.,Murray, G. J., Parker, R. I., Argoff, C. E., Grewal, R. P., Yu KT (1991) Replacement therapy for inherited enzyme deficiency: macrophage targeted glucocerebrosidase for Gaucher's disease. *N. Eng. J. Med.* 324: 1464.

30. Rappeport, J. M., Barranger, J. A., Ginns, E. H. (1986) Bone marrow transplantation antation in Gaucher disease. *In: Birth Defects: Original Article Series.* March of Dimes Birth Defects Foundation, New York, Alan R. Liss Inc., 22: 101.

31. Te Reile, H., Maandag, E. R., Berns, I. (1992) Highly efficient gene targeting in embryonic stem cells through homologous recombination with isogenic DNA constructs. *Proc. Nat. Acad. Sci. USA* 89: 5128.

32. Evans, M. J., Kaufmann, M. H. (1981) Establishment in culture of pluripotential cells from mouse embryos. *Nature* 292: 154.

33. Bradley, A., Evans, M. J., Kaufmann, M. H., Robertson, E. J. (1984) Formation of germline chimeras from embryo-derived teratocarcinoma cell lines. *Nature* 309: 255.

34. Finger, S., Heavans, R. P., Sirinathsinghji, D. J. S., Kuehn, M. R., Dunnett, S. B. (1988) Behavioral and neurochemical evaluation of a transgenic mouse model of Lesch-Nyhan syndrome. *J. Neurol. Sci.* 86: 203.

35. Hartley, W. J., Blakemore, W. F. (1911) Neurovisceral glucocerebroside storage (Gaucher's disease) in a dog. *Vet. Pathol.* 10: 973.

36. Adachi, M., Volk, B. W. (1977) Gaucher disease in mice induced by conduritol β epoxide: morphologic features. *Arch. Pathol. Lab. Med.* 101: 255.

37. Wu CL, Melton DW (1993) Production of a mouse model for Lesch-Nyhan syndrome in hypoxanthine phosphoribosyltransferase-resistant mice. *Nat. Genet.* 3: 235.

38. Sidransky, E., Sherer, D. M., Ginns, E. I. (1992) Gaucher disease in the neonate: a distinct Gaucher phenotype ia analogous to a mouse model created by targeted disruption of the glucocerebrosidase gene. *Pediat. Res.* 32(4): 494.

39. LaMarca, M. E., Yoshikawa, H., McKinney, C. E., Stubblefield, B. K., Winfield, S., Carmon, L., Martin, B. M., Sidransky, E., Ginns, E. I. (1993) Targeting the common N370S Gaucher point mutation to the glucocerebrosidase gene in murine embryonic stem cells. *Am. J. Hum. Genet.* 53(3): Abs. 917.

40. Bornstein, P., McKinney, C. E., LaMarca, M. E., Winfield, S., Shingu, T., Devaralu, S., Vos, H. L., Ginns, E. I. (1995) Metaxin, a gene contiguous to both thrombospondin 3 and glucocerebrosidase, is required for embryonic development in the mouse: Implications for Gaucher disease. *Proc. Nat. Acad. Sci. USA* 92: 4547.

41. O'Neill, R., Toshihou, T., Kozak, C. A., Brady, R. O. (1989) Comparison of the chromosomal localization of murrine and human glucocerebrosidase genes and of deduced amino acid sequences. *Proc. Nat. Acad. Sci. USA* 86: 5049.

42. Wei, F. S., Vallor, M. J., Wei, J. F., Barranger, J. A. (1994) Mouse model of Gaucher disease created by targeted disruption of murine glucocerebrosidase (GC) gene and transgenic expression of of a mutant human GC gene. Abs, *Cold Spring Harbor,* NY, September.

43. Neufeld, E. F., Muenzer, J. (1995) *In:* C. R. Scriver, A. L. Beaudet, W. S. Sly, D. Valle (eds), *The Metabolic Basis of Inherited Disease.* McGraw-Hill, Inc., New York, p. 2465.

44. Birkenmeier, E. H., Barker, J. E., Vogler, C. A., Kyle, J. W., Sly, W. S., Gwynn, B., Levy, B., Pegors, C. (1991) Increased life span and correction of metabolic defects in in murine mucopolysaccharidosis type VII after syngeneic bone marrow transplantation. *Blood* 78: 3081.

45. Miller, A. D. (1992) Human gene therapy comes of age. *Nature* 357: 455.

46. Wolfe, J. H., Sands, M. S., Barker, J. E., Gwynn, B., Rowe, L. B., Vogler, C. A., Birkenmeier (1992) Reversal of pathology in murine mucopolysaccharidosis type VII by somatic cell gene transfer. *Nature* 360: 749.

47. Maréchal, V., Naffakh, N., Danos, O., Heard, J. M. (1993) Disappearance of lysosomal storage in spleen and liver of mucopolysaccharidosis VII mice after transplantation of genetically modified bone marrow cells. *Blood* 82(4): 1358.

48. Auborg, P., Blanche, S., Jambaque, I., Rocchiccioli, F., Kalifa G., Naud-Saudreau, C., Rolland, M. O., Debre M., Chaussain JL. (1990) Reversal of early neurologic and neuroradiologic manifestations of X-linked adrenoleukodystrophy by bone marrow transplantation. *N. Engl. J. Med.* 332(26): 1860.

49. Wolfe, J. H., Deshmane, S. W. L., Fraser, N. W. (1992) Herpesvirus vector gene transfer and expression of β-glucuronidase in the central nervous system of MPS VII mice. *Nat. Genet.* 1: 379.

50. Snyder, E. Y., Taylor, R. M., Wolfe, J. H. (1995) Neural progenitor cell engraftment corrects lysosomal storage throughout the MPS VII mouse brain. *Nature* 374: 367.

51. Moullier, P., Bohl, D., Heard, J. M., Danos, O. (1993) Correction of lysosomal storage in the liver and spleen of MPS VII mice by implantation of genetically modified skin fibroblasts. *Nat. Genet.* 4: 154.

52. Moullier, P., Bohl, D., Heard, J. M., Danos (1993) Correction of lysosomal storage in the liver and spleen of MPS VII mice by implantation of genetically modified skin fibroblasts. *Nat. Genet.* 4: 154.

3 Gene Therapy for Cystic Fibrosis

D. J. Porteous, J. A. Innes

Summary

It is unarguable that results reported in the early trials of somatic gene therapy for CF are equivocal. However, the inappropriately negative response which these studies have created in some quarters should be qualified and countered. Remarkable progress has been made and independent strategies have been tested in the very few years since the *CFTR* gene was cloned. The inadequacies of the very early generation of adenoviruses were easily predicted, as were the inefficiencies of the first generation cationic liposomes. What is encouraging is, that these early studies have stimulated rapid re-evaluation and further development of second and third generation vectors, which can be tested in laboratory mice engineered to have the *CFTR* defect as a prelude to clinical trials. As with any new development of clinical medicine, the development of somatic gene therapy protocols for cystic fibrosis is likely to be incremental. Only when safe and effective methods of gene delivery have been developed, will it be appropriate to apply this form of treatment to young CF individuals prior to the onset of irreversible lung damage. Under these circumstances, there is every reason to believe that the rationale underlying CF somatic gene therapy will be realised.

3.1   Background

3.1.1   The Clinical Impact of Cystic Fibrosis

Cystic fibrosis (CF) is an autosomal recessive disease affecting 1 in 2500 live births in caucasian populations (reviewed in [1–3]). The CF gene was first identified in 1989. In CF patients both copies of the gene are mutated, with any of several hundred identified mutations, resulting in functional abnormalities affecting the epithelial tissues of the lung, pancreas, liver, intestine and sweat glands. The major clinical manifestations of cystic fibrosis are exocrine pancreatic insufficiency and chronic lung infection leading to bronchiectasis, pulmonary scarring and ultimately to respiratory failure. Improved nutrition, pancreatic enzyme supplements, physiotherapy and antibiotics have extended life expectancy from under 10 years of age in 1960 to a median of 28 in 1991. Despite these advances, the majority of patients become burdened with respiratory disability, the need for regular physiotherapy and multiple drug therapies for much of their shortened lives.

3.1.2 Cellular Pathophysiology of CF

The cystic fibrosis gene codes for a membrane protein, known as the Cystic Fibrosis Transmembrane Conductance Regulator (CFTR), which is expressed physiologically in a variety of epithelial tissues. The protein acts as a chloride channel in the apical (lumenal) membrane of epithelial cells and is activated by intracellular cAMP. In respiratory epithelium, CFTR-dependent chloride secretion is part of the complex and incompletely understood regulation of the water and electrolyte composition of epithelial lining fluid.

Cystic fibrosis respiratory epithelium, tested *in vitro,* shows both failure to secrete chloride and hyperabsorption of sodium. Sodium hyperabsorption occurs through epithelial sodium channels which appear, *in vitro,* to be inhibited by the presence of CFTR [4]. The anatomical distribution of expression of the CF gene, in healthy human airways has been studied by examining unused donor lung tissue [5]. In this study, CF gene mRNA was found in approximately 5–10% of airway epithelial cells, including ciliated and non-ciliated cells. Expression occurred in large and small airways down to the terminal bronchi, with less expression at the alveolar level.

Pathologically, studies of lungs from neonates and children who died with CF suggest that an inflammatory bronchiolitis of peripheral airways is the earliest lesion, with some airway dilatation [6]. Within a few years, however, small airway narrowing with fibrosis and strictures appear. In addition, microbial airway colonization, initially with *Staphylococcus aureus* and later with a variety of gram negative bacilli, supervenes with progressive inflammatory scarring of airways.

In advanced CF lung disease, there is extensive destruction of lung parenchyma resulting from chronic sepsis, with many dilated, pus-filled and fibrous-walled cavities within the lung substance. The large airways are thickened, scarred and distorted by chronic infection and microscopically there is extensive mucosal inflammation and loss of the epithelial cilia. Small airways are frequently destroyed by infective scarring or obstructed by thickened mucus, and loss of supporting elastic recoil in remaining airways leads to expiratory airflow limitation on pulmonary function testing.

3.2 Somatic Gene Therapy for Cystic Fibrosis

3.2.1 Rationale

In the majority of cases, progressive lung damage is the main life-threatening manifestation of CF, although, occasionally cirrhosis with portal hypertension and liver failure dominate. Pancreatic and intestinal manifestations of the disease are managed by careful attention to diet and pancreatic enzyme supplementation. Early attempts at gene therapy have, therefore, concentrated on the potential for airway-delivered topical treatment of the respiratory epithelium [7–10]. The aim is to transfer, into epithelial cells, a wild-type *CFTR* cDNA construct to make possible the expression of normal CFTR protein and thus, to restore normal chloride transport. A fundamental yet unproven belief underlying this approach is that the restoration of CFTR function will result in improvements in epithelial lining fluid physiology which will be beneficial to patients. When assessing the effects of gene replacement therapy, intended to reverse the primary defect, it is important to bear in mind the secondary damage inflicted on the CF lung by chronic infection and the associated immune response, which inevitably limits the scope for improvement in established CF lung disease.

3.2.1.1 Detecting Success in CF Gene Therapy Experiments

To test CF gene therapy strategies, it is necessary to develop a range of measurable end-points by which to judge success. As therapy moves from the lab to the clinical setting, these end points become both more important and harder to achieve.

3.2.1.2 Detecting Gene Transfer

Gene constructs currently under investigation contain sequences additional to the CFTR cDNA (eg. exogenous promoters and antibiotic resistance genes). Using specific oligonucleotide primers, it is therefore possible to detect transfected DNA by PCR in the presence of a variety of endogenous mutant CF genes. Vector specific sequences can be detected by PCR, DNA, *in situ* or Southern hybridisation.

3.2.1.3 Detecting Gene Expression

Many CF mutations express endogenous mRNA and manufacture mutant CF proteins. The use of reverse transcriptase PCR to detect transgene mRNA must therefore also involve the identification of transgene-specific sequences and appropriate controls to distinguish endogenous from transgene expression. Depending upon the specific nature of the construct, transgene specific detection by RNA *in situ* hybridisation may also be possible.

3.2.1.4 Detecting CFTR Generated as a Result of Therapy

The use of monoclonal antibodies to demonstrate CFTR synthesis is greatly complicated by the frequent occurrence of cross-reactions with endogenously expressed mutant CFTR protein and related cross-reacting proteins [11]. The commonest mutation in caucasian populations, ΔF508, generates a mutant protein which accumulates in the Golgi apparatus instead of inserting into the apical membrane, therefore the microscopic location of staining may be used to distinguish mutant from wild-type gene product [12]. Even then, there is evidence that a small proportion of ΔF508/ΔF508 cells can show apical localisation [13] so this in itself can not be used to conclude that (inefficient) somatic gene transfer has been successful [9]. Other mutations, (e.g. G551D) produce proteins which reach the cell membrane but remain dysfunctional, and cannot be identified by their subcellular localisation. In practice, the wealth of different mutations and resulting gene products greatly limits the usefulness of antibody staining for CFTR [14].

3.2.1.5 Testing CFTR Function in vitro

CFTR-dependent chloride currents may be detected in cell culture and in tissue samples by measuring short-circuit currents while stimulating cAMP-dependent CFTR activity using drugs such as forskolin [15]. These methods give a global answer for a large number of cells but may be hard to apply to small tissue samples (eg bronchial biopsies): In these samples, the effect of the relatively large-cut tissue edge may obscure the behaviour of the intact epithelium. Single cell recordings of Cl channel activity can also be made [16] but are labour-intensive and may not be representative of the overall behaviour of a treated epithelium. Halide-sensitive fluorescent dyes, such as SPQ, may be used to indicate chloride channel activity in cell cultures and tissue samples (biopsies or brushings) [17]. Using this technique, a field of cells harvested (eg. by brushing) from a treated respiratory epithelium may be examined for the percentage of cells showing Cl channel activity.

3.2.1.6 Transgenic Mouse Models of Cystic Fibrosis

Gene targeting or deletion have been used to create strains of mice which express either no CFTR or very low levels of CFTR(reviewed in [18]). Notwithstanding interspecies differences, these animal models give invaluable insights into the pathogenesis of disease and offer the opportunity to test gene therapy strategies for toxicity and efficacy *in vivo*. Null mutations tend to be lethal early in life, from intestinal complications. Mutations which allow some wild-type expression give prolonged survival [19, 20] and there is evidence that these animals are less able to clear CF associated pathogens than control mice and develop pathogen-specific lung disease in response to repeated exposure [21]. Tracheal tissue from CF mutant mice also shows characteristic defects of chloride channel function, which can be partially or wholly corrected following instillation [22] or inhalation [23] of suitable exogenous CF gene constructs.

3.3 Developing Vectors for Cystic Fibrosis Gene Therapy

In the past five years, a number of viral and non-viral delivery systems have been developed which can transfer genes to respiratory epithelia.

3.3.1 Adenoviruses

Adenoviruses evolved to infect mammalian respiratory epithelia and insert genetic material into the cytoplasm of epithelial cells. While this has resulted in the virus being equipped with capsid proteins which facilitate epithelial binding, it has also resulted in the evolution of complex host antibody and cytotoxic T-cell mediated immune defences which resist gene transfer and destroy transfected cells. In this respect, viral-mediated gene therapy has important differences to conventional pharmacotherapy and non-viral gene therapy (see below).

The development of adenoviral vectors has involved progressive attenuation of the wild-type viral genome by deletion of various early antigen genes. First generation vectors had E1A and E1B deletions and have been shown to be capable of producing efficient transfection of respiratory epithelium with marker genes and human *CFTR* in a variety of animal species (reviewed in [24]). These vectors, however, were also capable of provoking inflammatory responses in lung tissue, with T-cell activation and also neutralising antibodies [24]. Additional deletions in the E2 and E4 regions further reduce antigenicity [25], however, to date no adenoviral vector has been shown to be entirely free of the potential to cause a T cell-mediated acute inflammatory response in experimental animals. Efforts continue to remove further non-essential elements of the viral particle to reduce further its antigenicity, however, it is likely that even the fundamental capsid proteins (fiber, penton and hexon) will stimulate a host immune response. Attempts to modulate the host response to viral vectors to reduce toxicity appear attractive at first sight but carry the additional potential hazard of impairing host defences against wild-type viruses in this compromised group of patients.

Recent evidence from mouse studies and resected human bronchial tissue has demonstrated relative resistance of healthy columnar bronchial epithelium to transfection, in contrast to the efficient transfection seen in cultured epithelium or in basal cell layers exposed by abrasion [26]. Further refinement of adenoviral vectors is needed to overcome this inefficient and selective transfection *in vivo*.

3.3.2 Adeno-associated Viruses

Adeno-associated virus (AAV) is a parvovirus which is capable of site-specific integration in human chromosome 19 [27, 28]. In theory, stable integration carries the potential for long-term transgene expression following a single treatment [29]. Problems, which are required to be addressed before this vector can be used for practical CF gene therapy, include (a) its limited carrying capacity (b) low transfection efficiency and (c) its long-term safety profile *in vivo*. Deletion of Rep and Cap segments of the genome can free enough space for the CFTR gene but at present the co-infection with adenovirus is required to yield practical levels of transfection efficiency. The immunogenicity of this agent in human lung, under conditions resulting in potentially therapeutic levels of transduction and expression, remains unknown.

3.3.3 Liposomes

Cationic liposomes, in various formulations, have been in common use in the research laboratory for transient and stable transfection of cultured cells [30, 31]. The biophysics of cationic liposome mediated gene transfer is still poorly understood but the majority of cell types will, to varying extents, be transfectable by DNA/liposome complexes of net negative charge. The major reason for the relative inefficiency of transfection, *in vitro,* compared with viruses such as adenovirus is that there is no mechanism for the DNA/liposome complexes to escape the endosomal compartment. Thus, although the majority of cells take up DNA/liposome complexes, the DNA reaches the nucleus in only a very small proportion. Despite this limitation of the current generation of liposome DNA constructs, they have several powerful attributes: hypoimmunogenicity, capacity to transfect large DNA molecules and suitability for industrial production, quality control and formulation.

3.4 Controlling Expression; The Choice of Promoter

Ideally, the *CFTR* cDNA would be under the transcriptional control of the endogenous elements but the *CFTR* promoter is poorly defined and little is known about other possible local and distant control elements [32]. Consequently, the vectors to date have utilised either the endogenous adenoviral promoter, mammalian house-keeping promoters or, in the case of cationic liposomes, ubiquitous viral promoters. The hope is, that these promoters will drive *CFTR* expression in the appropriate cells at a level sufficient for therapeutic benefit, while avoiding possible negative effects of overexpression or inappropriate expression in cells not normally expressing *CFTR*.

3.5 Delivering Gene Therapy to Airway Epithelium

3.5.1 Direct Instillation

Early animal studies [22, 24] and human trials [9] have employed intra-tracheal or bronchoscopic, intra-lobar instillation of gene therapy. While this allows high efficiency of delivery and localization of treatment for these phase 1 trials, it is clearly not a practical method for routine clinical use, since bronchoscopy itself carries the risk of causing infections in CF patients.

3.5.2 Aerosol Delivery

The challenge for aerosol delivery is to achieve the appropriate airway distribution of gene therapy without degradation of the drug and with minimum wastage. It is known that commercial jet nebulizers can effectively deliver liposome-mediated gene therapy to the respiratory epithelium of mice [23], however, most commercial systems for human use have relatively low deposition efficiency (approx 10–15% of drug delivered to the airways).

The physiological distribution of CF gene expression extends from the upper airway to the respiratory bronchioles [5]. Together with the observation that CF lung disease begins in peripheral bronchioles [1, 6], this suggests that aerosol administration should be customized to deliver, efficiently, not only to major airways, but also to the smaller peripheral bronchioles. In practice, this means aerosols containing a high proportion of small particles (mass median diameter 3 mm or less) should be used. This introduces a further potential source of inefficiency, since small particles, although penetrating to distal airway generations, are more easily exhaled without deposition. Deposition fraction may be enhanced and efficiency improved by imposing a slow deep breathing pattern, ideally, with inspiratory pauses [33]. Further gains in efficiency can be achieved using nebulizers which generate aerosol only during inspiration.

Airway plugging by retained secretions and maldistribution of inspired gas in established CF lung disease will make uniform airway deposition difficult to achieve.

Physiotherapy, mucolytic and antibiotic therapy will be needed to optimize the condition of airways prior to treatment.

The ideal aerosol delivery device, yet to be designed, would allow for stable storage of the gene therapy agent and self-administration of repeated doses by inhalation.

3.6   Safety and Ethical Considerations

The ethics of introducing exogenous genetic material into human subjects are governed by the extent of the likely effects and the potential for clinical benefit. Topical administration of a vector, designed to generate temporary transfection of individual cells, is less contentious that systemic therapy, designed to induce permanent stable transfection in stem cells and their descendants. Similarly, as the lung disease of cystic fibrosis progresses in an individual, risks of unexpected adverse effects of therapy may be perceived to be acceptable when balanced against the reduced life expectancy of the advanced disease. Scientifically, however, the treatment is most likely to succeed in early childhood when lung disease is minimal but the patient is neither in immediate danger from disease, nor able to take part in discussions about the merits or dangers of treatment.

In the United Kingdom, the Gene Therapy Advisory Committee (GTAC) of the Department of Health has approved three related, but independent phase 1 liposome-based trials designed to test topical application of gene therapy in the nasal mucosa. Experimental data from animals, showing lack of pulmonary toxicity and the absence of germ line transfer of genetic material, was required before approval was granted for the second and third clinical trials. At the time of writing, approval had been given in the USA for a similar DNA/liposome trial and also human use of a stable integrating AAV vector for CF somatic gene therapy.

3.7   Detecting Toxicity in Human Studies

3.7.1   Nasal Studies

Symptoms and microscopy of nasal mucosal biopsies, taken under local anaesthesia are the usual end points for these early nasal trials. Using monoclonal antibodies, inflammatory cells in the mucosa and submucosa may be quantified per unit length or area but the sensitivity of this method is reduced by the paucity of data on inflammation in the untreated CF nose.

3.7.2   Lung Studies

Symptoms, signs, lung function tests, radiological changes (plain film and CT), serum markers (eg neutralising antibodies, cytokines, C-reactive protein) and examination of bronchoscopic biopsies are the principal methods by which pulmonary toxicity of gene therapy vectors are assessed in current studies. The indirect clinical measures are all relatively insensitive and mucosal biopsy is prone to sampling bias and very likely to show some inflammation even in untreated CF lung. Carefully controlled and blinded study design is required to avoid misleading results.

3.8   Detecting Successful Transfection in Human Studies

3.8.1   Nasal Studies

Apart from the measurement of protein, vector-derived DNA, mRNA and cAMP-dependent halide conductance in biopsies and brushings (see above), electrophysiological correction of the mucosa may be assessed *in vivo* from the nasal epithelial potential difference (PD) [8, 10, 15]. CF patients have high baseline nasal PD. After perfusion of the nasal cavity with amiloride to block sodium current, stimulation of the chloride current with low chloride and isoprenaline increases PD in normal subjects but not in CF. The response to this stimulation may be used to quantify any electrophysiological correction of the CF mucosa.

3.8.2   Lung Studies

Bronchoscopic biopsy and brushings may yield evidence of molecular correction as outlined above. Testing of the lower respiratory epithelium for functional correction is at present an unsolved problem. Restoring pulmonary function at rest and on exercise is the ultimate goal of therapy but it is very difficult to detect small improvements against the changing baseline of CF lung disease. Radioisotope measurements of muco-ciliary clearance and regional ventilation may conceivably show improvements in mucosal function, which are otherwise undetectable. There is a definite need for development of new functional measures of mucosal function which are applicable to the lower airway in life.

3.9   Early Clinical Trials

At the time of writing, four clinical studies of human CF gene therapy have been published [7–10].

Zabner and colleagues [7] performed an unblinded safety trial of local administration of an E1–deficient A2 adenoviral vector to a defined area of the inferior turbinate of the nose, in three CF patients. No clinical or biopsy evidence of adverse reactions was seen and in all three patients, both baseline and terbutaline-stimulated nasal epithelial potential difference showed changes towards the non-CF (normal) pattern, indicating a degree of functional correction. Two out of three nasal biopsies revealed *CFTR* mRNA.

Crystal and colleagues [9] treated the nasal mucosa and then instilled an adenoviral vector into one lower lobe in each of four CF patients. After nasal administration of viral vector, one subject showed evidence of mRNA transcription and positive staining for CFTR protein in some nasal epithelial cells. Another patient showed vector DNA in nasal brushings but no mRNA or protein. After lung administration, one patient showed vector DNA in brushings and a second showed evidence of CFTR protein expression in approximately 10% of the epithelial cells in one sample. One patient experienced transient leucocytosis fever and X-ray infiltrates and a second developed a febrile illness with tachycardia, hypotension and X-ray infiltrates in the treated lobe.

In the first placebo-controlled trial of CF gene therapy, in human subjects, Caplen and colleagues [8] used a nasal spray to apply the cationic liposome, DC-Chol, complexed to p*SV-CFTR* to the nasal mucosa of CF patients. Nine patients received active treatment and a further six received placebo to both nostrils. No adverse effects, attributable to treatment, were seen and small and variable changes in nasal PD towards normal were seen in the treated group: these reverted to pre-treatment values by day 7. Positive assays for vector-derived mRNA were observed in five of eight treated individuals and two false positive results for vector DNA in six control subjects.

Knowles et al. [10] tested adenoviral gene transfer, in a controlled trial, in the noses of 12 CF patients, with, in each case, one nostril receiving a variable dose of ad*CFTR*, the other a mock treatment. They found a vector DNA in two of the mock treated nostrils, low level of gene expression (mRNA) in the group receiving the highest doses (10^9 and 10^{10} pfu/ml) but also observed adverse effects (earache, jaw pain and nasal inflammation) in two out of three patients receiving the highest dose.

These results illustrate the need for further vector development to improve transfection efficiency and to reduce toxicity. It is too early to reach a general conclusion about the safety of non-viral vectors, however, both animal studies and these clinical studies indicate that the current generation of adenoviral vectors appears to carry the risk of toxic effects in patients.

3.10   Future developments

If adenoviral, AAV, or other animal viruses are to play anything, other than an early pathfinding role in gene therapy for CF, or indeed any other life-shortening but not life-threatening, disorder requiring long-term treatment, then the serious issues of safety (immunological reactivity, contamination with wild-type virus, or with other animal viruses), bulk production and formulation must be addressed. While these problems are largely avoided by use of cationic liposomes, they too require further development if they are to have a long term role. Perhaps most important will be to show that repeated dosing, which will likely be necessary to achieve a sufficient number of transfected cells to be of clinical benefit, can be achieved without toxic effects. Improvements in the efficiency of DNA translocation to the nucleus will be highly desirable. Likewise, efforts to develop vectors which maintain long-term physiological levels of gene expression will be valuable. In this regard, the currently available animal viruses, developed for gene therapy are, again, at a disadvantage because of their limited cloning capacity.

Yeast and bacterial artificial chromosomes have sufficient capacity to carry large fragments of exogenous DNA and,in theory, can yield stable and safe transfection of successive generations of cells [34]. Considerable further developmental work is still needed to ensure that unwanted recombination does not occur and that stable centromere function can be maintained.

As vectors improve and safety data emerge, a move towards treating patients earlier in life seems likely. Treating the underlying defect, before the onset of irreversible infective lung damage, should be the ultimate goal of CF gene therapy and the chances of success in relatively healthy lungs are correspondingly greater.

References

1. Welsh, M. J., Tsui, L., Boat, T. F. and Beaudet, A. L. (1995) Cystic fibrosis. *In*: C. R. Scriver, A. L. Beaudet, W. S. Sly, D. Valle (eds), *The Metabolic Basis of Inherited Disease.* McGraw-Hill, Inc., New York, pp. 3799–3876.
2. Collins, F. S. (1992) Cystic fibrosis: Molecular biology and therapeutic implications *Science* 256: 774–779.
3. Welsh, M. J. and Smith, A. E. (1995) Cystic fibrosis. *Sci. Amer.* 273(6): 52–59.
4. Stutts, M. J., Canessa, C. M., Olsen, J. C. et al. (1995) Cftr as a cAMP-dependent regulator of sodium channels. *Science* 269: 847–850.
5. Englehardt, J. F., Zepeda, M., Cohn, J. A. et al. (1994) *J. Clin. Invest.* 93: 737–749.
6. Sobonya, R. E. and Taussig, L. M. (1986) Quantitative aspects of lung pathology in cystic fibrosis. *Amer. Rev. Respir. Disease* 134: 290–295.
7. Zabner J., Couture L.A., Gregory R.J. (1993) Adenovirus-mediated gene transfer transiently corrects the chloride transport defect in nasal epithelia of patients with cystic fibrosis. *Cell,* 75: 207–216.

8. Caplen N.J., Alton E.W.F.W., Middleton, P. G. et al. (1995) Liposome-mediated *CFTR* gene transfer to the nasal epithelium of patients with cystic fibrosis. *Nat. Med.* 1: 39–46.

9. Crystal, R. G., McElvaney, N. G., Rosenfeld, M. A. et al. (1994) Administration of an adenovirus containing human CFTR cDNA to the respiratory tract of individuals with cystic fibrosis. *Nat. Genet.* 8: 42–51.

10. Knowles, M. R., Hohneker, K. W., Zhaoqing, Z. et al. (1995) A controlled study of adenoviral-vector-mediated gene transfer in the nasal epithelium of patients with cystic fibrosis. *N. Engl. J. Med.* 1995;333: 823–831.

11. Walker J., Watson, J., Holmes, C. et al. (1995) Production and characterisation of monoclonal and polyclonal antibodies to different regions of the cystic fibrosis transmembrane conductance regulator (CFTR): detection of immunologically related proteins. *J. Cell Sci.* 108: 2433–2444.

12. Kartner, N., Augustinas, O., Jensen, T. J. et al. (1992) Mislocalization of deltaF508 CFTR in cystic fibrosis sweat gland. *Nat.Genet.* 1: 321–327.

13. Dupuit, F., Kalin, N., Brezillon, S. et al. (1995) CFTR and differentiation markers expression in non-CF and delta F508 homozygous CF nasal epithelium. *J. Clin. Invest.* 96(3): 1601–1611.

14. Welsh, M. J. and Smith, A. E. (1993) Molecular mechanisms of CFTR chloride channel dysfunction in cystic fibrosis. *Cell* 73: 1251–1254.

15. Knowles, M., Gatzy, J. and Boucher, R. (1981) Increased bioelectric potential difference across respiratory epithelia in cystic fibrosis. *N. Engl. J. Med.* 305(25): 1489–1495.

16. Frizzell, R. A., Rechkemmer, G. and Shoemaker, R. L. (1986) Altered regulation of airway epithelial cell chloride channels in cystic fibrosis. *Science* 233: 558–560.

17. Chao, A. C., Dix, J. A., Sellers, M. C. et al. (1989) Fluorescence measurement of chloride transport in monolayer cultured cells: Mechanisms of chloride transport in fibroblasts. *Biophys. J.* 56(December): 1071–1081.

18. Dorin, J. R., Alton, E. W. F. W. and Porteous, D. J. (1994) Mouse models for cystic fibrosis. *In*: J. A. Dodge, D. J. H. Brock and J. H. Widdicombe (eds), *Cystic Fibrosis Current Topics.* John Wiley and Sons, Chichester, pp. 3–31.

19. Dorin, J. R., Dickinson, P., Alton, E. W. F. W. et al. (1992) Cystic fibrosis in the mouse by targeted insertional mutagenesis. *Nature* 359: 211–215.

20. Dorin, J. R., Stevenson, B. J., Fleming, S. et al. (1994) Long-term survival of the exon 10 insertional cystic fibrosis mutant mouse is a consequence of low level residual wild-type *CFTR* gene expression. *Mamm. Genome* 5: 465–472.

21. Davidson, D. J., Dorin, J. R., McLachlan, G. et al. (1995) Lung disease in the cystic fibrosis mouse exposed to bacterial pathogens. *Nat. Genet.* 9: 351–357.

22. Hyde, S. C., Gill, D. R., Higgins, C. F. et al. (1993) Correction of the ion transport defect in cystic fibrosis transgenic mice by gene therapy. *Nature* 362: 250–255.

23. Alton, E. W. F. W., Middleton, P. G., Caplen, N. J. et al. (1993) Non-invasive liposome-mediated gene delivery can correct the ion transport defect in cystic fibrosis mutant mice. *Nat. Genet.* 5: 135–142.

24. Wilson, J. M. (1995) *J. Clin. Invest.* 96: 2547–2554.

25. Yang, Y., Nunes, F. A., Berencsi, K. et al. (1994) Inactivation of *E2a* in recombinant adenoviruses improves the prospect for gene therapy in cystic fibrosis. *Nat. Genet.* 7: 362–369.

26. Grubb, B. R., Pickles, R. J., Ye, H. et al. (1994) Inefficient gene transfer by adenovirus vector to cystic fibrosis airway epithelia of mice and humans. *Nature* 371: 802–806.

27. Kotin, R. M., Siniscalo, M., Samulski, R. J. et al. (1990) Site-specific integration by adeno-associated virus. *Proc. Natl. Acad. Sci. USA* 87: 2211–2215.
28. Samulski, R. J., Zhu, X., Xiao, X. et al. (1991) Targeted integration of adeno-associated virus (AAV) into human chromosome 19. *EMBO J.* 10: 3941–3950.
29. Flotte, T. R., Afione, S. A., Conrad, C. et al. (1993) Stable *in vivo* expression of the cystic fibrosis transmembrane conductance regulator with an adeno-associated virus vector. *Proc. Natl. Acad. Sci. USA* 90: 10613–10617.
30. Felgner, P. L., Gadek, T. R., Holm, M. et al. (1987) Lipofection: A highly efficient, lipid-mediated DNA-transfection procedure. *Proc. Natl. Acad. Sci. USA* 84: 7413–7417.
31. Gao, X. and Huang, L. (1995) Cationic liposome-mediated gene transfer. *Gene Ther.* 2: 710–722.
32. Chou, J., Rozmahel, R. and Tsui, L. -C. (1991) Characterisation of the promoter region of the cystic fibrosis transmembrane conductance regulator gene. *J.Biol.Chem.* 266(36): 24471–24476.
33. Woolman, P. S., Coutts, C. T., Mole, D. R. et al. (1989) Sites of deposition of aqueous aerosols: a study of efficiency of delivery systems for lung ventilation imaging in man. *Nucl. Med. Commun.* 10: 171–180.
34. Cohen, P. (1995) Creators of the forty-seventh chromosome. *New Sci.* (11 November): 34–37.

4 Gene Therapy of Familial Hypercholesterolemia

G. Cichon, M. Strauss

4.1   Introduction

Familial hypercholesterolemia (FH) is one of the most common autosomal dominant diseases which affects about 1 in 500 individuals, in the heterozygous form.
The reason for elevated serum cholesterol levels in these patients is a defect of the
low density lipoprotein (LDL) receptor gene. The absence of functional LDL-receptors in the liver prevents clearance of LDL from the circulation, leaving serum cholesterol levels constantly elevated. In heterozygous carriers total serum cholesterol
ranges between 260–500 mg/100 ml (2–4 fold increased LDL-cholesterol compared
with normal patients). Patients are at high risk for coronary artery disease (CAD)
and about 85% up to the age of 60 experience myocardial infarction. Heterozygous
FH patients account for about 5% in the whole group of myocardial infarctions.
There are several sites for pharmacological intervention in the heterozygous group
but only a few therapeutic options are available for the treatment of homozygous
FH, which occurs only once in a million individuals. These patients suffer from
serum cholesterol levels between 500–1200 mg/100 ml (6–8 fold increased LDL-
cholesterol). They develop severe atherosclerosis, experience myocardial infarctions
during childhood and have a markedly reduced life expectancy.

In June 1992, the first patient suffering from homozygous FH was treated by
gene therapy [1, 2]. This was the second trial for somatic gene therapy, after the initial experiment in two patients with an inherited form of severe combined immunodeficiency 18 months earlier [3]. To date, a total of five patients with FH ranging in
age from 7 to 41 years, have been treated by retroviral mediated gene transfer. The
treatment resulted in a prolonged reduction of serum cholesterol in 3 of 5 patients
[5]. In the most successful case, serum cholesterol was lowered by about 20%. This
pilot study proved the general feasibility of this strategy but the lack of a therapeutic effect and its impact on the progression of atherosclerosis raised controversial
discussions about its efficacy.

This article will discuss strengths and drawbacks of the most common gene
transfer systems and their possible application to the development of a safe and efficient therapy for familial hypercholesterolemia.

T. Blankenstein (ed.) Gene Therapy
©1999, Birkhäuser Verlag Basel

4.2   Familial Hypercholesterolemia

4.2.1   The Genetic Defect

The gene for the LDL-receptor is located on the short arm of chromosome 19. There are over 100 different mutations characterized at the DNA-level including deletions, insertions, nonsense- and missense-mutations and exon duplications. The largest group are deletions of varying size along the whole gene. The broad heterogeneity of molecular defects suggests that most of the affected families have their own genetic defect. Nevertheless, there are some isolated populations, for example a French-Canadian group around Montreal, carrying the same mutation (>10 kb spanning deletion). This group descends from French immigrants who left France during the 17. and 18. century. Similarly, populations have been described in Finland, Greece and Italy. According to functional properties, Goldstein and Brown distinguished five different types of defects regarding: 1. receptor protein synthesis, 2. glycosylation (transport defect from the endoplasmic reticulum to the Golgi apparatus), 3. LDL binding, 4. internalization and 5. intracellular disintegration of LDL/LDL-receptor complex [6]. All different types of receptor malfunction lead to the same pathophysiological picture of hypercholesterolemia.

4.2.2   The Clinical Picture

Early diagnosis for the FH heterozygote could be performed by analysis of umbilical cord blood, directly after birth. Postnatal serum LDL-cholesterol is already 2–3 fold increased compared with normal newborns and remains elevated during early childhood and adolescence. The most important clinical symptom is the early onset and the fast progression of severe atherosclerosis. In heterozygous FH patients, coronary artery disease becomes apparent in the 3rd and 4th decade and the event of myocardial infarction occurs most often between the 4th and 5th decade. Physical examination reveals, very often, tendinous xanthomas especially around the Achilles tendon, tendons of the knee, the elbow and the back of the hands. The morphological basis for these knotty swellings are tissue macrophages, which have incorporated large amounts of cholesterol esters and form conglomerates of, so called, foam cells. Xanthomas around the eyes and an arcus lipoides cornea are often observed, but these findings are not specific for FH.

Sometimes, patients who are homozygous for FH reveal xanthomas in the postnatal period. The lipid infiltration appears as plain, yellowish plaques especially on sites which are often exposed to mild mechanical trauma like elbows, knees and the back. Clinical onset of atherosclerosis is very early and many patients die from myocardial infarction before the age of twenty.

4.2.3   Sources of Serum Cholesterol

Since cholesterol is an important structural component of cellular membranes and is the main precursor for steroid synthesis in adrenals, ovaries and testes, nearly all tissues possess the enzymatic equipment for cholesterol synthesis. Nevertheless, about 97% of endogenous cholesterol biosynthesis is located in the liver, the intestine and the skin. But only the hepatic cholesterol synthesis (500–1000 mg/day) is under control of a feed-back mechanism by nutritional chylomicron remnants, which render the suppression of endogenous cholesterol synthesis ineffective. The percentage of exogenous cholesterol in total serum cholesterol does not exceed 40%, even after excessive intake of exogenous cholesterol, because of this incomplete feed-back mechanism. This problem limits the efficiency of dietetic strategies. The situation is even more complicated in homozygous FH patients. The hepatic cholesterol production is thought to be increased, despite the high serum cholesterol levels, because the intracellular cholesterol level is decreased due to the receptor defect.

4.2.4   Treatment

Currently there are four strategies for conservative therapeutical interventions in hypercholesterolemia: 1. Reduction of exogenous intake by dietary regimes, 2. reduction of endogenous cholesterol synthesis by blocking synthetic pathways, namely the β-hydroxy-3-methyl-glutaryl CoA reductase by clofibrinic acid derivates, 3. blocking the reabsorption of bile acids in the gut by resins like colestyramin or cholesterol analogues like sitosterin and 4. induction of the expression of the remainder of receptors by agents like dextrothyroxin. Several clinical trials have been performed to ensure the efficacy of combined dietary and pharmacological treatment. Combination of diet and colestyramin could lower the total cholesterol in a group of homozygous children by 33% [7] and the combination of bezafibrate (clofibrinic acid derivative) and sitosterin could lower the total serum cholesterol by 40% over the whole period of observation (24 months) [8]. Although the long term treatment with clofibrinic acid is accompanied with a number of gastrointestinal side effects, severe problems like hepatomegaly with raised transaminases are rare and the general acceptance is good. Despite the impressive results of pharmacological intervention and the early onset of treatment in homozygous FH patients, it is difficult to estimate, to what extent, onset of severe atherosclerotic lesions could be delayed and more effective therapies are still desirable.

   Normalization of serum cholesterol in homozygous FH patients, after liver transplantation, suggests the importance of a liver directed gene therapy [9]. Orthotopic liver transplantation is currently the most effective treatment for FH but often requires life long immune suppression and a compatible donor liver is not available in all cases. Other surgical procedures, such as a portocaval shunt and an ileal bypass, yielded only partial and transient reduction of serum cholesterol [10, 11].

4.2.5 Animal Models

There are two animal models for familial hypercholesterolemia: the Watanabe heritable hyperlipidemic rabbit (WHHL) [42] and a LDL-receptor knock-out mouse [44]. The Watanabe rabbit carries a 12 base in-frame deletion in the LDL-binding domain which renders the LDL-receptor inactive. At the age of 6 months, the cholesterol level reaches about 1000 mg/100 ml and the animals exhibit early onset of progressive atherosclerosis. The Watanabe rabbit represents an excellent animal model for FH. The LDL-receptor knock-out mouse requires a cholesterol and coconut oil rich diet for hypercholesterolemia to occur since there is a fundamental difference in cholesterol metabolism between mouse and man. Mice are able to edit the mRNA of the ApoB 100, the LDL core protein, to a truncated form, the Apo B 48. Lipid micelles with an Apo B 48 core can be internalized and cleared from the circulation by a remnant receptor, which results in only a slight total serum cholesterol increase on a normal diet.

4.3 Gene Therapy of FH

4.3.1 Principles

The aim of gene therapy is the reconstitution of a lost gene function by gene transfer into mutant cells. There are two options for intervention: 1. the exchange of the defective part of a gene or 2. introduction of an additional gene into the nucleus. The first approach would keep the integrity of the gene and in particular the physiological control of gene expression would be preserved. Unfortunately, homologous recombination only occurs in 1 of 10^4–10^5 transduced cells and is even less efficient in nondividing cells. There is currently no perspective for homologous recombination as a therapeutic principle for somatic gene therapy. Thus, the introduction of an additional gene is the option of choice in current gene therapeutic strategies.

4.3.2 Gene Transfer Systems

4.3.2.1 Retroviral Vectors

Retroviruses are RNA tumor viruses which are widespread among mammals. A unique property of retroviruses is their efficient mechanism of integration into the host genome [45]. Integration happens, by chance, with a preference for transcriptional active areas. Retroviral vectors which are employed in gene therapy are derived from a murine retrovirus (Moloney mouse leukemia virus, MMLV). Essential gene functions (gag,pol,env) have been removed from the recombinant virus which leaves about 7000 bases insertional space for therapeutic genes. Viral functions which are required in the vector are the 'long terminal repeats' (LTRs), on both ends, providing promoter function, the sites for integration into host chromosomes, the primer sites for initiation of reverse transcription and a packaging signal (Fig. 4-1).

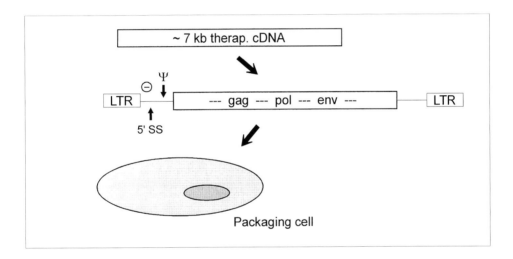

Figure 4-1. Construction of a recombinant retrovirus. ψ = packaging signal, 5' SS = tRNA primer binding site, LTR = long terminal repeats.

In vivo employment of MMLV-derived vectors is difficult because the murine virus is rapidly inactivated by human complement [13]. The second restriction results from the inability of MMLV to integrate into non proliferating tissues such as the liver [14]. With regard to these drawbacks, an *ex vivo* protocol has been developed with partial hepatectomy followed by isolation and transduction of hepatocytes in culture and re-infusion of transduced hepatocytes over the splenic vein. The efficiency of this is procedure is low and seems to be limited to replacement of not more than 2–5% of the liver mass by functional hepatocytes.

An alternative strategy for application of retroviral vectors *in vivo* has been developed [46–48]. This takes advantage of the regeneration-inducing effect of hepatectomy, which involves replication in most of the hepatocytes. Following hepatectomy, retroviral vectors are infused into the portal vasculature while normal blood circulation is temporarily blocked. However, despite the high degree of replication of hepatocytes, not more than 5% of them can be transduced [46, 47]. Optimized delivery protocols may help to increase this success rate [49]. Kay et al. [50] introduced the gene for canine factor IX into dogs with hemophilia B. Serum factor IX levels of 3–10 ng/ml, corresponding to 0,1% of the normal level, have been maintained constitutively for at least 9 months. The whole blood clotting time was reduced to less than 50%. This result suggests that the increase in expression, even in a low percentage of transduced hepatocytes, may lead to therapeutic benefits in diseases like hemophilia A and B. Recently, a new, less invasive strategy was developed for inducing regeneration which is based on the function of urokinase induced hepatocyte proliferation [51]. If retroviral vectors could infect resting cells they would be the vectors of choice for liver-directed gene transfer *in vivo*. Since it was shown recently that HIV infects resting cells due to its unique ability of targeting its pre-integration complex to the nucleus [52], the modification of existing retroviral vector systems seems feasible. Using VSV-pseudotyped retroviral vectors, it was possible to inject high concentrations of vector directly into neonatal mouse liver which resulted in 25–30% transduction of β-gal gene and to long term expression of HBV surface antigen [53]. If the latter turns out to be the result of integration of the vector genome in a comparably high percentage of hepatocytes, this would be the easiest way of liver-directed gene transfer with great prospects for application in gene therapy.

4.3.2.2 Adenoviral Vectors

The most efficient gene transfer vehicles known so far are recombinant adenoviruses. Adenoviruses do not integrate their genes into the host chromosome as do retroviruses. In fact, chromosomal integration does not seem to be absolutely required for long term gene expression in non proliferating tissues like the liver. Reporter gene expression from adenoviral vectors was stable for at least 3 months in nude mice and for 6 months in C57 Bl/6 mice after transfection [16, 17].

From fourty-seven already known different serotypes, only type 5 and type 2 vectors are employed in gene transfer experiments [18]. After systemic application (tail vein injection) of a recombinant Ad type 5 vector in mice over 90% of reporter gene expression was found in the liver [19]. The high transgene expression in the liver was surprising since adenoviruses were expected to have a high affinity to respiratory endothelial cells.

The receptor on the surface of hepatocytes, mediating internalization, remains unknown but a role for alpha v integrins is discussed [20]. The initial interpretation of liver specificity had to be modified after it was shown that intracardiac application leads not only to high gene transfer into the myocardium, but also to increased transfer rates into brain, lungs, kidneys and gonads compared with systemic i.v. application [21, 22]. After systemic application of about 2×10^9 virus particles in mice, about 50% of hepatocytes can be transduced. This rate can be increased to nearly 100% after portal vein infusion.

The major drawback in the employment of adenoviral vectors is the limited duration of transgene expression. One single systemic application of recombinant adenoviruses (1×10^{13} virus particles), carrying the human LDL-receptor gene, leads to 60% reduction of total serum cholesterol in the Watanabe rabbit but the therapeutic effect is only transient and does not exceed 10–14 days (Fig. 4-2). The repeated administration of recombinant viruses remains inefficient since production of neutralizing antibodies, following the first injection inactivates circulating viruses rapidly. In this setting, human LDL-receptor protein induces rapid antibody production which contributes to the limited therapeutic effect. However, even the use of the rabbit LDL-receptor gene did not extend the therapeutic effect by more than 21 days [22].

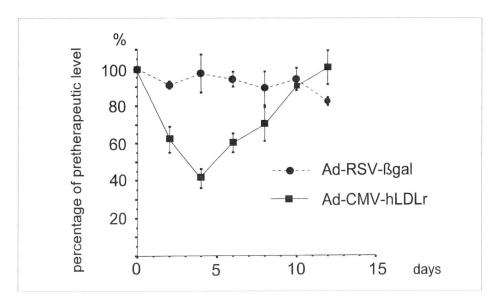

Figure 4-2. Course of serum cholesterol in the Watanabe rabbit after one single injection of 1×10^{13} virus particles via the ear vene.

Reporter gene expression in mice varies between 2–3 weeks and 6 months depending on the mouse strain and the reporter gene [17]. The reason for these differences is thought to be the ability of antigenic presentation based on strain related morphological specificity of MHC class I-molecules (respectively H 2). For long term expression studies, small soluble antigens with low antigenicity (like human alpha-1-antitrypsin) are thought to be better reporter genes compared with strong antigens like bacterial β-galactosidase. However, another important reason for the limited transgene expression seems to be the expression of adenoviral genes in the recipient cells [27]. The reason for viral antigenicity becomes apparent by looking at the structure of recombinant adenoviruses. The whole adenoviral genome spans about 35000 bases harboring at least 7 viral promoter sites [24]. The recombinant virus has been crippled by removing the essential E1 region. The removed gene function is supplemented by the 293 packaging cell. To increase the capacity for insertion of foreign DNA, the non-essential E3 region has been removed as well which generates space for about 8.000 bases for therapeutic cassettes [25, 26]. These vectors still bear nearly 80% viral genes and some of these genes seems to express viral antigens which mediate a strong cytotoxic attack to transgene cells after *in vivo* application [27].

1. There are four alternative strategies to overcome the immunological defense mechanisms: general immune suppression
2. induction of immune tolerance
3. reduction of antigenic presentation in transgenic cells and
4. reduction of vector antigenicity

Application of cyclosporin A (serum level 300–600 ng/ml) in factor IX deficient dogs, after adenoviral gene transfer with a recombinant virus, prolonged the clotting factor IX expression but did not prevent production of neutralizing antibodies rendering the repeated administration inefficient [27].

Another strategy targeted the costimulatory pathway in T-lymphocytes. Clonal expansion of T-cells does require antigen presentation by MHC class II and at least one costimulatory signal, which is mediated by interaction of the B7-receptor (on antigen presenting cells) and the CTLA4-receptor on naive T-cells. The administration of an antibody, against the CTLA4-receptor, blocks the costimulatory pathways and renders the T-lymphocytes inactive. One single administration of anti-CTLA4 in mice, prior to viral challenge, provided a constant reporter gene expression in 90% of treated individuals over a period of 180 days [29]. The serum level of anti-CTLA4 dropped from 100 µg/ml initially to levels averaging 1 µg/ml during the first 3–5 weeks. Serum levels of anti-CTLA4 of 1 µg/ml can induce an nonspecific immune suppressive effect, which means that the cellular immune system is unspecifically impaired at least for the first two weeks after the administration. Challenge with wild type HSV type I, during this phase, resulted in increased mortality of animals. This strategy would pose a definite health risk for the patient and needs further evaluation. Safer and more specific strategies are required.

Direct postnatal exposure to an adenoviral antigen can induce prolonged gene expression, most likely by a tolerance mechanism. However, it remains without clinical consequence, since, it is impossible to estimate the ability of the immune system to react in an adequate way against wild type virus [21].

A third strategy utilizes overexpression of genes of the adenoviral E3 region to escape from immune recognition (Tab. 4-1). There are nine predicted E3 proteins in group C adenoviruses, six of which have been identified in infected cells. The functions of four of these proteins have been characterized. One of the E3 products, a 19 kd glycoprotein (gp 19K), was shown to bind to MHC class I molecules, preventing transport to the cell surface [30]. Affinities of gp 19K to MHC class I molecules vary with MHC types. Cytolysis studies have revealed good interaction with murine Kd, Ld, Db, modest interaction with Dd and no interaction with Dk, Kk, Kb class I antigens [31–34].

Table 4-1. Adenoviral proteins interfering with host immune functions [43]

Protein	Function *in vitro*
E1A-289R/243R	induce cellular susceptibility to TNF cytolysis
E1B-19K	prevents cytolysis of human but not mouse cells by TNF
E3-gp19K	prevents cytolysis by CTL
E3-14.7K	prevents cytolysis of mouse cells by TNF
E3-10.4K/14.5K	prevents cytolysis of mouse cells by TNF down-regulates EGF receptor

The interaction with human HLA-A2 is different than with HLA-B7 [35]. The E3-14.7K protein as well as E3-10.4 K and E3-14.5 K proteins are required to prevent TNF mediated lysis (36–38). The overexpression of E3-gp 19K in target cells (3T3), after transduction with a recombinant adenoviral vector, significantly reduced the cytolytic activity of CTLs in cell culture but experiments *in vivo* have not been carried out so far [39].

All strategies which contribute to weakening of the immune system in a non specific fashion or interfere with immune recognition bear a certain health risk. The solution for these immunological problems would be the reduction of vector antigenicity. Two attempts have been made to reduce the expression of viral antigens after transduction. The first one uses a temperature sensitive adenovirus mutant which expresses reduced amounts of E2a proteins. The E2a complex codes for the DNA polymerase, the terminal protein and the DNA binding protein [24]. The DNA binding protein, expressed in large amounts after infection, is thought to account for an important part of immune recognition. Application of the recombinant mutant virus led to a reduction of antigenicity and a slight prolongation of reporter gene expression but additional antigens seem to be involved in immune recognition [40].

A second attempt to reduce the antigenicity has been developed only recently. Open reading frame 6 and, to a lesser extent, ORF 3 of the adenoviral E4 region are known to transactivate adenoviral late gene expression partially, through activation of adenoviral IVa2 genes [41]. The E4 region has been removed from the vector and stably transfected into the 293 packaging cell line. Since the simultaneous presence of E4 and E1 gene products in 293 cells is not tolerated, the expression of E4 is directed by an inducible promoter. Apart from increased transport capacity this third generation vector led to significantly prolonged reporter gene expression [54]. These advances in adenoviral vector construction and their impact on liver gene therapy need further evaluation but have potential as part of new important therapeutic strategies for the treatment of FH and many other inborn metabolic diseases.

4.3.2.3 Other Viral Vector Systems

Many attempts have been made to employ other viruses for gene therapy. Adeno-associated viruses appeared to be particularly interesting. Wild type adeno-associated viruses (AAV) integrate their genes into a specific locus on chromosome 19 [55]. Integration does not seem to be accompanied with any pathology. Unfortunately, recombinant AAV lose this unique property of site specific integration [56]. The advantage of AAV, compared with MMLV retroviruses, is the possibility of *in vivo* application. The major obstacle in their employment is the achievement of sufficiently high viral titers. For successful liver gene therapy in humans, a number of virus particles between 5×10^{13} and 1×10^{14} will be required. With the current methods of virus propagation, it is impossible to produce such high titers with AAV. Another problem is contamination with helper virus which is very difficult to avoid (for review see [57]).

Herpes viruses can be utilized as gene transfer vehicles as well. Vectors propagated in an amplicon technique have a theoretical transport capacity of about 150 kb [58]. Herpes virus mediated gene transfer into the CNS by retrograde transneuronal flow has been reported [59–61] and also investigations regarding their ability for liver gene transfer have been communicated [62]. After herpes virus mediated gene transfer, factor IX expression in mouse liver was stable for 3–5 weeks [62]. The major drawback with this type of vector is toxicity (J.Glorioso, pers. communication).

To overcome the immunological problems, which arise in the employment of adenoviruses, the therapeutic gene has been linked to the surface of inactivated adenoviruses [63]. These viruses are still able to infect cells but there is no more viral gene expression after transduction. These complexes are efficient gene transfer vehicles in cell culture but are very ineffective *in vivo*. Most likely, they are too big and most of the complexes end up in the reticulo-endothelial system of the liver.

4.3.2.4 Non-Viral Vector Systems

For some time, much attention has been drawn to non-viral systems like polylysine-DNA complexes [64] and cationic liposomes [65]. These synthetic particles can be produced in large amounts, they do not carry risk of recombination like viruses do and they could be directed to specific organs by a mechanism which is called receptor targeting. For liver targeting, transferrin or orosomucoid molecules are linked to the complex and they bind to the corresponding receptors on the surface of hepatocytes. After internalization they are not able to escape from the endosomes into the cytosol which results only in a marginal therapeutic *in vivo* effect. Several attempt have been made to provide them with mechanisms for endosomal disruption, for example with influenza hemagglutinin [12], but none of these attempt have led to satisfactory results, *in vivo*, so far. In general, all non-viral delivery systems are far too inefficient for *in vivo* gene transfer into the liver.

4.3.3   The Therapeutic Cassette

Regardless of the gene transfer system used, the organization of the therapeutic cassette remains the same. A strong viral promoter (e.g. CMV,RSV,LTR) or a minimal mammalian promoter is normally used to direct expression of the cDNA of the therapeutic gene, followed by a polyadenylation site. Considering the limited transport capacity of retro- and adenoviruses, much attention has been drawn to elements which are relatively small. The most critical functional feature of these cassettes is the potential loss of expression control in the viral context. The therapeutic gene is usually expressed constitutively, as long as the transgenic promoter is not inactivated by methylation. As long as hepatocytes are not harmed by a constantly high intake of LDL, this approach seems reasonable. There are no reports evaluating the long term viability of LDL-receptors overexpressing transgenic hepatocytes in an LDL-rich environment. Thus, it remains to be clarified whether or not a physiological mode of LDL receptor gene modulation would be desirable.

4.4   The First Clinical Trial

The first and only clinical trial for treatment of FH has been carried out by Wilson and coworkers. The initial step in the *ex vivo* gene transfer protocol was partial hepatectomy. Liver tissues ranging from 84 to 221 g were resected and hepatocytes were isolated by collagenase perfusion and plated in some 1000 culture plates (a total $3- \times 10^9$ cells). On the second day the hepatocyte cultures were infected with a recombinant retrovirus carrying the cDNA of the human LDL-receptor gene under control of a chicken β-actin promoter and an enhancer element derived from the cytomegaly virus promoter. Sixteen hours later, cells were washed, harvested and re-infused over an indwelling catheter into the portal vein. The number of harvested cells were in the range of $1.1–3.2 \times 10^9$ cells and fluorescence microscopy revealed, on average, 21% transduced cells. Thus, the number of re-infused, genetically corrected hepatocytes amounted to about 3×10^8. The total number of hepatocytes in the liver can be estimated to be about 5×10^{11} suggesting that in the case of a successful engraftment, less than 1 in a thousand of the total number of hepatocytes can be actively transduced and regrafted. Pharmacological treatment was stopped two months before gene therapy and resumed about 4 months later. During this post treatment period, total serum cholesterol levels dropped, in the best case, by 20%, in a second case by 19%, in a third patient by 6% and it remained unchanged in the two other patients. However, it is not clear yet if this is indeed a specific effect of the gene transfer. This physiological effect could also be a consequence of physical manipulation and partial hepatectomy, which could have reduced the VLDL production of the liver and could have increased LDL removal [15]. Another explanation might be some replication of transduced cells after engraftment. This is even more likely since re-infusion happened during the regeneration phase of the liver.

A convincing therapeutic effect, in the treatment of hypercholesterolemia in homozygous FH patients, requires a high percentage of successfully transduced hepatocytes. The limits of a retroviral *ex vivo* approach had already become, to some extent, obvious in the animal model. Diseases which require lower expression rates of a therapeutic gene product, like factor IX expression in hemophilia B, would possibly be better candidates for this kind of retroviral-mediated therapy. Therefore, the search for alternative strategies for LDL receptor gene transfer is justified.

4.5 Conclusions

The first clinical trial for gene therapy of familial hypercholesterolemia has proved the general feasibility of a retroviral *ex vivo* gene transfer but the limited effect on total serum cholesterol levels suggests, that a careful evaluation of alternative strategies is required. Over 60% reduction of serum cholesterol in animal models had been achieved by adenoviral gene transfer but the expression of the therapeutic gene is only short term and is abolished by T-cell mediated inactivation of hepatocytes expressing viral antigens. Since immunosuppression does not seem to be an ideal answer to the immunological problem, the reduction of viral antigenicity is the most important goal. Several promising attempts have been made but need further evaluation. Non-viral vectors are not efficient enough for *in vivo* gene transfer, at least in the current state of development.

Gene therapy of familial hypercholesterolemia is an ideal model for inherited metabolic diseases of the liver and a successful strategy could pave the way for treatment of a number of currently untreatable diseases.

References

1. Grossman, M., Raper, S. E., Kozarsky, K., Stein, E. A., Engelhardt, J. F., Muller, D., Lupien, P. J. and Wilson, J. M. (1994) Successful *ex vivo* gene therapy directed to liver in a patient with familial hypercholesterolemia. *Nat. Genet.* 6: 335–341.
2. Wilson, J. M. et al. (1992) Clincal Protocol: *Ex vivo* gene therapy of familial hypercholesterolemia. *Hum. Gene Ther.* 3 (2): 179–22.
3. Anderson, W. (1990) The ADA human gene therapy protocol. *Hum. Gene Ther.* 4 (4): 521–527.
4. Blaese R.M. et al. (1990) Clinical Protocol: Treatment of severe combined immune deficiency (SCID) due to adenosine deaminase (ADA) with autologues lymphocytes transduced with human ADA gene. *Hum. Gene Ther.* 1(3): 327–362.
5. Grossman, M., Rader, D. J., Muller DWM., Kolansky, D. M., Kozarsky, K., Clark, B. J., Stein, E. A., Lupien, P. J., Brewer, H. B., Raper, S. E., Wilson, J. M. (1995) A pilot study of *ex vivo* gene therapy for homozygous familial hypercholesterolemia. *Nat. Med.* 1 (11): 1148–1154.
6. Goldstein, J. L., Brown, M. S. (1984) Progress in understanding the LDL-receptor and HMD-CoA reductase, two membrane proteins that regulate the plasma cholesterol. *J. Lipid Res.* 25: 1450–61.
7. Koletzko, B., Kupke, I., Wendel, U. (1992) Treatment of hypercholesterolemia in children and adolescents. *Acta Paediatr.* 81: 682–5.
8. Becker, M., Staab, D., Von Bergmann, K. (1992) Long term treatment of severe familial hypercholesterolemia in children; effect of sitosterol and bezafibrate. *Pediatrics* 89: 138–42.
9. Bilheimer, D., Goldstein, J. L., Grundy, S., Starzl, T. (1984) Liver transplantation to provide low density lipoprotein receptors and lower plasma cholesterol in a child with homozygous familial hypercholesterolemia. *N. Engl. J. Med.* 296: 1658–1664.
10. Starzl, T. et al. (1983) Portocaval shunts in patients with familial hypercholesterolemia. *Ann. Surg.* 198: 273–283.
11. Deckelbaum, R., Lees, R., Small, D., Hedberg, S., Grundy, S. (1977) Failure of complete bile diversion and oral bile acid therapy in the treatment of homozygous familial hypercholesterolemia. *N. Engl. J. Med.* 296: 465–470.
12. Wagner, E., Plank, C., Zatloukal, K., Cotten, M., Birnstiel, M. L. (1992) Influenza virus haemagglutinin HA-2 N-terminal fusogenic peptides augment gene transfer by transferrin-polylysine-DNA complexes: towards a synthetic virus-like gene-transfer vehicle. *Proc. Natl. Acad. Sci. USA* 89: 7934–7938.
13. Russel, D. W., Berger, M. S., Miller, A. D. (1995) The effects of human serum and cerebrospinal fluid on retroviral vectors and packaging cell lines. *Hum. Gene Ther.* 6: 635–41.

14. Bukrinsky, M. I., Haggerty, S., Dempsey, M. P., Sharova, N., Adzubei, A., Spitz, L., Lewis, P., Goldfarb, D., Emerman, M., Stevenson, M. (1993) A nuclear localization signal within HIV-1 matrix protein that governs infection of non-dividing cells. *Nature* 365: 666–669.

15. Brown, M. S., Goldstein, J. L., Havel, R., Steinberg, D. (1994) Gene therapy to cholesterol. *Nat. Genet.* 7: 349–50.

16. Yang, Y., Nunes, F., Berencsi, K., Gönczöl, E., Engelhardt, J. F., Wilson, J. M. (1994) Inactivation of E2a in recombinant adenoviruses improves the prospect for gene therapy in cystic fibrosis. *Nat. Genet.* 7: 362–369.

17. Barr, D., Tubb, J., ferguson, D., Scaria, A., Lieber, A., Wilson, C., Perkins, H. J., Kay, M. A. (1995) Strain related variations in adenovirally mediated transgene expression from mouse hepatocytes *in vivo*: comparison between immunocompetent and immunodeficient inbred strains. *Gene Ther.* 2: 151–155.

18. Kremer, E. J., Perricaudet, M. (1995) Adenovirus and adeno-associated virus mediated gene transfer. *Brit. Med. Bull.* 51(1): 31–44.

19. Herz, J., Gerard, R. (1993) Adenovirus-mediated transfer of low density lipoprotein receptor gene acutely accelerates cholesterol clearance in normal mice. *Proc. Natl. Acad. Sci. USA* 90: 2812–2816.

20. Mathias, P., Wickham, T., Moore, M., Nemerow, G. (1994) Multiple adenovirus serotypes use alpha v integrins for infection. *J. Virol.* 1994: 6811–6814.

21. Kass-Eisler, A., Falck-Pederson, E., Elfenbein, D., Alvira, M., Buttrick, P., Leinwand, L. (1994) The impact of developemental stage, route of administration and the immune system on adenovirus gene transfer. *Gene Ther.* 1: 395–402.

22. Huard, J., Lochmüller, H., Acsadi, G., Massie, B., Karpati, G. (1995) The route of administration is a major determinant of the transduction efficiency of rat tissues by adenoviral recombinants. *Gene Ther.* 2: 107–115.

23. Li, J., Fang, B., Eisensmith, C., Hong, X., Li, C., Nasonkin, I., Woo SC (1995) *In vivo* gene therapy for hyperlipidemia: Phenotypic correction in watanabe rabbits by hepatic delivery of the rabbit LDL receptor gene. *J. Clin. Invest.* 95: 768–773.

24. Horwitz, M. (1990) Adenoviridae and their replication. *In*: B. N. Fields and D. M. Knipe (eds), *Virology*. Raven Press, New York, pp. 1679–1721.

25. McGrory, W., Bautista, D., Graham, F. (1988) A simple technique for the rescue of early region 1 mutations into infectious human adenovvirus type 5. *Virology* 163: 614–617.

26. Bett, A., Haddara, W., Prevec, L., Graham, F. (1994) An efficient and flexible system for the construction of adenovirus vectors with insertions or deletions in early regions 1 and 3. *Proc. Natl. Acad. Sci. USA* 91: 8802–8806.

27. Yang, Y., Ertl, H. C. J., Wilson, J. M. (1994) MHC class I-restricted cytotoxic T lymphocytes to viral antigens destroy hepatocytes in mice infected with E1 deleted recombinant adenovirusese. *Immunity* 1: 433–442.

28. Fang, B., Eisensmith, R., Wang, H., Kay, M., Cross, R., Landen, C., Gordon, G., Bellinger, D., Read, M., Hu, P., Brinkhous, K., Woo SC (1995) Gene therapy for Hemophilia B: Host immunosuppression prolongs the therapeutic effect of adenovirus-mediated Factor IX expression. *Hum. Gene Ther.* 6: 1039–1044.

29. Kay, M. A., Holterman, A., Meuse, L., Gown, A., Ochs, H. D., Linsley, P. S., Wilson, C. B. (1995) Long-term hepatic adenovirus-mediated gene expression in mice following CTLA 4 Ig administration. *Nat. Genet.* 11: 191–197.

30. Dähllof, B., Wallin, M., Kvist, S. (1991) The endoplasmatic reticulum retention signal of the E3/19 K protein of Adenovirus 2 is microtubule binding. *J. Biol. Chem.* 266: 1804–1808.

31. Burgert, H., Kvist, S. (1987) The E3/19K protein of adenovirus type 2 binds to the domains of histocompatibility antigens required for CTL recognition. *EMBO J.* 6: 2019–2026.

32. Tanaka, Y., Tevethia, S. (1988) Differential effect of adenovirus 2 E3/19K glycoprotein on the expression of H-2K^b - and H-2D^b-restricted SV 40-specific CTL-mediated lysis. *Virology* 165: 357–366.

33. Rawle, F., Tollefson, A., Wold, W. and Gooding, L. (1989) Mouse anti-adenovirus cytotoxic T-Lymphocytes. Inhibition of lysis by E3 gp 19K but not E3 14.7K. *J. Immunol.* 143: 2031–2037.

34. Cox, J., Yewdell, J., Eisenlohr, P. and Bennik, J. (1990) Antigen presentation requires transport of MHC class I molecules from the endoplasmatic reticulum. *Science* 247: 715–718.

35. Severinsson, L., Martens, I. and Peterson, P. (1986) Differential association between two human MHC class I antigens and an adenoviral glycoprotein. *J. Immunol.* 137: 1003–1009.

36. Gooding, L., Elmore, L., Tollefson, A., Brady, H., Wold, W. (1988) A 14.7 K protein from the E3 region of adenovirus inhibits cytolysis by tumor necrosis factor. *Cell* 53: 341–346.

37. Gooding, L., Sofola, I., Tollefson, A., Duerksen-Hughes, P., Wold, W. (1990) The adenovirus E3-14.7 K protein is a general inhibitor of tumor necrosis factor-mediated cytolysis. *Immunology* 145: 3080–3086.

38. Gooding, L., Ranheim, T., Tollefson, A., Aquino, L., Duerksen-Hughes, P., Horton, T., Wold, W. (1991) The 10.4 and 14.5 dalton proteins encided by region E3 of adenovirus function together to protect many but not all mouse cell lines against lysis by tumor necrosis factor-mediated cytolysis. *J. Virol.* 65 (8): 4114–4123.

39. Lee, M., Abina, M., Haddada, H., Perricaudet, M. (1995) The constituive expression of the immunomodulatory gp 19K protein in E1–, E3– adenoviral vectors strongly reduces the host cytotoxic T cell response against the vector. *Gene Ther.* 2: 256–262.

40. Yang, Y., Nunes, F., Berencsi, K., Gönczol, E., Engelhardt, J., Wilson, J. M. (1994) Inactivation of E2a in recombinant adenoviruses improves the prospect for gene therapy in cystic fibrosis. *Nat. Genet.* 7: 362–369.

41. Tribouley, C., Lutz, P., Staub, A., Kedinger, C. (1994) The product of the adenovirus intermediate gene IVa2 is a transcriptional activator of the major late promoter. *J. Virol.* 68 (7): 4450–4457.

42. Watanabe, Y., Itpo, T., Shiomi, M. (1985) The effect of selective breeding on the developement of coronary artherosclerosis in WHHL rabbits. An animal model for familial hypercholesterolemia. Artherosclerosis 56: 71–97.

43. Wold, W., Gooding, L. (1991) Region E3 of adenovirus: A cassette of genes involved in host immunosurveillance and virus-cell interactions. *Virology* 184: 1–8.

44. Ishibashi, S., Brown, M. S., Goldstein, J. L., Gerard, R., Hammer, R., Herz, J. (1993) Hypercholesterolemia in low density lipoprotein receptor knockout mice and its reversal by adenovirus-mediated gene delivery. *J. Clin. Invest.* 92: 883–893.

45. Gilboa E. (1990) Retroviral gene transfer: Applications to human therapy. *In*: *The biology of hematopoesis*. Wiley-Liss, Inc., pp. 301–311.

46. Ferry, N., Duplessis, O., Houssin, D. et al. (1991) Retroviral-mediated gene transfer into

hepatocytes *in vivo. Proc. Natl. Acad. Sci. USA* 88: 8377–8381.

47. Kay, M. A., Li, Q., Liu, T. J. et al. (1992) Hepatic gene therapy: persistent expression of human alpha 1-antitrypsin in mice after direct gene delivery *in vivo. Hum. Gene Ther.* 3: 641–647.

48. Cardoso, J. E., Branchereau, S., Jeyaraj et al. (1993) *In situ* retrovirus-mediated gene transfer into dog liver. *Hum. Gene Ther.* 4: 411–418.

49. Brancherau, S., Calise, D., Ferry, N. (1990) Factors influencing retroviral-mediated gene transfer into hepatocytes *in vivo. Hum. Gene Ther.* 5: 803–808.

50. Kay, M. A., RothenbergS, Landen, C. N. et al. (1993) *In vivo* gene therapy of hemophilia B: Sustained partial correction in factor IX-deficient dogs. *Science* 262: 117–119.

51. Lieber, A., Vrancken Peeters J. F. T. F. D., Gowen, A., Perkins, J., Kay, M. A. (1995) A modified urokinase plasminogen activator induces liver regeneration without bleeding. *Hum. Gene Ther.* 6: 1029–1037.

52. Bubrinsky MI, Haggerty S, Dempsy MP et al. (1993) A nuclear localization signal within HIV-1 matrix protein that governs infection of non-dividing cells. *Nature* 365: 666–669.

53. Miyanohara, A., Yee, J. K., Bouic, K., La Porte, P., Friedmann, T. (1995) Efficient *in vivo* transduction of the neonatal liver with pseudotyped retroviral vectors. *Gene Ther.* 2: 138–142.

54. Yeh, P., Dedieu, J. F., Orsini, C., Vigne, E., Denette, P., Perricaudet, M. (1996) Efficient dual transcomplementation of adenovirus E1 and E4 regions from a 293-derived cell line expressiong a minimal E4 functional unit. *J. Virol.* 70: 559–565.

55. Samulski, R. J., Zhu, X., Xiao, X. et al. (1991) Targeted integration of adeno-associated virus (AAV) into human chromosome 19. *EMBO J.* 10(12): 3941–50 (erratum in: *EMBO J.* 1992, 11(3): 1228).

56. Muzyczka, N. (1992) Use of adeno-associated virus as ageneral transduction vector for mammalian cells. *Curr. Top. Microbiol. Immunol.* 158(97): 97–129.

57. Kremer, E. J., Perricaudet, M. (1995) Adenovirus and adeno-associated virus mediated gene transfer. *Brit. Med. Bull.* 51 (1): 31–44.

58. Frenkel N, Singer O, Kwong AD. Minireview: The herpes simplex virus amplicon – a versatile defective virus vector. *Gene Ther.* 1: 40–46.

59. Dobson, A. T., Sedarati, F., Devi-Rao, G. et al. (1989) Identification of the latency associated transcript promoter by expression of rabbit beta-blobin mRNA in mouse sensory nerve ganglia latently infected with a recombinant herpes simplex virus. *J. Virol.* 63: 3844–3851.

60. Ho DY, Mocarski ES (1989) Herpes simplex virus latent RNA (LAT) is not required for latent infection in the mouse. *Proc. Natl. Acad. Sci. USA* 86: 7596–7600.

61. Goins, W. F., Sternberg, L. R., Croen, K. D. et al. (1994) A novel latency-active promoter is contained within the herpes simplex virus type 1 U_L flanking repeats. *J. Virol.* 68: 2239–2252.

62. Miyanohara, A., Johnson, P. A., Elam, R. L. et al. (1992) Direct gene transfer to the liver with herpes simplex virus type I vectors: transient production of physiologically relevant levels of circulating factor IX. *New. Biol.* 4: 238–246.

63. Cotten, M., Wagner, E., Zatloukal, K., Phillips, S., Curiel, D., Birnstil, J. (1992) High-efficiency receptor-mediated delivery of small and large (48 kb) gene constructs using the endosome-disruption activity of defective or chemically-inactivated adenovirus particles. *Proc. Natl. Acad. Sci. USA* 89: 6094–6098.

64. Cotten, M., Wagner, E. (1993) Non-viral approaches to gene therapy. *Curr. Opin. Biotechnol.* 4: 705–710.
65. Schofield, J. P., Caskey, C. T. (1995) Non-viral approaches to gene therapy *Brit. Med. Bull.* 51(1): 56–71.

III. Gene Marking

1 Gene Marking in Bone Marrow and Peripheral Blood Stem Cell Transplantation

K. Cornetta, E. F. Srour, C. M. Traycoff

1.1   Introduction

In this chapter we will evaluate the use of genetic marking in autologous bone mar-row transplantation. After identifying the controversial issues in the field, clinical studies using gene-marked bone marrow will be reviewed and discussed in the con-text of these controversies. Autologous bone marrow transplantation and peripher-al blood cell transplantation (AuBMT and AuPBPC, respectively) are becoming an increasingly common procedure for the treatment of solid tumors and hematologi-cal malignancies. The International Bone Marrow Transplantation Registry esti-mates that over 10,000 autologous transplants are performed each year and the number is growing at an annual rate of 20% [1]. For hematological malignancies, autologous transplantation is an attractive alternative to allogeneic transplantation because it does not require a histocompatible donor, it lacks the risk of graft-versus-host disease and can be used in patients up to 70 years of age. In solid tumors, au-tologous transplantation is utilized as a method of dose intensification, which side-steps the dose limiting hematological toxicity of most cancer treatment regimens.

Traditionally, transplanted cells have been obtained by harvesting bone marrow, a process usually requiring general anesthesia, in which multiple passes (100–200) are made into the posterior iliac crest with removal of approximately one liter of bone marrow. More recently, cells for transplantation have been obtained from pe-ripheral blood by "pheresis" in which marrow elements are "mobilized" into the pe-ripheral blood through the use of cytokines (such as GM-CSF or G-CSF), with or without additional chemotherapy. Pheresis is an outpatient procedure requiring ap-proximately 4 h, during which time, white blood cells are selectively removed from the bloodstream and the patient's red cells, platelets and plasma are returned to the circulation. Between 1 and 4 daily procedures are required normally to obtain suf-ficient cells for transplantation. A simplified schema for AuBMT and AuPBPC is shown in Figure 1-1. In most cases, the marrow or peripheral blood progenitor cells are frozen in DMSO and stored in liquid nitrogen until the patient has completed chemotherapy and/or radiation therapy. Following thawing and intravenous ad-ministration, the cells home to the bone marrow micro-environment where they re-

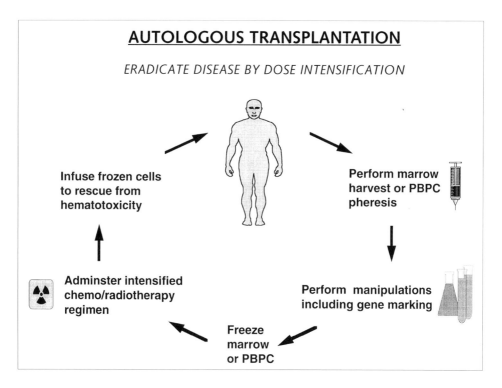

Figure 1-1. Schema for patients undergoing autologous bone marrow or peripheral blood stem cell transplantation. Cells for transplantation are harvested, manipulated, then frozen and stored in liquid nitrogen. After the patient has completed an intensive chemotherapy-radiation therapy preparative regimen, the frozen cells are thawed and used to rescue the patient from the myelosuppressive therapy.

store granulopoiesis in approximately two weeks, with platelet and red cell counts returning in approximately one month.

1.2   Controversies in Autologous Transplantation

1.2.1   Engraftment of Transplanted Autologous Cells

While AuBMT and AuPBPC are being performed in large numbers, the biology of engraftment is poorly understood. Specifically, do transplanted autologous cells lead to long-term engraftment or are they merely a pool of committed progenitor cells providing short-term recovery of blood counts until long-term hematopoiesis is restored by residual stem cells which survive the transplant chemotherapy regimen?

This is a valid concern since many chemotherapy regimens, used in autologous transplantation, are not truly myeloablative. If transplanted autologous cells do not contribute to long-term hematopoiesis, intensification of treatment regimens may result in marrow failure, months or years after autologous transplantation. In the case of peripheral blood progenitor cells, which are known to provide an enhanced rate of engraftment after transplantation compared with bone marrow, the question remains whether or not they also contribute to long-term hematopoiesis.

Interpretation of marrow engraftment kinetics is further complicated by contributions from residual marrow that survives the bone marrow transplantation preparative regimen. Comparisons between transplant studies also must consider variations in the types and strength of chemotherapy and radiation administered. Gene marking of transplanted cells is the only method, currently available, to help distinguish cells which arise from transplanted marrow from residual cells that survive the transplant chemotherapy-radiation regimen.

1.2.2   Purification of Marrow Stem Cells

The questions of short and long-term engraftment become increasingly important as new marrow purification methods are brought to clinical trial. CD34 selection is one such manipulation being implemented in autologous transplantation [2]. The CD34 protein is a marker of hematopoietic progenitor cells found on approximately 1% of marrow elements [3]. Committed marrow progenitors (such as colony-forming unit granulocyte-monocyte and the burst forming unit erythroid) as well as primitive progenitors (such as the long-term bone marrow culture initiating cells) express the CD34 molecule [4, 5]. CD34 is also believed to be expressed on cells responsible for long-term marrow engraftment, namely the stem cells. Although the human stem cell has yet to be fully characterized and isolated to purity, it is defined as a cell with self-renewal and multi-lineage potentiality capable of differentiating into a variety of cell types including erythroid, granulocytic, megakaryocytic and lymphoid elements. Since many cancer cells, such as breast cancer cells, do not express CD34, selection for CD34 expressing cells is being evaluated as a means of decreasing contamination of the graft with malignant cells. Gene-marking of CD34-selected grafts provides a means of assessing the long-term engraftment potential and engraftment kinetics of cells exposed to the selection process.

1.2.3   Cytokines and *ex vivo* Manipulations in Autologous Transplantation

The use of exogenous cytokines in autologous transplantation represents another manipulation which has the potential to improve the rate of marrow engraftment but whose effect on long-term hematopoiesis will likely require gene marking.

Cytokines comprise an increasing number of recently identified hematopoietic growth factors and interleukins, believed to be important in the normal growth and regulation of hematopoiesis. G-CSF and GM-CSF are cytokines which have proven useful in increasing the number of committed progenitor cells prior to marrow harvest, mobilizing marrow progenitor cells into the blood prior to pheresis and enhancing the rate of granulocyte engraftment when administered after marrow or peripheral blood progenitor cells infusion [6–9]. Numerous new cytokines have been identified and are being evaluated for their effect on engraftment and mobilization. Cytokines have also opened the possibility for *ex vivo* expansion of marrow cells [10–12]. The hope of *ex vivo* expansion is the *in vitro* generation of sufficient numbers of cells for transplantation, from small numbers of marrow or blood cells, eliminating the need for a full bone marrow harvest or peripheral blood progenitor cells pheresis.

Gene marking can make critical contributions in assessing the effect of stem cell selection, cytokine therapy and *ex vivo* expansion of engraftment by permitting us to study the contribution that transplanted marrow makes to short and long-term hematopoiesis. Since cytokine therapy and *ex vivo* manipulation may actually promote differentiation of hematopoietic stem cells, their clinical use could lead to the development of late marrow failures. Unfortunately, an *in vitro* assay of marrow stem cells does not exist. There are a number of systems which detect very early human progenitor cells, such as the Long-term Bone Marrow-Culture Initiating Cell [13], Colony Forming Unit-Blast [4], and High Proliferative Potential-Colony Forming Unit [14] assays, but there is still great debate regarding their relevance to the cell responsible for marrow repopulation and long-term engraftment. Gene marking is the only method currently available to evaluate the effects of marrow manipulation on long-term engraftment.

1.2.4 Understanding Disease Relapse after Autologous Transplantation

The final controversy we will consider in our discussion will be the source of disease relapse after autologous transplantation. For those not cured by autologous transplantation, the major cause of treatment failure is recurrent disease. Malignancy after transplantation can result from the proliferation of residual neoplastic cells in the host, which survived the transplantation preparative regime, from malignant cells contained within the transplanted marrow, or both. Understanding the contribution of each of these potential sources in disease relapse will be important in designing more effective transplant protocols.

1.3 Retroviral Vectors as Markers

1.3.1 Advantages of Retroviral Mediated Gene Transfer

One approach in determining the kinetics of marrow reconstitution and evaluating the role of transplanted cells in disease relapse is to mark the transplanted cells used for autologous transplantation. To date, clinically applicable procedures for marking long-lived autologous cells *in vivo* are not available. Radionuclides have been evaluated extensively as labels but their long-term *in vivo* use is limited by either rapid decay or radiation exposure [15, 16]. In addition, the loss, re-utilization or sequestration of label by cells not related to the original labeled cells is a significant problem for long-term *in vivo* studies. These limitations can be overcome by the use of retroviral-mediated gene transfer (RMGT).

Retroviral vectors are extremely well suited to labeling autologous cells since they possess a number of advantages: (1) They insert stably into the target cell genome; (2) The label dies with the cell. It is not lost or sequestered; (3) If the cell divides, all of its progeny also contain the marker; (4) Very sensitive methods of vector detection, namely the polymerase chain reaction, exist which can detect approximately 1 vector-containing cell in 10^5 unmarked cells; (5) If a selectable marker gene is used, it may be possible to select out marked cells from a population of cells which do not contain the vector and (6) RMGT is a simple technical procedure which does not expose the marked cell to toxic compounds or radioactivity which might alter the function of the marked cells.

RMGT has been used extensively *in vitro* to mark many types of cells including leukemia cell lines and bone marrow. Human leukemic cell lines are easily transduced with retroviral vectors, with an efficiency of 10–60%, using supernate transduction protocols [17–19]. RMGT has also been used extensively in murine syngeneic bone marrow transplantation with stable integration and expression of exogenous genes [20–24]. Rosenberg et al. were the first to demonstrate the use of this technology in human clinical trials [25]. Tumor Infiltrating Lymphocytes, or TIL, were marked with the neomycin gene and were shown to persist in the circulation for up to two months.

Based on these encouraging studies, the use of retroviral vectors for gene marking have been applied in the context of autologous transplantation. Most vectors have utilized amphotropically packaged retroviral vectors containing the neomycin resistance gene. Since this gene is of bacterial origin it serves as a unique genetic marker in mammalian cells. Also, the gene product, neomycin phosphotransferase, will protect mammalian cells from the toxic effect of the neomycin analogue G418, permitting selection of vector transduced cells *in vitro*. To date, two similar vectors have been utilized: the G1N retroviral vector created by Genetic Therapy Incorporated (Gaithersburg Maryland, USA) and the LNL6 vector [13] created by Dr. Dusty Miller of the University of Washington (Fig. 1-2). The vectors are based on the Moloney murine leukemia virus, which was altered by the removal of most of the viral genes and the insertion of the bacterial neomycin resistance gene. The vectors have a number of modifications which decrease the likelihood that transplant recipients will be exposed to replication-competent virus. These include the introduction of a stop codon at the *gag* start codon and substitution or deletion of 5' and 3' sequences to minimize homology between vector and packaging cell line genome [27]. Differences in noncoding regions between LNL6 and G1N permit the design of vector specific PCR primers enabling the vectors to be used in double marking studies.

1.3.2   Disadvantage of Retroviral Mediated Gene Transfer

Integration of retroviral vectors into target cells requires cell division. Metabolically active cells (cells residing in the active phases of cell cycle, S and G_2+M) integrate retroviral vectors more efficiently than resting cells, such as those in G_0 or G_1 [28].

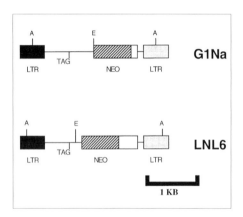

Figure 1-2. Retroviral vectors utilized in clinical gene marking studies. The G1Na and LNL6 retroviral vector containing the neomycin resistance gene (NEO). Abbreviations used: LTR, long terminal repeat region, Moloney murine sarcoma virus LTR (black box), Moloney murine leukemia virus LTR (open box), non-coding region of the neomycin resistance gene (hatched box), and replacement of the gag start codon with a stop codon (TAG). Restriction enzyme sites; Asp718 (A) and EcoRI (E).

This requirement likely poses as the major disadvantage of the use of RMGT in bone marrow transplantation, since stem cells are characteristically a very quiescent population. What factors induce a stem cell to undergo self renewal, without irreversibly committing these cells to unilineage differentiation, are unknown. Identifying these factors or culture conditions remains a major area of research, both in regards to gene therapy of hematopoietic disorders and for *ex vivo* expansion of marrow cells.

1.4 Clincal Gene Marking Studies

1.4.1 Introduction

In the setting of bone marrow transplantation, over 20 gene marking studies targeting bone marrow, peripheral blood progenitor cells and peripheral blood T cells have been proposed. Since T cell marking will be discussed elsewhere in this book, our discussion will be limited to studies involving marrow and peripheral blood progenitor cells. Protocols which have accrued patients or are currently open are listed in Table 1-1. All but one utilize neomycin resistance gene vector(s) as the marker gene. A recent study by Steward et al. proposes to mark cells with a vector containing both the neomycin resistance gene and the thymidine kinase gene [29]. Interested investigators can obtain information regarding the status of clinical gene therapy protocols in the December issue of the journal, Human Gene Therapy. Selected clinical protocols, including many of those discussed in this chapter, can be found in this publication.

1.4.2 Engraftment of Gene Marked Marrow

As illustrated in Table 1-1, there are still relatively few patients who have undergone gene marking and only a few centers have reported their results. Table 2 contains a summary of studies in autologous bone marrow transplantation from four centers. Three protocols evaluated the patients undergoing autologous transplantation for leukemia, [30–32] while another three studies evaluated patients with neuroblastoma [33–34], breast cancer [35–36], or multiple myeloma [35–36]. All centers utilized the G1Na vector or the LNL6 vector (Fig. 1-2). Since Genetic Therapy Incorporated supplied vector to all four centers, the retroviral vectors themselves were not a variable in the findings described in Table 1-2.

Gene marking studies have utilized two distinct transduction protocols. The three centers evaluating leukemic patients utilized short term supernate incubation of target cells without cytokines, since cytokines can stimulate leukemic cell growth and theoretically increase the chance of disease relapse. In these studies, gene transfer at the time of transduction ranged from 1–10% of committed progenitors (Tab. 1-2). Although CD34 selection did not appear to increase the transduction efficacy [37], CD34 selection does, however, decrease the amount of vector supernate required per patient by at least ten fold.

Pre-clinical studies would suggest that gene transfer into committed progenitors can be increased if cytokines are included in the transduction protocol [38–40]. Since cytokines can also stimulate leukemic cells, they were utilized only in the study of Dunbar et al. in patients transplanted for breast cancer (stem cell factor, IL-3 and IL-6) and multiple myeloma (stem cell factor and IL-3). At the time of transduction,

Table 1-1. Clinical gene marking studies in autologous transplantation

Principle investigator	Institution	RAC protocol number	Date first patient entered	Protocol status	Patient accrual	Marking target	Single or double marking	Patient population
Brenner, MK	St Jude, Memphis	4	09.09.1991	closed	12	BM	S	Acute myeloid leukemia in first remission
Brenner, MK	St Jude, Memphis	5	22.01.1992	closed	6	BM	S	Stage D Neuroblastoma in first relapse
Brenner, MK	St Jude, Memphis	6	16.01.1992	closed	3	BM	S	Neuroblastoma without marrow involvement
Deisseroth, AB	MD Anderson	7	31.07.1992	closed	5	BM	S	Chronic myeloid leukemia
Cornetta, K	Indiana Univ	14	15.05.1992	closed	5	BM	S	Acute myeloid and lymphoid leukemia
Deisseroth, AB	MD Anderson	20	17.12.1993	open	4	BM/PBPC	D	Chronic Myeloid Leukemia
Dunbar, C	NIH, Bethesda	23	18.03.1993	open	9	CD34+BM/PBPC	D	Multiple Myeloma
Dunbar, C.	NIH, Bethesda	24	07.12.1992	open	6	CD34+BM/PBPC	D	Breast Cancer
Schuening, FG	Univ of Washington	27	10.04.1994	open	1	CD34+ PBPC	S	Breast and Hodgkin's Diseases
Brenner, MK	St Jude, Memphis	32	21.03.1995	open	1	BM	D	Purging in Stage D Neuroblastoma
Brenner, MK	St Jude, Memphis	39	14.10.1993	open	7	BM	D	Purging in acute myeloid leukemia

Table 1-1. (continued)

Principle investigator	Institution	RAC protocol number	Date first patient entered	Protocol status	Patient accrual	Marking target	Single or double marking	Patient population
Heslop, H	St Jude, Memphis	76	12.07.1994	open	U	CD34+ BM	D	Cytokine treatment of CD34+ cells
Dunbar, C	NIH, Bethesda	25	pending	open	U	CD34+BM/PBPC	D	Chronic Myeloid Leukemia
Deisseroth, AB	MD Anderson	30	pending	open	U	CD34+ BM/PBPC	D	Purging in Indolent Lymphomas
Douer, D	Univ of S. California	92	pending	open	U	BM/PBPC	D	Lymphoma and Breast Cancer
Verfaillie, C	Univ of Minnesota	106	pending	open	U	PBPC	U	Chronic Myeloid Leukemia
Bjorkstrand, B	Karolinska Institute	13	pending	open	U	CD34+ BM/PBPC	U	Multiple Myeloma
Stewart, AK	Univ of Toronto	14	pending	open	U	cultured BMT	S	Multiple Myeloma (neo-thy. kinase vector)

U=unknown, D=double, S=single, BM=bone marrow, PBPC=peripheral blood progenitor cells

Table 1-2. Initial results of clinical gene marking studies

Institution	Disease	Number of patients treated	Initial % Gene Marked Cells	Vector Supernate	Trans-duction time	Cell selection	Cytokines in incubation	Long-term Marking	Longest positive signal post-transplant (months)	# marked relapses/ # total relapses	Reference
St.Jude	AML	12	6.5%	G1Na	6 h	none	none	5%	18+	2/2	30,42
St. Jude	NB	8	4.0%	G1Na	6 h	none	none	"	15+	-	30,42
Indiana	AML/ALL	5	3.6%	G1Na	4 h	none	none	rare	12+	0/2	43
MD Anderson	CML	4	1–10%	LNL6	6 h	CD34	none	-	23+	2/2	37
NIH	Breast	5	18.4%	G1Na/LNL6	72 h	CD34	SCF,IL3,IL6	0.1–0.01%	15+	0/2	36
NIH	Myeloma	6	23.7%	G1Na/LNL6	72 h	CD34	SCF,IL3	"	19+	-	36

Institutions: St. Jude's Hospital, Memphis, TN; Indiana University, Indianapolis, IN; MD Anderson, Houston, TX; NIH, National Institutes of Health, Bethesda, MD. Disease: AML = Acute myeloid leukemia; ALL = Acute lymphoid leukemia; CML = Chronic myeloid leukemia; Breast = Breast carcinoma; Myeloma = Multiple myeloma. Cytokines: SCF = Stem cell factor; IL3 = Interleukin 3; IL6 = Interleukin 6.

the gene transfer rates were significantly higher in this study (18.4% in breast cancer and 23.7% in myeloma patients) compared with the less than 10% noted in those studies which did not utilize cytokines during transduction. Other modifications used by Dunbar may also have increased gene transfer, including the use of a multiple transduction regimen over a three day period and *in vivo* chemotherapy priming prior to harvesting and marking the graft cells. The rational for *in vivo* priming is based on an observation in mice that gene transfer can be increased significantly if performed during recovery from myelosuppressive chemotherapy [41]. While cytokines, multiple transduction and *in vivo* priming are all likely to increase gene transfer into committed progenitors, the relative importance of the individual interventions is currently unknown.

Interestingly, the gene transfer rates noted in committed progenitors immediately after transduction did not always correlate with the long-term engraftment of gene-marked cells. Brenner noted a gene marking rate of 5% immediately after transplantation which appeared to persist long-term [42]. In contrast, our study documented low levels of long-term marking [43], despite having a gene transfer rate of approximately 5% at the time of transduction (Tab. 1-2). This discrepancy was even more notable in the study of Dunbar *et al.* who noted an immediate gene transfer rate of about 20%, although marked cells represented only 0.1–0.01% of marrow elements in those patients with persistent evidence of gene marked cells [36]. There are a number of possible explanations for these findings: Firstly, the highest long-term gene transfer rates, in the study of Brenner *et al.*, were seen in gene marking studies utilizing pediatric patients [42], in contrast with the Dunbar and Cornetta studies which marked adult patients [36, 43]. For the Dunbar study, it is also possible that the addition of cytokines, or the three days of *in vitro* culture used in their transduction protocol, may have had a negative effect upon gene transfer into primitive stem cells.

Since patients involved in the gene marking studies differed in the type and stage of disease, the amount of previous therapy, and the transplant preparative regime which they received, there may have been differences in the proportion of residual marrow contributing to long-term hematopoiesis. For example, all of our acute myeloid leukemia (AML) patients were transplanted in second remission [43],while all but one of the children in the Brenner study were first remission patients [42]. As expected, heavy pre-treatment of our AML patients resulted in a considerably longer time to engraftment, with a mean of 58 days compared with 35 days for the Brenner study. It is possible that very few stem cells contribute to reconstitution after autologous transplantation in heavily pretreated adults. Reconstitution after induction chemotherapy [44] and allogeneic transplantation [45] appears to arise from multiple stem cells in humans, although syngeneic transplantation studies in mice suggest that very few stem cells contribute to engraftment and do so possibly in a clonal succession fashion [46]. If the number of stem cells transplanted in heavily pretreated patients is small, residual stem cells that survive the transplantation conditioning regimen may comprise a greater component of long-term hematopoiesis.

To date, only one study in which double marking of bone marrow versus peripheral blood cells has been reported [36]. Immediately after transduction, there appeared to be decreased marking in the peripheral blood progenitor cell pool (14.5%) compared with bone marrow (29.2%) but preliminary results indicate that peripheral blood cells appear to have greater rates of long-term engraftment. This study appears to contradict findings reported by Bregni et al. [47], which noted peripheral blood progenitor cells to be superior targets for retroviral mediated gene transfer.

While important questions remain to be answered, these studies have made important contributions to our understanding of the biology of autologous transplantation. While not all patients studied demonstrate the presence of gene marked cells long-term, all four investigators noted that some of their patients did have gene marked cells detectable one year or more after transplantation. These studies are the first to show that transplanted autologous cells contribute to long-term hematopoiesis. Persistence of gene marked cells, along with the demonstration of marking in multiple lineages (i.e. granulocytes, T cell and B cells) suggest that very primitive multipotential hematopoietic progenitor cells and possibly stem cells, can be marked with retroviral vectors.

1.4.3 Gene Marking and Disease Relapse

The studies described in Table 1-2 also sought to determine whether or not transplanted marrow contributes to disease relapse. Autologous transplantation for acute leukemias utilizes complete remission marrows, i.e. marrow with no visible leukemia when analyzed by pathological examination. Unfortunately, this method may only detect leukemia if it is present at levels greater than 0.1–1%. Since most transplants often utilize over 10^{10} cells, it is possible that up to 10^8 leukemic cells could, inadvertently, be infused along with the transplanted marrow progenitors. Since leukemia is a disease of the bone marrow, contaminating leukemic cells, in transplanted marrow, would intuitively seem to be an important cause of disease relapse. While methods to rid the transplanted marrow of residual acute leukemia cells ("purging") are theoretically attractive, most clinical studies have failed to show a survival advantage in patients transplanted with purged marrow [48–51]. Without an understanding of the biology of disease relapse, we do not know whether purging methods failed to increase survival because (1) they are ineffective in eliminating disease from the marrow; (2) disease relapse results from residual disease and purging is not needed or (3) purging is effective and required because the transplanted marrow contributes to relapse, but residual disease also contributes to relapse and therefore negates the advantage of purging on overall survival.

Therefore, a method such as gene marking is required if we are to make educated decisions when assessing the need and efficacy of marrow purging.

Support for marrow purging was first demonstrated by Brenner *et al.* [30]. In this study, two children with AML relapsed after transplantation with a portion of leukemic cells containing the neomycin resistance gene. In our study, marked leukemic cells were not detected at relapse in adult patients transplanted for AML. Interestingly, the marked leukemic relapses reported by Brenner and colleagues occurred 2 and 6 months after transplantation [30] while our relapses occurred relatively late: 11 and 18 months after marrow infusion [43]. It is possible that late relapses may arise from residual disease. However, until higher gene transfer rates are attained, the significance of unmarked relapses can not be interpreted with certainty.

In contrast to acute leukemias, chronic myeloid leukemia (CML) is a disease with a significant relapse rate even after allogeneic transplantation, suggesting residual disease is an important contributor to disease relapse in this patient population, even with marrow ablative therapies. Deisseroth and colleagues were successful in demonstrating that infused marrow also plays a significant role in disease relapse, at least for patients with advanced stages of CML [37].

Contamination of the bone marrow graft with tumor cells is also a concern for patients with solid tumors. For example, breast cancer has been found to have occult bone marrow involvement in up to 50% of patients with primary disease [52]. To deal with this problem, bone marrows have been treated *ex-vivo* using drug purging and/or monoclonal antibodies [53–54]. However, the need and effectiveness of *ex-vivo* purging has not be confirmed. In the Dunbar study, three breast cancer patients relapsed at the site of original disease and biopsies in two invaluable patients failed to detect vector sequences in the tumor mass, suggesting relapse resulting from residual tumor [36].

1.5   Conclusion

Gene marking studies have made important contributions to our understanding of autologous transplantation. They have confirmed the contribution of transplanted autologous marrow to long-term hematopoiesis. The finding of gene-marked cells in multiple hematopoietic lineages, for over one year post-transplant, supports the enthusiasm for RMGT as an attractive gene therapy approach for bone marrow disorders. However, the finding of marked relapses in children with AML and adults with CML confirms the need for marrow purging in this population.

Studies that are currently underway should further increase our understanding of hematopoiesis and transplantation. Double marking studies will define the contribution of bone marrow and peripheral blood progenitor cells to short and long-term engraftment. As new cytokines are developed, gene marking will permit us to demonstrate the ability, or inability, of these agents to stimulate proliferation while maintaining the self-renewal potential of marrow stem cells. As attempts to expand marrow elements *ex vivo* are brought to clinical trial, gene marking will help demonstrate the long-term engraftment potential of cells maintained *in vitro*. Increasing gene transfer rates remains the major obstacle to RMGT in marrow elements. Recent manipulations, such as the use of fibronectin fragments to increase the efficiency of supernate transduction, hold promise for providing efficient gene transfer into the stem cell population [55]. Gene marking should continue to provide important information regarding the biology of autologous transplantation, keeping our understanding in pace with the rapid implementation of this procedure in cancer therapy.

References

1. Horowitz, M. M. (1995) New IBMTR/ABMTR slides summarize current use and outcome of allogeneic and autologous transplants. *IBMTR Newsletter* 2(1): 1–8.
2. Berenson, R., Heimfeld, S., Hallagan, J., Jones, R., Shpall, E. (1993) Human stem cell transplantation. *J. Hematother.* 2: 114–115.
3. Strauss, L. C., Rowley, S. D., LaRussa, V. F., Sharkis, S. J., Stuart, R. K., Civin, C. I. (1986) Antigenic analysis of hematopoiesis. V. Characterization of My-10 antigen expression by normal lymphohematopoietic progenitor cells. *Exp. Hematol.* 14: 878.
4. Brandt, J., Baird, N., Lu, L., Srour, E., Hoffman, R. (1988) Characterization of a human hematopoietic progenitor cell capable of forming blast cell containing colonies *in vitro*. *J. Clin. Invest.* 82: 1017.
5. Terstappen LWMM, Huang, S., Safford, D. M., Lansdrop, P. M., Loken, M. R. (1991) Sequential generations of hematopoietic colonies derived from single nonlineage committed CD34+ CD38– progenitor cells. *Blood* 77: 1218.
6. Bregni, M., Siena, S., Magni, M., Bonadonna, G., Gianni, A. M. (1991) Circulating hemopoietic progenitors mobilized by cancer chemotherapy and by rhGM-CSF in the treatment of high-grade non-Hodgkin's lymphoma. *Leukemia* 1: 123–127.
7. Cashman, J. D., Eaves, A. C., Eaves, C. J. (1992) Granulocyte-macrophage colony-stimulating factor modulation of the inhibitory effect of transforming growth factor-beta on normal and leukemic human hematopoietic progenitor cells. *Leukemia* 6: 886–892.
8. Hogge, D. E., Cashman, J. D., Humphries, R. K., Eaves, C. J. (1991) Differential and synergistic effects of Human Granulocyte-Macrophage Colony-Stimulating Factor and Human Granulocyte Colony-Stimulating Factorr on hematopoiesis in human long-term marrow cultures. *Blood* 77: 493–499.
9. Siena, S., Bregni, M., Bonsi, L. et al. (1993) Increase in peripheral blood megakaryocyte progenitors following cancer therapy with high-dose cyclophosphamide and hematopoietic growth factors. *Exp. Hematol.* 21: 1583–1590.
10. Traycoff, C. M., Kosak, S. T., Grigsby, S., Srour, E. F. (1995) Evaluation of ex vivo expansion potential of cord blood and bone marrow hematopoietic progenitor cells using cell tracking and limiting dilution analysis. *Blood* 85: 2059–2068.
11. Verfaillie, C. M., Miller, J. S. (1995) A novel single-cell proliferation assay shows that

long-term culture-initiating cell (LTC-IC) maintenance over time results from the extensive proliferation of a small fraction of LTC-IC. *Blood* 86: 2137–2145.

12. Moore, M. A. S. (1993) *Ex vivo* expansion and gene therapy using cord blood CD34+ cells. *J. Hematother.* 2: 221–224.

13. Sutherland, H. J., Eaves, C. J., Eaves, A. C., Dragowska, W., Lansdorp, P. M. (1989) Characterization and partial purification of human marrow cells capable of initiating long-term hematopoiesis *in vitro*. *Blood* 74: 1563–1570.

14. McNiece, I. K., Stewart, F. M., Deacon, D. M. et al. (1989) Detection of a human CFC with a high proliferative potential. *Blood* 74: 609–612.

15. Griffith, K. D., Read, E. J., Carrasquillo, J. A. (1989) *In vivo* distribution of adoptively transfered Indium-111-labeled tumor infiltrating lymphocytes and peripheral blood lymphocytes in patients with metastatic melanoma. *J. Nat. Cancer Inst.* 81: 1709–1717.

16. Fisher, B., Packad, B. S., Read, E. J. (1989) Tumor localization of adoptively transferred Indium-111 labeled tumor infiltrating lymphocytes in patients with metastatic melanoma. *J. Clin. Oncol.* 7: 250–261.

17. Smith, L. J., Benchimol, S. (1987) Introduction of new genetic material into human myeloid leukemic blast stem cells by retroviral infection. *Mol. Cell. Biol.* 8: 974–977.

18. Hogge, D. E., Humphries, R. K. (1987) Gene transfer to primary normal and malignant human hematopoietic progenitors using recombinant retroviruses. *Blood* 69: 611–617.

19. Cornetta, K., Nguyen, N., Morgan, R. A., Muenchau, D. D., Hartley, J., Anderson, W. F. (1993) Infection of human cells with murine amphotropic replication-competent retroviruses. *Hum. Gene Ther.* 4: 579–588.

20. Belmont, J. W., Henkel-Tigges, J., Chang, S. M. W. et al. (1986) Expression of human adenosine deaminase in murine haematopoietic progenitor cells following retroviral transfer. *Nature* 322: 385–387.

21. Eglitis, M. A., Kantoff, P., Gilboa, E., Anderson, W. F. (1985) Gene expression in mice after high efficiency retroviral-mediated gene transfer. *Science* 230: 1395–1398.

22. Keller, G., Paige, P., Gilboa, E., Wagner, E. F. (1985) Expression of a foreign gene in myeloid and lymphoid cells derived from multipotent haematopoietic precursors. *Nature* 318: 149–154.

23. Miller, A. D., J., E. R., Jolly, D. J. (1984) Expression of a retrovirus encoding human HPRT in mice. *Science* 230: 1395–1398.

24. Williams, D. A., Orkin, S. H., Mulligan, R. C. (1986) Retrovirus-mediated transfer of human adenosine deaminase gene sequences into cells in culture and into murine hematopoietic cells *in vivo*. *Proc. Natl. Acad. Sci. USA* 83: 2566–2570.

25. Rosenberg, S. A., Aebersold, P. M., Cornetta, K. et al. (1990) Gene transfer into humansimmunotherapy of patients with advanced melanoma, using tumor infiltrating lymphocytes modified by retroviral gene transduction. *N. Engl. J. Med.* 323: 570–578.

26. Bender, M. A., Palmer, T. D., Gelinas, R. E., Miller, A. D. (1987) Evidence that the packaging signal of Moloney murine leukemia virus extends into gag region. *J. Virol.* 61: 1639–1646.

27. Miller, A D., Buttimore, C. (1986) Redesign of retrovirus packaging cell lines to avoid recombination leading to helper virus production. *Mol. Cell. Biol.* 6: 2895–2902.

28. Miller, D. G., Mohammed, A. D., Miller, A. D. (1990) Gene transfer by retrovirus vector occurs only in cells that are actively replicating at the time of infection. *Mol. Cell. Biol.* 8: 4239–4242.

29. Stewart, A. K. (1995) A phase I study of autologous bone marrow transplantation with

stem cell gene marking in multiple myeloma. *Hum. Gene Ther.* 6: 107–119.

30. Brenner, M. K., Rill, D. R., Moen, R. C. et al. (1993) Gene-marking to trace origin of re-lapse after autologous bone-marrow transplantation. *Lancet* 341: 85–86.

31. Cornetta, K., Tricot, G., Broun, E. R. et al. (1992) Clinical Protocols: Retroviral mediat-ed gene transfer of bone marrow cell during autologous bone marrow transplantation for acute leukemia. *Hum. Gene Ther.* 3: 305–318.

32. Deisseroth, A. B. (1991) Autologous bone marrow transplantation for chronic myeloge-nous leukemia in which retroviral markers are used to discriminate between relapse which arises from systemic disease remaining after preparative therapy versus relapse due to residual leukemic cells in autologous marrow: A pilot study. *Hum. Gene Ther.* 2(4): 359–376.

33. Brenner, M. K. (1991) A phase I trial of high dose carboplatin and etoposide with autol-ogous marrow support for treatment of relapsed/refractory Neuroblastoma without ap-parent bone marrow involvement. *Hum. Gene Ther.* 2: 273–286.

34. Brenner, M. K. (1991) A phase I trial of high dose carboplatin and etoposide with autol-ogous marrow support for treatment of Stage D Neuroblastoma in first remission. *Hum. Gene Ther.* 2: 257–272.

35. Dunbar, C. E. (1993) Genetic marking with retroviral vectors to study the feasibility of stem cell gene transfer and the biology of hematopoietic reconstitution after autologous transplantation in multiple myeloma, chronic myelogenous leukemia or metastatic breast cancer. *Hum. Gene Ther.* 4: 205–222.

36. Dunbar, C. E., Cottler-Fox, M., O'Shaughnessy, J. A. et al. (1995) Retrovirally marked CD34– enriched peripheral blood and bone marrow cells contribute to long-term en-graftment after autologous transplantation. *Blood* 85: 3048–3057.

37. Deisseroth, A. B., Zu, Z., Claxton, D., Hanania EG, Fu, S., Ellerson, D. (1994) Genetic marking shows that Ph+ cells present in autologous transplants of chronic myelogenous leukemia (CML) contribute to relapse after autologous bone marrow in CML. *Blood* 83: 3068.

38. Bodine, D. M., Karlsson, S., Nienhuis, A. W. (1989) Combination of interleukin-3 and 6 preserves stem cell function in culture and enhances retrovirus-mediated gene transfer in-to hematopoietic stem cells. *Proc. Natl. Acad. Sci. USA* 86: 8897–8901.

39. Nolta, J. A., Kohn, D. B. (1990) Comparison of the effects of growth factors on retrovi-ral vector-mediated gene transfer and the proliferative status of human hematopoietic progenitor cells. *Hum. Gene Ther.* 1: 257–268.

40. Luskey, B. D., Rosenblatt, M., Zsebo, K., Williams, D. A. (1992) Stem cell factor, IL-3 and IL-6 promote retroviral-mediated gene transfer into murine hematopoietic stem cells. *Blood* 80: 396–402.

41. Bodine, D. M., McDonagh, K. T., Seidel, N. E., Nienhuis, A. W. (1991) Survival and retrovirus infection of murine hematopoietic stem cells *in vitro*: Effects of 5-FU and method of infection. *Exp. Hematol.* 19: 206–212.

42. Brenner, M. K., Rill, D. R., Holladay, M. S. et al. (1993) Gene marking to determine whether autologous marow infusion restores long-term haemopoiesis in cancer patients. *Lancet* 342: 1134–1137.

43. Cornetta, K., Srour, E. F., Berebitsky, D. et al. (1994) Retroviral marking of autologous bone marrow grafts in adult acute leukemia. *Blood* 10 (Suppl 1): 401a.

44. Fialkow, P. J., Janssen, J. W. G., Bartram, C. R. (1991) Clonal remissions in acute non-lymphocytic leukemia: Evidence for a multistep pathogenesis of the malignancy. *Blood*

77: 1415–1417.

45. Nash, R., Storb, R., Neiman, P. (1988) Polyclonal reconstitution of human marrow after allogeneic bone marow transplantation. *Blood* 72: 2031.

46. Lemischka, I. R., Raulet, D. H., Mulligan, R. C. (1986) Developmental potential and dynamic behavior of hematopoietic stem cells. *Cell* 45: 917–927.

47. Bregni, M., Magni, M., Siena, S., Di Nicola, M., Bonadonna, G., Gianni AM (1992) Human peripheral blood hematopoietic progenitors are optimal targets of retroviral-mediated gene transfer. *Blood* 80: 1418–22.

48. Chopra, R., Goldstone, A. H., McMillan, A. K. (1991) Successful treatment of acute myeloid leukemia beyond first remission with autologous bone marrow transplantation using bysulfan/cyclophosphamide and unpurged marrow: The British Autograft Group Experience. *J. Clin. Oncol.* 9: 1840–1847.

49. Ramsay, N., LeBien, T., Nesbit, M. (1985) Autologous bone marrow transplantation for patients with acute lymphoblastic leukemia in second or subsequent remission: Results of bone marrow treated with BA-1, BA-2, and BA-3 with complement. *Blood* 66: 508–513.

50. Zittoun, R. A., Mandelli, F., Willemze, R. et al. (1995) Autologous or allogeneic bone marrow transplantation compared with intensive chemotherapy in acute myelogenous leukemia. *N. Engl. J. Med.* 332: 217–223.

51. Gulati, S. C., Romero, C. E., Ciavarella, D. (1994) Is bone marrow purging proving to be of value? *Oncology* 8: 19–24.

52. Cote, R. J., Rosen, P. P., Hakes, T. B. et al. (1988) Monoclonal antibodies detect occult breast carcinoma metastases in the bone marrow of patients with early stage disease. *Amer. J. Surg. Pathol.* 12: 333–340.

53. Jones, R. J., Miller, C. B., Zehnbauer, B. A., Rowley, S. D., Colvin, O. M., Sensenbrenner, L. L. (1990) Bone Marrow *Transplantation* 5: 301–307.

54. Cheson, B. D., Lacerna, L., Leyland-Jones, B., Sarosy, G., Wittes, R. E. (1989) Autologous bone marrow transplantation. Current status and future directions. *Ann. Internal Medicine* 1: 51–65.

55. Moritz, T., Patel, V. P., Williams, D. A. (1994) Bone marrow extracellular matrix molecules improved gene tranfer into human hematopoietic cells via retroviral vectors. *J. Clin. Invest.* 93: 1451–1457.

2 Gene Marking of T Lymphocytes

C. Bonini, C. Bordignon

2.1   Introduction

In the past years, a number of clinical trials involving the adoptive transfer of genetically modified T lymphocytes have been reported [1–3]. Gene marking studies
were the first gene transfer protocols to enter clinical practice. The principal objective of a gene marking study is to introduce, in the target cells, a gene which does
not modify the function of the cells but allows them to be detected, providing information on survival, distribution, and function of the infused genetically-modified
cells. Moreover, gene marking studies provided crucial informations about the feasibility, safety, and efficacy of genetically-modified cells, an important pre-requisite
for future gene therapy trials. To date, two general groups of marking studies have
been conducted. The first group of protocols focuses on the transduction of lymphocytes with potential antitumor or antiviral activity. Target cells of these studies
include tumor infiltrating lymphocytes (TIL), virus-specific cytotoxic T cells (anti-
Epstein-Barr virus specific CTLs and anti-HIV specific CTLs), and donor-derived
lymphocytes infused in the context of allogeneic bone marrow transplantation
(BMT). The second group of gene marking studies focuses on the transduction of
autologous bone marrow cells from patients with neoplastic diseases. These studies
provided important information concerning the biology of BMT and the source of
post-BMT relapse of neoplastic diseases. The principal purpose of these studies was
to determine whether neoplastic cells, present in unpurged autologous bone marrow, contribute to relapse following autologous BMT. In all disease settings (acute
leukemia, chronic myelogenous leukemia, and neuroblastoma) gene marked tumor
cells were found in relapsed patients [4, 5]. This observation represents the basis for
ongoing second generation studies, which focus on the comparison of different
purging techniques performed prior to transplantation. The second purpose of these
studies was to investigate *in vivo* the possibility of introducing a gene into normal
hematopoietic progenitors. The presence of the marker gene in hematopoietic progenitor cells was confirmed *ex vivo* by clonogenic assays. The marker gene continued
to be detected and expressed for up to 4 years in the mature progeny of marrow precursor cells, suggesting that a relatively immature hematopoietic cell population had
been transduced.

T. Blankenstein (ed.) Gene Therapy
©1999, Birkhäuser Verlag Basel

Although marking studies have been useful, it is becoming apparent that the marker genes used have a number of undesirable characteristics. Future applications of marking, in the hematopoietic system and elsewhere, will require the use of marker elements that will not produce any modification of the cells' behavior and potentially escape host immune surveillance.

2.2   Gene Transfer Technology and Safety Issues

Retroviral vectors based on the Moloney murine leukemia virus were the first utilized in clinical trials and remain the most effective approach for the introduction of genes into human T lymphocytes. In fact, such vectors provide some important advantages compared with other gene transfer techniques. These advantages are represented mainly by the ability to incorporate up to 9 Kb of cDNA [6], a high efficiency of gene transfer in replicating cells (more than 50% on human lymphocytes) [7], and the ability to integrate in the DNA of the target cell, leading to long-term gene expression and avoiding dilution of the transferred gene by cell replication [8]. Being the first clinical trials of gene transfer, safety of gene marking protocols was the major concern. For that reason, three generations of packaging cell lines were developed in order to decrease the risk of recombination events, resulting in the production of helper viruses. The combination of PA317 packaging cell lines [9] and LNL6 retroviral vectors was utilized for the first clinical trial of gene marking. The ability of murine amphotropic retrovirus to infect and replicate in primate cells was assessed *in vitro*, and the safety of gene transfer was assessed extensively, in monkeys, documenting no clinical illness up to 43.9 months after the infusion of replication-competent retrovirus (RCR) [10]. In a subsequent study, the exposure of severely immunosuppressed monkeys to high titer RCR was followed by the development of T-cell lymphoma [11], suggesting that replication competent retrovirus could be pathogenic to primates under certain conditions. These animals had been irradiated and infused with T cell depleted marrow. The lymphoma occurred only in those animals that failed to mount an antiviral immune response. The presence of RCR was checked extensively before infusion of transduced cells, although a small number of replication-competent viral particles is not expected to be sufficient to produce any problem in humans [8].

2.3   The Clinical Application

2.3.1   Gene Marking of Tumor Infiltrating Lymphocytes

The first gene marking study was performed in 1990 by Rosenberg et al. [12] at the National Institute of Health in Bethesda. This was a continuation of previous studies, begun in 1986 and involving a clinical protocol for the treatment of advanced malignant melanoma with a newly discovered class of immune cells called TIL [13]. TIL are lymphocytes directly isolated from the tumor and that are then grown and expand *in vitro* culture in the presence of IL-2. This study documented the ability of TIL plus IL-2 to mediate the regression of metastatic melanoma (35–40% of objective responses). Since TIL seem to accumulate in tumor deposits, they could be used as vehicles to deliver molecules, to the tumor-site, that would increase their anti-tumor effectiveness (such as IL-2, IL-4, IL-12, gamma-IFN, or TNF). Using gene transfer techniques, new properties could be introduced into TIL to enhance existing functions or to generate a kind of lymphocyte that normally did not exist in nature. This could either increase the efficacy of the treatment or reduce the toxicity, since high dose systemic administration of cytokine could be replaced by local delivery. Thus, gene marking studies were designed with two different purposes: first, to answer important questions concerning the survival and the *in vivo* distribution of TIL; second, to demonstrate the safety and efficacy of using retroviral-mediated gene transduction to introduce genes into TIL, providing the essential basis for future transfer of therapeutic genes. The marker utilized in this trial was the gene coding for neomycin phosphotransferase (neoR) that would mark cells by expression of a bacterial enzyme conferring resistance to the antibiotic neomycin and to its analogue G418. The use of a gene marking strategy provided many advantages that did not exist with previous radio-labeled marking studies, such as those utilizing Indium-111. Indium-111 has a half life of 2.8 days and the combination of the natural decay of the isotope and the spontaneous release of the label from the cell severely limited the time in which these cells could be used to study cell survival *in vivo*. Furthermore, cell division resulted in dilution of the label, making the detection of high-proliferating cells extremely difficult. Finally, autoirradiation of the cell, resulting from indium-111 incorporation, led to potential damage of TIL which could alter their function. Transfer of the neomycin resistance gene appeared to represent an ideal label that could overcome all these problems. In fact, it could be incorporated into the genome of the cell without altering its function and would replicate along with the cell so that progeny would have an equal amount of label. Furthermore, this marker could be used to re-isolate the infused cells. Moreover, it was possible to detect as few as one modified cell in 10^5 normal cells thanks to the polymerase chain reaction technique. The first five patients were treated in 1989 and received between 3.3 and 14.5×10^{10} cells. The estimated percentage of transduced cells ranged between 1 and 11%. In the first five treated patients, genetically-modified cells could be detected in peripheral blood up to 189 days after the first infusion, although, in general, transduced cells were found in the circulation for the first 3 weeks after infusion, which corresponded to the time of IL-2 administration. Moreover, transduced-TIL could be detected in tumor sites up to 64 days after infusion. No side effects due to the gene transfer procedure were observed in this trial. Two of the first 5 treated patients had objective responses after the infusion of gene-modified TIL, documenting that the gene transfer procedure did not alter the *in vivo* function of TIL. [12, 1]. This study demonstrated the feasibility and safety of using retroviral-mediated gene transduction into lymphocytes as a method of introducing new genes into humans, providing crucial information for future gene therapy trials, either for the treatment of cancer or genetic diseases. In a subsequent gene marking study, conducted by Cai et al. [14], four melanoma patients received genetically modified TIL expressing the neoR gene. The marker gene could be detected by PCR in peripheral blood samples of all treated patients and in tumor biopsies of some patients for up to 3 months after infusion. A similar gene marking protocol was initiated at the Centre Leon Berard in Lyon, France. Other gene marking studies were developed to investigate the characteristics of TIL therapy. Using a similar gene marker protocol, investigators at the University of UCLA Medical Center, Los Angeles studied whether TIL specifically homed to tumor deposits, comparing the amount of marked TIL and IL-2 stimulated peripheral blood lymphocytes (PBL) present in tumor tissue versus biopsies of normal skin, fat, and muscle using a semiquantitative PCR method. Nine patients affected by melanoma or renal cancer received between 4.5×10^8 and 1.24×10^{10} total cells (both TIL and PBL transduced with two retroviral vectors distinguishable by PCR and both containing the neoR gene). Both TIL and PBL could be detected by PCR analysis in peripheral blood and tumor biopsies up to 99 days after infusion. No convincing pattern of preferential trafficking of TIL versus PBL to tumor sites was noted. Moreover, concurrent biopsies of muscle, fat, and skin demonstrated the presence of TIL and PBL in comparable or greater numbers than in tumor sites in five patients [15]. Other studies investigating the role of different TIL subsets are now in progress [14].

2.3.2 Gene Marking of T Lymphocytes in the Context of Allo-BMT

Allogeneic bone marrow transplantation (allo-BMT) is the treatment of choice for many hematologic malignancies [16, 17]. It is now established that the allogeneic immune advantage, in addition to the effect of high dose chemoradiotherapy, is responsible for the curative potential of allo-BMT. Although the nature of the effector cells has not yet been fully elucidated, the allogeneic advantage in bone marrow transplantation is well documented by the superior results produced by transplants from allogeneic donors compared with autologous or syngeneic transplants [18]. However, the therapeutic impact of the allogeneic advantage is limited by the risk of graft versus host disease (GvHD), a potential life-threatening complication of allo-BMT. Severe GvHD can be circumvented by the removal of T lymphocytes from the graft [17]. However, T-depletion increases the incidence of disease relapse, graft rejection, and reactivation of endogenous viral infections [19]. For all these reasons, the delayed administration of donor lymphocytes has recently become a new tool for treating leukemic relapse after BMT. Patients affected by post-BMT recurrence of chronic myelogenous leukemia, acute leukemia, lymphoma, or multiple myeloma could achieve complete remission after the infusion of donor leukocytes without re-quiring cytoreductive chemotherapy or radiotherapy [20–22]. Moreover, the infu-sion of donor lymphocytes after allogeneic bone marrow transplantation is a promising therapeutic tool for the treatment of other complications related to the severe immunosuppressive status of transplanted patients, such as Epstein Barr virus induced lymphoproliferative disorders (EBV-BLPD) [23] or reactivation of CMV in-fection [24]. In fact, the primary defect contributing to the pathogenesis of progres-sive infection in these patients appears to be the inability to generate adequate virus-specific alpha-beta T cell responses. Thus, the adoptive transfer of T lymphocytes to prevent and treat these complications seems a very interesting and efficacious tool. Despite all these beneficial effects, severe GvHD represents a frequent and poten-tially lethal complication [20]. No specific treatment exists for established GvHD. Different strategies have been utilized in order to circumvent this complication, and gene marking has been used in order to investigate the safety and efficacy of im-munotherapy in the context of allo-BMT.

2.3.2.1 Adoptive Transfer of Virus-Specific T Cells

When the targets of T lymphocytes are known, it is possible to generate and expand antigen-specific CTLs *in vitro*. Adoptive transfer of virus-specific T cells has been extensively studied in animal models providing important information on the role of specific effector subsets and demonstrating the therapeutic role of T cells. At the present time, adoptive immunotherapy with T cells specific for viral antigens is be-ing explored as treatment for human diseases associated with CMV, EBV, and HIV-1 infections [25]. In the context of allogeneic BMT, the transfer of these specific ef-fector cells rather than unselected lymphocytes should significantly reduce the risk of GvHD.

The first study in humans, investigating adoptive immunotherapy with virus spe-cific T cells was performed in allogeneic bone marrow recipients at high risk for de-veloping CMV disease. The development of CMV pneumonia in immunosuppressed transplanted patients is a life-threatening complication which occurs exclusively in individuals without post-BMT recovery of CD8+ CMV specific T cell responses [26]. This observation is consistent with previous studies in a murine CMV (mCMV) model which provided principles to guide the development of adoptive im-munotherapy for human CMV infection. In fact, the importance of subsets of mCMV-specific T cells in protection from lethal disease has been demonstrated by mCMV infection in immunosuppressed BALB/c mice. Adoptive transfer of mCMV-specific CD8+ T cells into such mice was sufficient for protective immunity to mCMV [27], whereas adoptive transfer of CD4+ was ineffective in preventing pro-gression of infection, although CD4+ cells were required to resolve persistent sali-vary gland infection [28]. These observations were the basis for clinical application.

At the Fred Hutchinson Research Institute, specific anti-CMV CD8+ donor-derived CTL clones were infused to allograft recipients as prophylaxis for CMV disease. No toxicity was observed and short-term reconstitution of anti-CMV specific activity could be documented. Moreover, the adoptive transfer of T cells appeared to confer protective immunity, since no patient developed CMV viremia or disease after having received T cell therapy [24].

However, the long-term value of CTL therapy might depend on the durability and antigen-responsiveness of the transferred lymphocytes, as it will be important to have antigen-specific cell populations that expand during periods of viral or tumor challenge and contract after viral or tumor load is eliminated. In order to investigate the long-term value of CTL therapy, investigators of the St. Jude Children's Research Hospital designed a clinical protocol involving the infusion of neoR-marked donor specific anti-EBV CTLs to allograft recipients [29]. EBV-induced lymphoproliferation in the immunocompromised host provides an ideal model to analyze the potential of this immunotherapeutic strategy both for virus-induced disease and for cancer. In fact, EBV has been found in a number of malignant diseases, such as immunoblastic lymphomas arising in allograft recipients and HIV-infected patients, in the oral leukoplakia of AIDS, and in Reed-Sternberg cells. EBV is controlled in the immunocompetent host by virus-specific cytotoxic T lymphocytes which lyse EBV-infected cells when they recognize fragments of viral protein presented in the context of class I MHC molecules [30]. Although unable to completely eliminate EBV from the body, CTLs seem to be essential in maintaining control of latently infected cells. Specific anti-EBV CTLs can be easily obtained *in vitro* by usage of autologous lymphoblastoid cell lines as stimulators, since the majority of immunocompetent individuals have a strong immune response to EBV. To determine whether adoptive transfer of donor derived EBV specific CTLs can reconstitute the immune response to EBV and prevent EBV-lymphoma in allograft recipients, both CD4+ and CD8+ anti-EBV T-cells were transduced with a retroviral vector containing neoR. This study provided important informations related to the persistence and immunologic activity of CTLs. Fourteen patients received $4 \times 10^7 - 1.2 \times 10^8$ genetically-modified T cells per square meter. Marked cells could be detected in the peripheral blood for a median of 10 weeks. The transferred cells showed specific anti-EBV cytotoxic activity. Three patients with very low or undetectable levels of anti-EBV CTL precursor before the infusion, could obtain a 5–100 fold increase in the frequency of specific CTLp after infusion, entering the normal range of anti-EBV CTLp. Although the number of neoR-CTLs fell under the threshold of PCR detection by 4 to 5 months after infusion, stimulation of PBMCs with appropriate EBV-LCL could enrich both CD4+ and CD8+ cell populations in marked cells to a level detectable by PCR analysis up to 18 months after infusion. The physiological relevance of this phenomenon was observed in a patient with undetectable levels of marked cells 18 months after infusion. In this patient, the rise in the EBV-DNA level, indicating reactivation of latent virus, was followed by an increase in the frequency of marked cells, which could be detected again in freshly isolated PBMCs by PCR analysis [31]. All these observations suggest a wider use of CTL therapy in the future, even for the treatment of solid tumors and leukemia.

Gene marking studies were also performed to investigate the role of HIV-1 specific CTL in the treatment of HIV infection. HIV-1 infection elicits host immune responses in the immunocompromised host, leading to a period of clinical latency characterized by active virus replication, which causes a progressive destruction of the host immune system and the development of AIDS [32]. Although the immune response elicited by HIV-1 does not result in complete resolution of infection, there is evidence that the CD8+ T cell response to HIV-1 contributes significantly to the temporary control of HIV-1 replication. During the early phase of infection, a vigorous polyclonal CD8+ response develops and individuals commonly have readily detectable CD8+ in the peripheral blood reactive with many HIV-1 proteins [32]. Coincident with the development of HIV-1 specific CD8+ T cells, CD4+ T cell counts often recover to near normal levels, and plasma viremia and p24 antigen levels decline dramatically [33]. Moreover, the persistence of high levels of CD8+ HIV-1-specific activity is associated with a longer asymptomatic phase and more gradual decline of CD4+ circulating cells. The terminal phase of HIV infection is characterized by severe immunodeficiency, high levels of HIV viremia, and the development of malignancies and infections with opportunistic pathogens [34]. This stage of disease is characterized by undetectable levels of CD8+ HIV-1-specific T cells. Since the transient control of HIV replication during the early clinical latent phase seems to be mediated by CD8+ HIV-specific T cells, the adoptive transfer of autologous CD8+ HIV-specific T cells might be of therapeutic benefit [35]. However, safety issues must be considered in the context of adoptive transfer of T cells to HIV patients. Microglial cells in the CNS are known to be reservoirs of the virus and could be targets for transferred HIV-1 specific T cells, potentially leading to fatal immunopathologic lesions. Moreover, the migration of activated CD8+ HIV-1 specific T cells into the lungs has been postulated to result in the lymphocytic alveolitis occasionally observed during the natural host immune response to HIV-1. Therefore, at the Fred Hutchinson Research Institute, autologous specific anti-HIV-1 CD8+ CTL clones used in adoptive immunotherapy were modified by retrovirus-mediated gene transfer to express a suicide gene coding for a bifunctional protein carrying both hygromycin phosphotransferase activity and thymidine kinase activity of the Herpes Simplex virus (HSV-TK) and thus conferring sensitivity to gancyclovir. Genetically modified cells were infused at an escalating cell dose to six HIV-1 sieropositive patients. No significant toxicity was observed and no patient required ablation with gancyclovir. However, five out of six patients developed cytotoxic T-lymphocytes response against the novel protein, leading to the elimination of transduced CTL and limiting the therapeutic potential of the strategy [36]. This observation underlines the need to improve gene transfer technology and the choice of marker genes in order to escape immunosurveillance.

2.3.2.2 Gene Marking of Donor Lymphocytes for Graft Versus Leukemia and Anti-Viral Activity

A different approach of gene marking was used by our group at the H.S. Raffaele in Milano. In allogeneic marrow transplantation, donor lymphocytes play a central therapeutic role for both graft versus leukemia (GvL) and immune reconstitution. Patients who developed leukemic relapse or EBV lymphoma after BMT received donor PBL transduced by a retroviral vector. The use of a heterogeneous T cell population, with a wide range of antigen-specificity carries some advantages over the use of specific CTLs. First, such a population can provide a more complete immunological reconstitution to immunocompromised transplanted patients. Second, since the antigen specificity of the GvL effector cells is not completely clear, the use of the entire T cell repertoire is the best option to obtain a GvL effect to date. However, the therapeutic use of donor T cells is limited by severe graft vers host disease. Although the delay in the administration of T lymphocytes is expected to reduce the risk of severe GvHD, this risk is still present at higher doses of donor T-cells. For this reason, in this study, donor T cells were transduced with a retroviral vector for the transfer and expression of a suicide gene into human PBLs, which would allow the *in vivo* selective elimination of cells responsible for severe GvHD. Additionally, this approach might allow *in vivo* modulation of donor anti-tumor responses, and separation of GvL from GvHD. Finally, the presence of gene-marked cells allows us to address important and unresolved questions concerning the survival and function of donor lymphocytes used in this way [37].

The vector utilized in this study carries two genes: the first coding for the truncated form of the low affinity receptor for the nerve growth factor (ΔLNGFR); the second coding for a bifunctional protein carrying both the HSV-TK activity, conferring gancyclovir sensitivity to transduced cells, and neoR. In previous studies, the possibility of using the LNGFR as a cell surface marker was investigated [38]. The use of a surface molecule for cell marking represents an important advantage over conventional strategies, which typically involve the use of nuclear markers, e.g. the gene coding for neomycin resistance (neoR). The cell surface molecule, in fact, allows a rapid *in vitro* selection of transduced cells by the use of magnetic immunobeads. This selection requires a shorter culture time, an important variable in order to preserve the immune repertoire and to achieve cell numbers suitable for clinical application [38]. In addition, a surface marker allows an easy *ex vivo* detection and characterization of the transduced cells by FACS analysis.

In order to eliminate any possible functional activity of the cell surface marker, a modified form of the NGFR gene, coding for a protein deleted of its intracellular domain, was utilized. This deletion completely abrogated the signal transduction activity of the receptor. Moreover, being a human protein, this marker is not expected to be a target for a specific immune response. In the pilot study, 8 patients affected by hematologic malignancies who developed severe complications (EBV lymphoproliferative disease, CML relapse, and AML relapse) following an allogeneic T-cell depleted BMT, received escalating doses (beginning at 1×10^5/kg up to 2×10^7/kg) of genetically-modified donor PBL. In this study, the *in vivo* persistence of transduced cells in high proportions (up to 13.4% of circulating PBL) and long term (more than 5 months) could be easily documented by FACS analysis; an anti-tumor response was observed in five patients, with three complete responses; three patients developed GvHD that required gancyclovir treatment; gancyclovir-mediated selective elimination of transduced cells resulted in near resolution of all clinical and biochemical signs of acute GvHD in all affected patients [39]. These data indicate that genetically modified cells maintain their *in vivo* potential to develop both anti-tumor and GvHD effects, and might represent a new potent tool for providing a specific mean for elimination of acute GvHD, making allogeneic marrow transplantation more efficacious, safer, and available to a larger number of patients.

References

1. Anderson, W. F. (1992) *Hum. Gene Ther. Science* 256: 808–813.
2. Brenner, M. (1996) Gene marking. *Hum. Gene Ther.* 7: 1927–1936.
3. Hege, K. M., Roberts, M. R. (1996) T-cell gene therapy. *Curr. Opin. Biotechnol.* 7: 629–634.
4. Brenner, M. K., Rill, D. R., Moen, R. C. et al. (1993) Gene-marking to trace origin of relapse after autologous bone-marrow transplantation. *Lancet* 341: 85–86.
5. Deisseroth, A. B., Zu, Z., Claxton, D. et al. (1994) Genetic marking shows that Ph+ cells present in autologous transplants of chronic myelogenous leukemia (CML) contribute to relapse after autologous bone marrow in CML. *Blood* 83: 3068–3076.
6. Crystal, R. G. (1995) Transfer of genes to humans: early lessons and obstacles to success. *Science* 270: 404–410.
7. Bunnell, B. A., Muul, L. M., Donahue, R. E. et al. (1995) High-efficiency retroviral-mediated gene transfer into human and nonhuman primate peripheral blood lymphocytes. *Proc. Natl. Acad. Sci. USA* 92: 7739–7743.
8. Blease, M., Blankenstein, T., Brenner, M. et al. (1995) Vectors in cancer therapy: how will they deliver? *Cancer Gene Ther.* 2: 219–297.
9. Miller, A. D., Buttimore, C. (1986) Redesign of retrovirus packaging cell lines to avoid recombination leading to helper virus production. *Mol. Cell. Biol.* 6: 2895–2902.
10. Cornetta, K., Morgan, R. A., Anderson, W. F. (1991) Safety issues related to retroviral-mediated gene transfer in humans. *Hum. Gen. Ther.* 2: 5–14.
11. Donahue, R. E., Kessler, S. W., Bodine, D. et al. (1992) Helper virus induced T cell lymphoma in nonhuman primates after retroviral mediated gene transfer. *J. Exp. Med.* 176: 1125–1135.
12. Rosenberg, S. A., Aebersold, P., Cornetta, K. et al. (1990) Gene transfer into humans – immunotherapy of patients with advanced melanoma, using tumor-infiltrating lymphocytes modified by retroviral gene transduction. *N. Engl. J. Med.* 323: 570–578.
13. Rosenberg, S. A., Packard, B. S., Aebersold, P. M. et al. (1988) Use of tumor-infiltrating lymphocytes and interleukin-2 in the immunotherapy of patients with metastatic

melanoma, special report. *N. Engl. J. Med.* 319: 1676–1680.

14. Cai, Q., Rubin, J. T., Lotze, M. T. (1995) Genetically marking human cells-results of the first clinical gene transfer studies. *Cancer Gene Ther.* 2: 125–136.

15. Economou, J. S., Belldegrun, A. S., Glaspy, J. et al. (1996) *In vivo* trafficking of adoptively transferred interleukin-2 expanded tumor-infiltrating lymphocytes and peripheral blood lymphocytes. Results of a double gene marking trial. *J. Clin. Invest.* 97: 515–521.

16. Thomas, E. D., Clift, R. A., Fefer, A. et al. (1986) Marrow transplantation for the treatment of chronic myelogenous leukemia. *Ann. Int. Med.* 104: 155–163.

17. O'Reilly, R. (1993) Bone Marrow Transplantation. *Curr. Opin. Hematol.* : 221–222.

18. Horowitz, M. M., Gale, R. P., Sondel, P. M. et al. (1990) Graft versus leukemia reactions after bone marrow transplantation. *Blood* 75: 555–562.

19. Kernan, N. A., Bordignon, C., Collins, N. H. et al. (1989) Bone marrow failure in HLA-identical T-cell depleted allogeneic transplants for leukemia: I. Clinical aspects. *Blood* 74: 2227–2236.

20. Kolb, H. J., Schattenberg, A., Goldman, J. M. et al. (1995) Graft-versus-leukemia effect of donor lymphocyte transfusions in marrow grafted patients. *Blood* 86: 2041–2050.

21. Slavin, S., Naparstek, E., Nagler, A. et al. (1996) Allogeneic cell therapy with donor peripheral bloood cells and recombinant human interleukin-2 to treat leukemia relapse after allogeneic bone marrow transplantation. *Blood* 87: 2195–2204.

22. Tricot, G., Vesole, D. H., Jagganath, S. et al. (1996) Graft-versus-myeloma effect: proof of principle. *Blood* 87: 1196–1198.

23. Papadopoulos, E. B., Ladanyi, M., Emanuel, D. et al. (1994) Infusions of donor leukocytes to treat Epstein-Barr virus-associated lymphoproliferative disorders after allogeneic bone marrow transplantation. *N. Engl. J. Med.* 330: 1185–1191.

24. Riddell, S. R., Watanabe, K. S., Goodrich, J. M. et al. (1992) Restoration of viral immunity in immunodeficient humans by adoptive transfer of T cell clones. *Science* 257.

25. Riddell, S. R., Greenberg, P. D. (1995) Principles for adoptive T cell therapy of human viral diseases. *Annu. Rev. Immunol.* 13: 545–586.

26. Reusser, P., Riddel, S. R., Meyers, J. D. et al. (1991) Cytotoxic T-lymphocyte response to cytomegalovirus after human allogeneic bone marrow transplantation: pattern of recovery and correlation with cytomegalovirus infection and disease. *Blood* 83: 1373–1380.

27. Reddehase, M. J., Mutter, W., Munch, K. et al. (1987) CD8-positive T lymphocytes specific for murine cytomegalovirus immediate-early antigens mediate protective immunity. *J. Virol.* 61: 3102–3108.

28. Lucin, P., Pavic, I., Polic, B. et al. (1992) Gamma interferon-dependant clearance of cytomegalovirus infection in salivary gland. *J. Virol.* 66: 1977–1984.

29. Rooney, C. M., Smith, C. A., Ng, C. Y. C. et al. (1995) Use of gene-modified virus-specific T lymphocytes to control Epstein-Barr-virus related lymphoproliferation. *Lancet* 345: 9–13.

30. Strauss, S. E., Cohen, J. I., Tosato, G. et al. (1992) Epstein-Barr virus infection: biology, pathogenesis and management. *Ann. Intern. Med.* 118: 45–58.

31. Heslop, H. E., Ng, C. Y. C., Li, C. et al. (1996) Long-term restoration of immunity against Epstein-Barr virus infection by adoptive transfer of gene-modified virus-specific T lymphocytes. *Nat. Med.* 2: 551–555.

32. McCune, J. M. (1991) HIV-1: the infective process *in vivo. Cell* 64: 351–363.

33. Borrow, P., Lewicki, H., Hahn, B. H. et al. (1994) Virus-specific CD8+ cytotoxic T-lymphocyte activity associated with control of viremia in primary human immunodeficiency

virus type 1 infection. *J. Virol.* 68: 6103–6110.

34. Fauci, A. S. (1988) The human immunodeficiency virus: infectivity and mechanism of pathogenesis. *Science* 239: 617–622.

35. Carmichael, A., Jin, X., Sissons, P. et al. (1993) Quantitative analysis of the human immunodeficiency virus type 1 (HIV-1)-specific cytotoxic T lymphocytes (CTL) response at different stages of HIV-1 infection: differential CTL responses to HIV-1 and Epstein-Barr virus in late disease. *J. Exp. Med.* 177: 249–256.

36. Riddell, S. R., Elliott. M., Lewinsohn, D. A. et al. (1996) T-cell mediated rejection of gene-modified HIV-specific cytotoxic lymphocytes in HIV-infected patients. *Nat. Med.* 2: 216–223.

37. Bordignon, C., Bonini, C. (1995) A clinical protocol for gene transfer into peripheral blood lymphocytes for *in vivo* immunomodulation of donor anti-tumor immunity in patients affected by recurrent disease after allogeneic bone marrow transplantation. *Hum. Gene Ther.* 2: 813–819.

38. Mavilio, F., Ferrari, G., Rossini, S. et al. (1994) Peripheral blood lymphocytes as target cells of retroviral vector-mediated gene transfer. *Blood* 83: 1988–1997.

39. Bonini, C., Ferrari, G., Verzeletti, S. et al. (1997) HSV-TK gene transfer into donor lymphocytes for control of allogenic graft-versus-leukemia. *Science* 276: 1719–1724.

3 *MDR*1 Gene Transfer to Hematopoietic Cells

T. Licht, M. M. Gottesman, I. Pastan

3.1   Indroduction

Gene therapy, which has the potential to cure hereditary or acquired diseases, is still hampered by low gene expression in target organs. This is particularly true for hematopoietic disorders because the efficiency of gene transfer is limited and stable expression of transgenes in bone marrow has been found difficult to accomplish (Miller, 1990). In preclinical primate studies as well as in clinical trials, the percentage of hematopoietic cells expressing transgenes long-term have been found to be disappointingly low (Bodine et al., 1993; Dunbar et al., 1994; Van Beusechem et al., 1995). One possible strategy for overcoming this problem employs the use of selectable drug resistance markers, that allow enrichment of transduced cells *in vivo*, if or when expression levels decrease. Selectable markers provide an advantage to transduced cells by conferring resistance to cytotoxic drugs and result in the selected elimination of non-transduced cells by drug treatment.

Several genes have been identified that render mammalian cells chemo-resistant. These genes were originally found in naturally occurring cancers that failed to respond to clinical treatment, or by stepwise drug selection of cells in tissue culture. Cloning of drug resistance genes and their introduction into eukaryotic expression vectors has made it possible to confer chemo-resistance to drug-sensitive target cells. Drug resistance genes either inactivate single classes of cytotoxic agents, or confer simultaneous resistance to multiple drugs. Examples of the latter type of cross-resistance include the multidrug transporter (P-glycoprotein) and the *MRP* transporter (Cole et al., 1992; Gottesman and Pastan, 1993).

While several genes that confer chemo-resistance have been cloned and characterized, not all of them are useful *in vivo* because of the pharmacology and toxicology of the respective drugs. Combination chemotherapy makes it possible to limit the dose levels of single drugs below the threshold for each of severe organ toxicity. Thus, it would be advantageous to use a flexible system that confers drug resistance to a range of different drugs. *MDR*1, the multidrug resistance gene which encodes P-glycoprotein, has been suggested as a suitable gene for use *in vivo* (Gottesman et al., 1991). P-glycoprotein confers resistance to a broad variety of natural toxic substances and their analogs, including clinically used anticancer drugs such as Vinca alkaloids, doxorubicin, daunorubicin, mitoxantrone, actinomycin D, etoposide and taxol.

Preclinical studies have indicated that overexpression of the *MDR1* gene in hematopoietic cells may protect them from the adverse effects of anticancer chemotherapy and clinical trials on retroviral transfer of *MDR1* into bone marrow progenitor cells of cancer patients have been approved and are in progress. In this paper we will focus on transfer of the *MDR1* gene to hematopoietic cells as well as on potential applications of this technology for gene therapy of cancer and hereditary metabolic disorders.

3.2   Functional Aspects and Physiologic Expression of P-glycoprotein

Human P-glycoprotein is a plasma membrane protein comprising 1280 amino acids. The protein is believed to contain twelve transmembrane domains and two ATP binding sites that form two highly homologous halves, each consisting of six transmembrane segments and one intracytoplasmic ATP-binding site. P-glycoprotein is a member of a superfamily of ATP-binding proteins, known as the ATP-binding cassette (ABC) superfamily. P-glycoprotein has been the subject of extensive research extended over more than 20 years (for reviews see Endicott et al., 1989; Germann, 1993; Gottesman et al., 1995).

The multidrug transporter binds and extrudes a multitude of compounds from cells. Its substrates are usually hydrophobic and often consist of aromatic rings. Although the detailed molecular mechanisms responsible for the extremely wide substrate specificity of this transporter are not yet completely understood, it is well known that the main mechanism of action is active drug efflux. In addition, decreased influx of substrates has been demonstrated, probably as a result of detection and efflux of substrates from within the plasma membrane (Raviv et al., 1990; Stein et al., 1994). Both mechanisms cause reduced intracellular drug accumulation by which cells expressing P-glycoprotein become chemo-resistant (reviewed in Gottesman et al., 1993). In addition to cytotoxic drugs, many non-toxic compounds bind to the multidrug transporter, e.g., calcium antagonists, cyclosporine, reserpine, neuroleptic and antidepressant drugs, and steroid hormones. In several preclinical and clinical studies, such compounds have been used as inhibitors of the pump. For instance, efficient chemosensitization with a P-glycoprotein inhibitor, cyclosporine, has been shown to eliminate multidrug-resistant cells in refractory multiple myeloma during chemotherapy (Sonneveld et al., 1992).

It should be mentioned that the protein encoded by the highly homologous human *MDR2*-gene (also designated as *MDR3*) is not involved in chemo-resistance, although, its predicted structure is similar (Schinkel et al., 1991). As determined by targeted disruption of the *MDR2* gene, the product of this gene functions in hepatocytes as a transporter for phosphatidylcholine which it secretes into the bile (Smit et al., 1993).

P-glycoprotein is detectable at significant levels in up to 50% of human neoplasms (Goldstein et al., 1990). Physiologically, P-glycoprotein is expressed in adrenal gland, kidney, jejunum and colon, in the capillaries of the brain and testes, in small biliary and pancreatic ductules, and on the biliary canalicular surface of hepatocytes (Fojo et al., 1987; Sugawara et al., 1988; Thiebaut et al., 1989). In normal hematopoietic cells, P-glycoprotein is expressed at low to moderate levels (reviewed in List et al., 1992; Licht et al., 1994). Functional P-glycoprotein has been described on hematopoietic stem cells (Chaudary et al., 1991), on the majority of T-lymphocytes (Neyfakh et al., 1989) and NK-cells as well as B-lymphocytes (Chaudary et al., 1992). In contrast, there is very little P-glycoprotein present on granulocytes and their myeloid progenitor cells (Drach et al., 1992; Marie et al., 1992; Klimecki et al., 1994). These low expression levels are not sufficient to render myeloid cells resistant to the toxicity of cytotoxic drugs. Thus, granulocytopenia caused by myelosuppression is a common adverse effect of anticancer chemotherapy.

3.3 Transgenic Models of P-glycoprotein Expression in Bone Marrow

The ability of P-glycoprotein to protect hematopoietic cells from myelosuppression was first demonstrated in a transgenic mouse model in which a human *MDR*1 cDNA was transcribed under the control of a chicken beta-actin promoter (Galski et al., 1989). These animals expressed human P-glycoprotein in bone marrow, and up to several-fold higher doses of taxol and daunomycin could be administered safely compared with normal animals of the respective background strains. The protective effect was specifically due to the multidrug transporter, since it could be overcome by inhibitors like verapamil. Normal bone marrow function remained unaltered in these mice (Mickisch et al., 1991). In subsequent experiments, bone marrow cells from these transgenic mice were transplanted into otherwise lethally irradiated normal host mice. In these studies, the recipients were also less susceptible to myelosuppression by chemotherapy (Mickisch et al., 1992). Although the expression of the human *MDR*1 transgene in these animals eventually proved unstable, these experiments demonstrated that the constitutive expression of P-glycoprotein was chemoprotective in bone marrow. That this protection could be transferred suggests that myeloprotection after *MDR*1 gene transfer would be possible.

3.4 Retroviral *MDR*1 Gene Transduction of Hematopoietic Progenitor Cells

Evidence that gene transfer to normal hematopoietic progenitor cells may protect them from cytotoxic drugs was first provided by *in vitro* studies in which primary bone marrow cells (McLachlin et al., 1990) or cultured erythroleukemia cells (DelaFlor-Weiss et al., 1992) were transduced retrovirally with the *MDR*1 gene. Retroviral transduction was performed with a vector containing a full-length *MDR*1 cDNA promoted by the long terminal repeats of the Harvey sarcoma virus (Pastan et al., 1988).

Based on these studies, *in vivo*-animal models for the use of the *MDR*1 gene in somatic gene therapy were developed: *MDR*1 cDNA was transferred retrovirally to murine bone marrow and transduced cells were transplanted into either anemic W/W^v mice (Sorrentino et al., 1992) or lethally irradiated normal mice (Podda et al., 1992). In both investigations, *MDR*1 gene expression was increased when mice were treated with taxol, a cytotoxic substrate of P-glycoprotein. Additional support for the concept of a selective advantage of hematopoietic cells expressing P-glyco-protein was provided by experiments in which *MDR*1-transduced bone marrow was first transplanted into recipient mice. After taxol treatment of recipient animals, their bone marrow was transplanted into a second generation of recipient mice. In several cycles of re-transplantation and taxol treatment of recipient mice, increas-ingly high levels of chemo-resistance were generated *in vivo* (Hanania et al., 1994).

Protection by *MDR*1 is not restricted to myeloid cells. Certain drugs exert high cytotoxicity to subpopulations of hematopoietic or lymphatic cells. For instance, bisantrene, an anthracycline compound and substrate of the multidrug transporter, has been found to be very toxic to B-lymphocytes. *MDR*1 gene transfer to murine bone marrow protects mainly lymphoid cells from bisantrene toxicity (Aksentijevich et al., 1996).

Recent studies were aimed at demonstrating *MDR*1 gene transfer to immature subpopulations of hematopoietic cells. Hematopoietic stem cells are ideal targets for gene therapy because they have a virtually indefinite life-span. Only stem cells have the capacity of self-renewal, whereas more mature progenitor cells become more dif-ferentiated with each step of cell duplication and eventually undergo terminal dif-ferentiation and apoptosis. Targeting of more mature, lineage-committed progenitor cells may, therefore, result in transient gene expression. However, while incidental targeting of stem cells has been confirmed in numerous investigations by gene trans-fer to whole bone marrow, transduction of hematopoietic stem cells has been found to be elusive (Miller, 1990; Nolta et al., 1996), mainly because most of these cells are not dividing *in vivo*: Cell proliferation is required for integration of retroviral se-quences into the genome. Activation of stem cells is accomplished by the use of hematopoietic growth factors and cytokines, and by chemotherapeutic pretreatment of bone marrow donors (Bodine et al., 1991; Ogawa, 1993). The capacity of cytokine-treated stem cells to engraft bone marrow after chemotherapy, which previously had been questioned, was demonstrated in a clinical study (Brugger et al., 1995).

Several groups have studied *MDR*1 gene transfer into subpopulations of hematopoietic progenitors enriched for stem cells. Ward et al. (1994) transduced human CD34+ cells using amphotropic retroviruses containing an *MDR*1-vector and cDNA of the transgene was detected in erythroid and myeloid clones derived from them. These experiments showed an increased resistance of cell clones derived from transduced CD34+ cells to high doses of taxol. For gene therapy purposes, immature hematopoietic cells can be obtained from various tissues besides bone marrow. For example, transfer of *MDR*1 to human CD34+ cells from cord blood was found to be more efficient compared with CD34+ bone marrow cells (Bertolini et al., 1994). In one study the *MDR*1 cDNA was transferred to murine fetal liver cells which contain hematopoietic precursor cells at very early developmental stages (Richardson et al., 1994). This study revealed that fetal liver cells express receptors only for ecotropic but not for amphotropic retroviruses.

We have transferred *MDR*1 to a small population of murine bone marrow cells that expressed no lineage-committed or MHC class II antigens but high levels of Sca-1 (stem cell antigen 1, also referred to as Ly6A/E) which is highly expressed on stem cells (Szilvassy et al., 1993). Stem cell characteristics of the fractionated cells were confirmed *in vivo* by transplantation experiments that showed the sustained presence of a *MDR*1 marker cDNA in transplanted SCID-mice, multilineage engraftment and presence of the marker gene after re-transplantation into a second generation of recipient mice (Licht et al., 1998). In this study, cells were expanded *ex vivo*, with the use of a combination of growth factors to recruit quiescent stem cells into cell cycle. Gene transfer by coculture resulted in expression of functional P-glycoprotein in more than 60% of stem cells. Following transplantation of the *ex vivo*-transduced cell population into recipient SCID-mice, P-glycoprotein was expressed in a high proportion of bone marrow cells of recipient animals at levels comparable to clinical multidrug-resistant cancers (Licht et al., 1998). Peripheral blood stem cells have also been found to be useful targets for *MDR*1 gene transfer. Bodine et al. (1994) transduced an *MDR*1 cDNA retrovirally to hematopoietic precursor cells mobilized from bone marrow of splenectomized mice by administration of granulocyte-colony stimulating factor and stem cell factor.

Transfer of the *MDR*1 gene into isolated stem cells may further improve treatment of disseminated cancers. Based on the findings of animal studies, that retroviral transfer of *MDR*1 cDNA appears to be safe and that P-glycoprotein may protect bone marrow from chemotherapy-induced toxicity, clinical trials on *MDR*1 transfer to bone marrow of cancer patients have been approved. The first trials in the U.S.A. are being performed on patients suffering from advanced breast and ovarian cancer (Hesdorffer et al., 1994; O'Shaughnessy et al., 1994; Deisseroth et al., 1994). Since bone marrow micrometastases are frequent in metastasized breast cancer, gene transfer into isolated pluripotent stem cells rather than unfractionated bone marrow should reduce the risk of unwanted transduction of P-glycoprotein in contaminating tumor cells.

3.5   Gene Therapy using the MDR1 Gene as a Selectable Marker

MDR1 may be useful for gene therapy, not only by virtue of rendering target cells chemo-resistant, but also by ensuring expression of a second passenger gene. Several approaches can be used to co-express two genes. The simplest approach is cotransfection or cotransduction of separate vectors but this does not necessarily ensure stable coexpression. Alternatively, two genes can be integrated into one vector. They can be driven from separate promoters, but this strategy does not ensure expression of both genes because, often, expression of the unselected gene is extinguished. A different approach is to construct gene fusions encoding for a chimeric protein. A fusion between MDR1 and the adenosine deaminase gene, which encodes a chimeric protein has been used to stably transduce fibroblasts (Germann et al., 1990). However, this approach cannot be used for proteins that need to be located in different cellular compartments (Aran et al., 1996). A preferred approach is the use of polycistronic mRNAs, that facilitate transcription of multiple genes by internal ribosomal entry sites (IRES). Such transcriptional fusion vectors have been developed containing the MDR1 cDNA and cDNAs that correct inherited metabolic disorders (Sugimoto et al., 1994). In these cassettes, both genes are transcribed from a single retroviral promoter but are translated from two open reading frames into two separate proteins. In spite of the large size of these vectors, in PA317 retroviral packaging cells, titers of 1×10^5/ml are achievable.

Aran et al. (1994) described a transcriptional fusion between MDR1 and the glucocerebrosidase gene which is defective in Gaucher's disease. This inherited metabolic disorder is characterized by accumulation of a glucosylceramide in glucocerebrosidase-deficient hematopoietic cells: Patients suffer from skeletal lesions and severe hepatosplenomegaly. The transcriptional fusion vector allowed expression of functional glucocerebrosidase in NIH 3T3 fibroblasts which was increased by selection with cytotoxic substrates of P-glycoprotein. Studies on this vector in hematopoietic cells are currently being performed. Similarly, a bicistronic vector that facilitates coexpression of MDR1 and alpha-galactosidase has been engineered (Sugimoto et al., 1995a). Defects of alpha-galactosidase are the cause of Fabry's disease, another metabolic disorder affecting hematopoietic organs. In these vectors, the non-selectable gene was cloned downstream from the internal ribosomal entry site. It was found that cDNAs in the downstream position were translated at three to five-fold lower levels than the upstream MDR1 gene. If the positions were reversed, i.e., if MDR1 was positioned downstream, fewer drug resistant clones were obtained, but almost all of them expressed the passenger gene at high levels (Gottesman et al., 1995). Vectors can thus be constructed to either confer high levels of drug resistance and a good yield of selectable cells, or to facilitate very high levels of expression of a non-selectable gene.

If bicistronic vectors containing *MDR*1 are introduced into hematopoietic cells of patients suffering from inherited diseases, drug treatment will eliminate untransduced cells whose genetic defect has not been corrected. All surviving cells should contain both *MDR*1 and the passenger gene. Moreover, chemotherapy should create space in bone marrow for the surviving progenitor cells, stimulating them to expand and repopulate the marrow. A marking vector, in which *MDR*1 is coexpressed with lacZ encoding beta-galactosidase, has been constructed and should be useful for analyzing expression in individual hematopoietic progenitor cells (Aran et al., 1995).

3.6   Safety Aspects

Transfer of *MDR*1 cDNA to hematopoietic progenitor cells of patients suffering from metastatic cancer bears the risk of unwanted transduction to previously undetected tumor micrometastases. This risk should be greatly reduced by the use of isolated stem cells. In addition, there is a risk of insertional mutagenesis when vectors are integrated into the genome and genome integration is a requirement for stable long-term gene expression. Because retroviral vectors integrate randomly into the genome, there is a very small possibility of activating oncogenes or disrupting anti-oncogenes if the integration site is located in close proximity to such a gene. In case of such an event, the action of P-glycoprotein in transduced cells has to be overcome and transduced cells must be eliminated. This may be accomplished by combined treatment with cytotoxic drugs and inhibitors of P-glycoprotein (Lum et al., 1993). Alternatively, multidrug resistant cells might be removed by the use of antibodies (Pearson et al., 1991) or immunotoxins (Mickisch et al., 1993), or the transcription of *MDR*1 may be interrupted by ribozymes (Kobayashi et al., 1993; Scanlon et al., 1994) or antisense cDNA (Corrias et al., 1992; Efferth et al., 1993). Except for the combination therapy with P-glycoprotein inhibitors, which has been used in many investigations, these approaches are still experimental although they might constitute future strategies to overcome multidrug resistance.

To increase the safety of gene therapy using *MDR*1, a bicistronic vector, containing thymidine kinase from herpes-simplex virus (HSV-TK), has been constructed as a passenger gene (Sugimoto et al., 1995b). HSK-TK is a suicide gene that confers hypersensitivity to the antiviral agent, ganciclovir. Treatment with ganciclovir would kill host cells that were inadvertently transduced.

Another strategy employs differences in substrate specificities of P-glycoproteins from different species or of mutated human P-glycoproteins. Human and mouse P-glycoprotein confer different levels of resistance to various drugs (Tang-Wai et al., 1993) and point mutatations have been found to alter the pattern of cross-resistance of P-glycoprotein (reviewed in Gottesman et al., 1995). For instance, substitution of valine for glycine at position 185 of *MDR*1 increases resistance to etoposide and decreases resistance to Vinca alkaloids (Safa et al., 1993). Many gene transfer studies with *MDR*1 have been performed with retroviral vectors containing the Val185 mutant *MDR*1. Conversely, a substitution of proline for alanine, in position 866, decreases colchicine resistance (Clarke et al., 1993) and a mutation of serine, in position 941, to phenylalanine decreases doxorubicin and colchicine resistance (Gros et al., 1991). Thus, vectors may be designed to make it possible to distinguish transduced cells from cells expressing endogenous P-glycoprotein. Such vectors may also prove useful to provide an additional selective advantage to transduced cells compared with the patient's cancer cells, thereby, allowing safe dose intensification of anticancer drugs. Vectors containing mutant P-glycoproteins might enhance the safety of *MDR*1 gene transfer if transduced cells were particularly sensitive to certain P-glycoprotein inhibitors. In the Val185 mutant, the amount of circumvention of drug resistance differs from that conferred by wildtype P-glycoprotein (Cardarelli et al., 1995).

3.7   Conclusions

Animal models and *in vitro* studies have shown that expression of P-glycoprotein in hematopoietic tissues can protect progenitor cells from the cytotoxicity of anticancer drugs and that transduced cells have a selective advantage over non-transduced cells. Transfer of *MDR*1 cDNA to hematopoietic progenitor cells may be useful for gene therapy in two ways: first, to protect bone marrow from myelosuppression following chemotherapy and second, to introduce and overexpress otherwise non-selectable genes that correct inherited disorders. New vectors containing mutated *MDR*1 might distinguish transduced cells from cells expressing endogenous P-glycoprotein, thereby giving an additional selective advantage.

References

Aksentijevich, I., Cardarelli, C. O., Pastan, I., Gottesman, M. M. (1996) Retroviral transfer of the human MDR1 gene confers resistance to bisantrene specific hematotoxicity. *Clin. Cancer Res.* 2: 973–980.

Aran, J. M., Gottesman, M. M., Pastan, I. (1994) Drug-selected coexpression of human glucocerebrosidase and P-glycoprotein using a bicistronic vector. *Proc. Natl. Acad. Sci. USA* 91: 3176–3180.

Aran, J. M., Licht, T., Gottesman, M. M., Pastan, M. (1995) Design of a multidrug-resistance – marking vector. *Proc. Amer Assoc. Cancer Res.* 36: 336.

Aran, J. M., Germann, U. A., Gottesman, M. M., Pastan, I. (1996) Construction and characterization of a selectable multidrug resistance – glucocerebroside fusion gene. *Cytokines*

Mol. Ther. 2: 47–57.

Bertolini, F., de Monte, L., Corsini, C., Lazzari, L., Lauri, E., Soligo, D., Ward, M., Bank, A., Malvasi, F. (1994) Retrovirus-mediated transfer of the multidrug resistance gene into human haematopoietic progenitor cells. *Brit. J. Haematol.* 88: 318–324.

Bodine, D. M., McDonagh, K. T., Seidel, N. E., Nienhuis, A. W. (1991) Survival and retrovirus infection of murine hematopoietic stem cells *in vitro*: effects of 5-FU and method of infection. *Exp. Hematol.* 19: 206–212.

Bodine, D. M., Moritz, T., Donahue, R. E., Luskey, B. D., Kessler, S. W., Martin, D. I., Orkin, S. H., Nienhuis, A. W. (1993) Long-term *in vivo* expression of a murine adenosine deaminase gene in rhesus monkey hematopoietic cells of multiple lineages after retroviral mediated gene transfer into CD34+ bone marrow cells. *Blood* 82: 1975–1980.

Bodine, D. M., Seidel, N. E., Gale, M. S., Nienhuis, A. W., Orlic, D. (1994) Efficient retrovirus transduction of mouse pluripotent hematopoietic stem cells mobilized into the peripheral blood by treatment with granulocyte colony-stimulating factor and stem cell factor. *Blood* 84: 1482–1491.

Brugger, W., Heimfeld, S., Berenson, R. J., Mertelsmann, R., Kanz, L. (1995) Reconstitution of hematopoiesis after high-dose chemotherapy by autologous progenitor cells generated ex vivo. *N. Engl. J. Med.* 333: 283–287.

Cardarelli, C. O., Aksentijevitch, I., Pastan, I., Gottesman, M. M. (1995) Differential effects of P-glycoprotein inhibitors on NIH 3T3 cells transfected with wild-type (G185) or mutant (V185) multidrug transporters. *Cancer Res.* 55: 1086–1091.

Chaudhary, P. M., Roninson, I. B. (1991) Expression and activity of P-glycoprotein, a multidrug efflux pump, in human hematopoietic stem cells. *Cell* 66: 85–94.

Chaudhary, P. M., Mechetner, E. B., Roninson, I. B. (1992) Expression and activity of the multidrug resistance P-glycoprotein in human peripheral blood lymphocytes. *Blood* 80: 2729–2934.

Clarke, D. M., Loo, T. W. (1993) Functional consequences of proline mutations in the predicted transmembrane domain of P-glycoprotein. *J. Biol. Chem.* 268: 3143–3149.

Cole, S. P. C., Bhardwaj, G., Gerlach, J. H., Mackie, J. E., Grant, C. E., Almquist, K. C., Stewart, A. J., Kurz, E. U., Duncan, A. M. V., Deeley, R. G. (1992) Overexpression of a transporter gene in a multidrug-resistant human lung cancer cell line. *Science* 258: 1650–1654.

Corrias, M. V., Tonini, G. P. (1992) An oligomer complementary to the 5' end region of the MDR1 gene decreases resistance to doxorubicin of human adenocarcinoma multidrug-resitant cells. *Anticancer Res.* 12: 1431–1438.

Deisseroth, A. B., Kavanagh, J., Champlin, R. (1994) Use of safety-modified retroviruses to introduce chemotherapy resistance sequences into normal hematopoietic cells for chemoprotection during the therapy of ovarian cancer: a pilot trial. *Hum. Gene Ther.* 5: 1507–1522.

DelaFlor-Weiss, E., Richardson, C., Ward, M., Himelstein, A., Smith, L., Podda, S., Gottesman, M., Pastan, I., Bank, A. (1992) Transfer and expression of the human multidrug resistance gene in mouse erythroleukemia cells. *Blood* 80: 3106–3111.

Drach, D., Zhao, S., Drach, J., Mahadevia, R., Gattringer, C., Huber, H., Andreeff, M. (1992) Subpopulations of normal and peripheral blood and bone marrow cells express a functional multidrug resistant phenotype. *Blood* 80: 2735–2739.

Dunbar, C. E., Bodine, D. M., Sorrentino, B., Donahue, R., McDonagh, K., Cottler-Fox, M., O'Shaughnessy, J. A. (1994) Gene transfer into hematopoietic cells: implications for cancer therapy. *Ann. N.Y. Acad. Sci.* 716: 216–224.

Efferth, T., Volm, M. (1993) Modulation of P-glycoprotein mediated multidrug resistance by monoclonal antibodies, immunotoxins or antisense oligodeoxynucleotides in kidney carcinoma and normal kidney cells. *Oncology* 50: 303–308.

Endicott, J. A., Ling, V. (1989) The biochemistry of P-glycoprotein-mediated multidrug resistance. *Annu. Rev. Biochem.* 58: 137–171.

Fojo, A. T., Ueda, K., Slamon, D. J., Poplack, D. G., Gottesman, M. M., Pastan, I. (1987) Expression of a multidrug resistance gene in human tumors and tissues. *Proc. Natl. Acad. Sci. USA* 86: 695–698.

Galski, H., Sullivan, M., Willingham, M. C., Chin, K. -V., Gottesman, M. M., Pastan, I., Merlino, G. T. (1989) Expression of a human multidrug resistance cDNA (MDR1) in the bone marrow of transgenic mice: resistance to daunomycin induced leukopenia. *Mol. Cell. Biol.* 9: 4357–4363.

Germann, U. A., Chin, K. -V., Pastan, I., Gottesman, M. M. (1990) Retroviral transfer of a chimeric multidrug resistance-adenosine deaminase gene. *FASEB J.* 4: 1501–1507.

Germann, U. A. (1993) Molecular analysis of the multidrug transporter. *Cytotechnology* 12: 33–62.

Goldstein, L., Galski, H., Fojo, A., Willingham, M., Lai, S. -I., Gadzar, A., Pirker, R., Green, A., Crist, W., Brodeur, G. M., Lieber, M., Cossman, J., Gottesman, M. M., Pastan, I. (1989) Expression of a multidrug resistance gene in human cancers *J. Nat. Cancer Inst.* 81: 116–124.

Gottesman, M. M., Pastan, I. (1991) The multidrug resistance (*MDR*1) gene as a selectable marker in gene therapy. *In*: O. Cohen-Haguenauer and M. Boiron (eds), *Human Gene Transfer*, Vol. 219. Coloque INSERM/John Libbey Eurotext, London, pp. 185–191.

Gottesman, M. M., Pastan, I. (1993) Biochemistry of multidrug resistance mediated by the multidrug transporter. *Annu. Rev. Biochem.* 62: 385–427.

Gottesman, M. M., Germann, U. A., Aksentijevitch, I., Sugimoto, Y., Cardarelli, C. O., Pastan, I. (1994) Gene transfer of drug resistance genes. *N.Y. Acad. Sci.* 716: 126–139.

Gottesman, M. M., Hrycyna, C. A., Schoenlein, P. V., Germann, U. A., Pastan, I. (1995) Genetic analysis of the multidrug transporter. *Annu. Rev. Genet.* 29: 607–648.

Gros, P., Dhir, R., Croop, J., Talbot, F. (1991) A single amino acid substitution strongly modulates the activity and substrate specificity of the mouse mdr1 and mdr3 drug efflux pumps. *Proc. Natl. Acad. Sci. USA* 88: 7289–7293.

Hanania, E. G., Deisseroth, A. B. (1994) Serial transplantation shows that early hematopoietic precursor cells are transduced by MDR-1 retroviral vector in a mouse gene therapy model. *Cancer Gene Ther.* 1: 21–25.

Hesdorffer, C., Antman, K., Bank, A., Fetell, M., Mears, G., Begg, M. (1994) Human MDR1 gene transfer in patients with advanced cancer. *Hum. Gene Ther.* 5: 1151–1160.

Klimecki, W. T., Futscher, B. W., Grogan, T. M., Dalton, W. S. (1994) P-glycoprotein expression and function in circulating blood cells from normal volunteer. *Blood* 83: 2451–2458.

Kobayashi, H., Dorai, T., Holland, J. F., Ohnuma, T. (1993) Cleavage of human MDR1 by a hammerhead ribozyme. *FEBS Lett.* 319: 71–74.

List, A. F., Spier, C. M. (1992) Multidrug resistance in acute leukemia: a conserved physiologic function. *Leuk. Lymphoma* 8: 9–14.

Licht, T., Pastan, I., Gottesman, M. M., Herrmann, F. (1994) P-glycoprotein-mediated multidrug resistance in normal and neoplastic hematopoietic cells. *Ann. Hematol.* 69:

159–171.

Licht, T., Aksentijevich, I., Gottesman, M. M., Pastan, I. (1995) Efficient expression of functional human *MDR*1 gene in murine bone marrow after retroviral transduction of purified hematopoietic stem cells. *Blood* 86: 111–121.

Licht, T., Gottesman, M. M., Pastan, I. (1998) Retroviral transfer of multidrug transporter to murine hematopoietic stem cells. *Methods Enzymol.* 292: 546–557.

Lum, B. L., Fisher, G. A., Brophy, N. A., Yahanda, A. M., Adler, K. M., Kaubisch, S., Halsey, J., Sikic, B. I. (1993) Clinical trials of modulation of multidrug resistance. Pharmacokinetic and pharmacodynamic considerations. *Cancer* 72 (suppl. 11): 3502–3514.

Marie, J. P., Brophy, N. A., Ehsan, M. N., Aihara, Y., Mohamed, N. A., Cornbleet, J., Chao, N. J., Sikic, B. I. (1992) Expression of multidrug resistance gene mdr1 mRNA in a subset of normal bone marrow cells. *Brit. J. Haematol.* 81: 145–152.

McLachlin, J. R., Eglitis, M. A., Ueda, K., Kantoff, P. W., Pastan, I., Anderson, W. F., Gottesman, M. M. (1990) Expression of a human complementary DNA for the human multidrug resistance gene in murine hematopoietic precursor cells with the use of retroviral gene transfer. *J. Nat. Cancer Inst.* 82: 1260–1263.

Mickisch, G. H., Licht, T., Merlino, G. T., Gottesman, M. M., Pastan, I. (1991) Chemotherapy and chemosensitization of transgenic mice which express the human multidrug resistance gene in bone marrow: Efficacy, potency and toxicity. *Cancer Res.* 51: 5417–5424.

Mickisch, G. H., Aksentijevich, I., Schoenlein, P. V., Goldstein, L. J., Galski, H., Staehle, C., Sachs, D. H., Pastan, I., Gottesman, M. M. (1992) Transplantation of bone marrow cells from transgenic mice expressing the human MDR1 gene results in long-term protection against the myelosuppressive effect of chemotherapy in mice. *Blood* 79: 1087–1093.

Mickisch, G. H., Pai, L. H., Siegsmund, M., Campain, J., Gottesman, M. M., Pastan, I. (1993) Pseudomonas exotoxin conjugated to monoclonal antibody MRK16 specifically kills multidrug resistant cells in cultured renal cell carcinomas and in MDR-transgenic mice. *J. Urol.* 149: 174–178.

Miller, A. D. (1990) Progress towards *Hum. Gene Ther. Blood* 76: 271–278.

Neyfakh, A. A., Serpinskaya, A. S., Chervonsky, A. V., Apasov, S. G., Kazarov, A. R. (1989) Multidrug-resistance phenotype of a subpopulation of T-lymphocytes without drug selection. *Exp. Cell* 185: 496–505.

Nolta, J. A., Dao, M. A., Wells, S., Smogorzewska, E. M., Kohn, D. B. (1996) Transduction of pluripotent human[211] hematopoietic stem cells demonstrated by clonal analysis after engraftment in immune-deficient mice. *Proc. Natl. Acad. Sci. USA* 19: 2414–2419.

Ogawa, M. (1993) Differentiation and proliferation of stem cells. *Blood* 81: 2844–2853.

O'Shaughnessy, J. A., Cowan, K. H., Nienhuis, A. W., McDonagh, K. T., Sorrentino, B. P., Dunbar, C. E., Chiang, Y., Wilson, W., Goldspiel, B., Kohler, D., Cottler-Fox, M., Leitman, S. F., Gottesman, M. M., Pastan, I., Denicoff, A., Noone, M., Gress, R. (1994) Retroviral mediated transfer of the human multidrug resistance gene (MDR-1) into hematopoietic stem cells during autologous transplantation after intensive chemotherapy for metastatic breast cancer. *Hum. Gene Ther.* 5: 891–911.

Pearson, J. W., Fogler, W. E., Volker, K., Usui, N., Goldenberg, S. K., Gruys, E., Riggs, C. W., Komschlies, K., Wiltrout, R. H., Tsuruo, T., Pastan, I., Gottesman, M. M., Longo, D. L. (1991) Reversal of drug resistance in a human colon cancer xenograft expressing MDR1 complementary DNA by *in vivo* administration of MRK-16 monoclonal antibody. *J. Nat.*

Cancer Inst. 83: 1386–1391.

Pastan, I., Gottesman, M. M., Ueda, K., Lovelace, E., Rutherford, A. V., Willingham, M. C. (1988) A retrovirus carrying an MDR1 cDNA confers multidrug resistance and polarized expression of P-glycoprotein in MDCK cells. *Proc. Natl. Acad. Sci. USA* 85: 4486–4490.

Podda, S., Ward, M., Himelstein, A., Richardson, C., de la Flor-Weiss, E., Smith, L., Gottesman MM, Pastan, I., Bank, A. (1992) Transfer and expression of the human multiple drug resistance gene into live mice. *Proc. Natl. Acad. Sci. USA* 89: 9676–9680.

Raviv, Y., Pollard, H. B., Bruggemann, E. P., Pastan, I., Gottesman, M. M. (1990) Photosensitized labeling of a functional multidrug transporter in living drug-resistant tumor cells. *J. Biol. Chem.* 265: 3975–3980.

Richardson, C., Ward, M., Podda, S., Bank, A. (1994) Mouse fetal liver cells lack amphotropic retroviral receptors. *Blood* 84: 433–439.

Safa, A. R., Stern, R. K., Choi, K., Agresti, M., Tamai, I., Mehta, N. D., Roninson, I. B. (1990) Molecular basis of preferential resistance to colchicine in multidrug-resistant human cells conferred by Gly-185–Val-185 substitution in P-glycoprotein. *Proc. Natl. Acad. Sci. USA* 87: 7225–7229.

Scanlon, K. J., Ishida, H., Kashani-Sabet, M. (1994) Ribozyme-mediated reversal of the multidrug-resistant phenotype. *Proc. Natl. Acad. Sci. USA* 91: 11123–11127.

Schinkel, A. H., Roelofs, M. E. M., Borst, P. (1991) Characterization of human MDR3 P-glycoprotein and its recognition by P-glycoprotein-specific monoclonal antibodies. *Cancer Res.* 51: 2628–2635.

Smit, J. J., Schinkel, A. H., Oude Elferink, R. P. J., Groen, A. K., Wagenaar, E., van Deemter, L., Mol CAAM, Ottenhoff, R., van der Lugt NMT, van Room, M. A., van der Valk, M. A., Offerhaus GJA, Berns AJM, Borst, P. (1993) Homozygous disruption of the murine mdr2 P-glycoprotein gene leads to a complete absence of phospholipid from bile and to liver disease. *Cell* 75: 451–462.

Sonneveld, P., Durie, B. D. M., Lokhorst, H. M., Marie, J. P., Solbu, G., Suciu, S., Zittoun, R., Löwenberg, B., Nooter, K. for the Leukemia Group of the EORTC and the HOVON (1992) Modulation of multidrug-resistant multiple myeloma by cyclosporine. *Lancet* 340: 255–259.

Sorrentino, B. P., Brandt, S. J., Bodine, D., Gottesman, M. M., Pastan, I., Cline, A., Nienhuis, A. W. (1992) Selection of drug-resistant bone marrow cells *in vivo* after retroviral transfer of human MDR1. *Science* 257: 99–103.

Srein, W. D., Cardarelli, C., Pastan, I., Gottesman, M. M. (1994) Kinetic evidence suggesting that the multidrug transporter differentially handles influx of its substrates. *Mol. Pharmacol.* 45: 763–772.

Sugawara, I., Kataoka, I., Morishita, Y., Hamada, H., Tsuruo, T., Itoyama, S., Mori, S. (1988) Tissue distribution of P-glycoprotein encoded by a multidrug-resistant gene as revealed by a monoclonal antibody, MRK 16. *Cancer Res.* 48: 1926–1929.

Sugimoto, Y., Aksentijevich, I., Gottesman, M. M., Pastan, I. (1994) Efficient expression of drug-selectable genes under control of an internal ribosome entry site. *Biotechnology* 12: 694–698.

Sugimoto, Y., Aksentijevich, I., Murray, G., Brady, R. O., Pastan, I., Gottesman, M. M. (1995a) Retroviral co-expression of a multidrug-resistance gene (*MDR1*) and human alpha-galactosidase A for gene therapy of Fabry disease. *Hum. Gene Ther* 6: 905–915.

Sugimoto, Y., Hrycyna, C. A., Aksentijevich, I., Pastan, I., Gottesman, M. M. (1995b) Co-expression of a multidrug resistance gene (*MDR1*) and Herpes Simplex Virus thymidine kinase gene as a part of a bicistronic mRNA in a retrovirus vector allows selective killing of MDR1-transduced cells. *Clin. Cancer Res.* 2: 447–457.

Szilvassy, S. J., Cory, S. (1993) Phenotypic and functional characterization of competitive long-term repopulating hematopoietic stem cells enriched from 5-fluorouracil-treated murine marrow. *Blood* 81: 2310–2320.

Tang-Wai, D. F., Kajaji, S., DiCapua, F., de Graaf, D., Roninson IB, Gros, P. (1995) Human (MDR1) and mouse (mdr1, mdr3) P-glycoproteins can be distinguished by their respective drug resistance profiles and sensitivity to modulators. *Biochemistry* 34: 32–39.

Thiebaut, F., Tsuruo, T., Hamada, H., Gottesman, M. M., Pastan, I., Willingham, M. C. (1989) Immunocytochemical localization in normal tissues of different epitopes in the multidrug transport protein P170: Evidence for localization in brain capillaries and cross-reactivity of one antibody with a muscle protein. *J. Histochem. Cytochem.* 37: 159–164.

Van Beusechem, V. W., Valerio, D. (1996) Gene transfer into hematopoietic stem cells of non-human primates. *Hum. Gene Ther.* 7: 1649–1668.

Ward, M., Richardson, C., Pioli, P., Smith, L., Podda, S., Goff, S., Hesdorffer, C., Bank, A. (1994) Transfer and expression of the human multiple drug resistance gene in human CD34+ cells. *Blood* 84: 1408–1414.

IV. Gene Therapy of Cancer

1 The c-myb Protooncogene: A Novel Target for Human Gene Therapy

A. M. Gewirtz

1.1 Introduction

Significant progress in the identification of genes, responsible for cellular transformation has been seen in recent years [1]. For example, rapidly transforming retroviruses are known to carry specific genes likely to be responsible for induction of the malignant phenotype in the cells they infect [2–5]. These genes, termed viral-oncogenes (v-onc), are derived from highly conserved, normal cellular genes which were almost certainly incorporated into the viral genome during its transit through the host cell. These normal cellular genes, called protooncogenes (c-onc), are thought to be intimately involved in the processes of cell proliferation and differentiation [6–12]. Accordingly, it is not difficult to envision a scenario where c-onc amplification, mutation, translocation leading to structural alteration, or change in transcriptional regulation might either lead to, or be associated with, induction of a malignant phenotype in the cell where these changes occurred [13–15].

The advent of innovative technologies for disrupting the expression of specific genes has also developed in recent years. All rely on some type of nucleotide sequence recognition for specificity but differ in how and where they perturb the flow of genetic information. Inhibition, prior to transcription, can be accomplished by two methods: One is the widely employed process of homologous recombination [16]. This approach leads to physical destruction of the gene target. Though quite effective, it is unlikely to have a clinical application, in the near future, because it is extremely inefficient, labor intensive and expensive [16]. An alternative "antigene strategy", can be effected by targeting polypurine-polypyrimidine sequences either within the gene or flanking the gene one wishes to disrupt [17]. This leads to "triplex" formation in the major groove of the DNA helix with consequent interference of transcription. Perturbation of gene function at the post-transcriptional level can also be effected by impairing utilization of RNA. This is the "antisense strategy" [18]. This technique relies on introducing or expressing, in the target cell, a nucleotide sequence which is the reverse complement (antisense) of the mRNA one is trying to disrupt. Hybridization between the target and the exogenous nucleotide sequence inhibits the target's function and may lead to its destruction. A variation of this technique relies on complementary base pairing, to target a catalytic RNA molecule or ribozyme [19]. In contrast to triplex forming molecules, antisense compounds do not have the same nucleotide sequence constraints which govern their ability to hybridize with an mRNA molecule. In contrast to ribozymes, they do not require viral vectors for entry into cells and therefore, are easier to design and perhaps utilize for disease therapy. Antisense (AS) oligodeoxynucleotide (ODN) molecules have already been used to block replication of several types of viruses, including HIV [20], and to block synthesis of a diverse array of proteins, including those encoded by cell cycle regulated genes [21–22], cell adhesion proteins [23–24], growth factors and their receptors [25–27], and elements of the signal transduction apparatus [26, 28–29]. A third and final approach is to destabilize the mRNA, not by direct interaction of the oligo and its mRNA, but rather by providing an alternative binding site, or decoy, for mRNA stabilizing proteins. All these approaches provide useful data for one another and may even lead to practical cooperations between these alternative methods.

A number of strategies have been employed to introduce ODN sequences into cells, including transfection with vectors carrying the antisense sequence [30], microinjection [31], encapsulation into liposomes [32] and incubation in high concentrations of short length synthetic ODN [33]. The latter approach, first utilized by Zamecnik and Stephenson [34], has now been successfully utilized by a number of groups in a variety of *in vitro* [21, 35–36] and *in vivo* [37–40] experimental systems. How ODN gain access to the cell remains uncertain. For some time, it was believed that the polyanionic charge on these molecules would deny them entrance [41] but this is clearly not the case. Most recent studies suggest that the process of ODN internalization may be receptor mediated [42–43] and is both energy, temperature and time dependent [42–45]. A putative receptor of ~80kD has been reported to be capable of specifically binding ODN [42, 43]. This receptor has been partially purified and some data suggest that, at least in hematopoietic cells, the receptor is similar to CD4 [46]. Once bound to the cell surface, the ODN may be internalized by endocytosis or pinocytosis [42, 44].

The finding that ODN gain access to their target cells is both useful and encouraging from several points of view in regard to therapeutic applications. Firstly, the use of AS-ODN for disrupting gene function avoids potential complications arising from the use of viral vectors [47]. This methodology may then prove to be safer with respect to the induction of malignancies and insertional mutagenesis. Secondly, the efficiency of cellular delivery appears to be much higher for AS-ODN than what is usually observed with retroviral vectors [47, 48]. While adeno and adeno-associated viral vectors do have high infection efficiency, their immunogenicity presently limits their utility. Finally, from a drug manufacturing point of view, it is likely that it will be easier to produce consistent lots of antisense DNA than the components of viral based delivery systems.

A stable compound must be employed if ODN are to be used for *in vivo* therapy. DNA with unmodified phosphodiester bonds is susceptible to 3' and 5' exonuclease attack. Such exonucleases are common in plasma and are found intracellularly [49]. Modification of the phosphodiester bonds between the bases also renders the molecules less sensitive to nuclease attack and therefore, significantly increases their survival. Though many such modifications have been synthesized [33, 50], the two most widely employed are the methylphosphonates [CH_3-O-P-] and the phosphorothioates (S-O-P) (Fig. 1-1). Methylphosphonates are highly nuclease resistant and tend to form RNA-DNA hybrids with high melting points [51]. Nevertheless, the lack of an ionizable hydrogen renders them relatively water insoluble. Furthermore, they are incapable of binding RNAse H thereby, eliminating a potentially important mechanism for disrupting gene expression. Phosphorothioates are also quite stable to nuclease attack, but in contrast to the methylphosphonates retain their charge and thus, their water solubility [51–52]. They also bind RNAse H. Some investigators have reported that the phosphorothioates may manifest significant non-sequence specific toxicity due to charge related protein binding [53]. This has not proven to be an insurmountable problem but ODN molecules that did not exhibit such effects might have a higher therapeutic index.

The potential therapeutic usefulness of AS-ODN has been demonstrated in many systems and against a number of different targets including viruses, hematopoietic cell genes and genes in solid malignancies [32, 51–53]. Examples in solid tumors include inhibition of tumor growth and oncogene expression using AS-ODN, complementary to K-ras RNA, to inhibit lung cancer cell growth [54], and c-myc in breast [55] and colon carcinoma [56]. These studies suggest that AS-ODN have the potential to become an important new therapeutic agent for the treatment of solid tumors. However, data needs to be gathered in several areas to develop these compounds as therapeutic, including identification of appropriate gene targets and knowledge of the ODN's pharmacokinetics and pharmacodynamics.

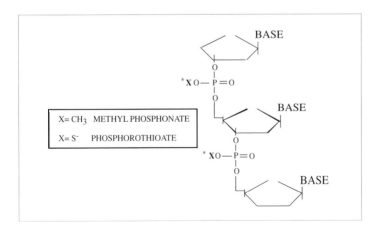

Figure 1-1. Common modifications of the phosphodiester linkage between nucleotides designed to increase nuclease resistance

A number of studies have addressed ODN pharmacokinetics in animal systems [57–59]. For example, Agrawal and co-workers studied the pharmacokinetics, biodistribution and excretion of ^{35}S-labeled $^{S-O-P}$ AS-ODN in mice [57]. After the intravenous or intraperitoneal administration of a single dose (30 mg/kg of body weight) of labeled ODN, material was found in most tissues for up to 48 h. About 30% of the dose was excreted in urine within 24 h, regardless of the mode of administration. Interestingly, the excreted material was found to be extensively degraded. In plasma, stomach, heart and intestine, the $^{S-O-P}$ AS-ODN was degraded by only ~15%, whereas, in the kidney and liver, degradation was about 50% at 48 h. Data have also been generated by Iversen et al. [58] in mice, rats and rabbits using ^{35}S-labeled compound. The single administration of a 27-mer into adult male rats, either by the intravenous or intraperitoneal route, revealed a biphasic plasma elimination. An initial half-life of 15–25 min, which likely represented distribution out of the plasma compartment, was followed by a second half life of 20 to 40 h representing elimination from the body. The second half-life was significantly longer than a variety of nucleic acids such as poly-IC suggesting that therapeutic administration of $^{S-O-P}$ AS-ODN would be practical. Repeated daily injections of the 27-mer provided steady state concentrations in 6–9 days, confirming the estimated long half-life from single injection studies. The drug appears to distribute itself into all organs and fluid compartments except the brain and testes. Chronic treatment studies revealed little toxicity; a finding confirmed in our own laboratories. Human studies are almost non-existent but Bayever et al. reported effects of administration of a ten day infusion of an $^{S-O-P}$ AS ODN to a 19 year old patient with AML [60]. The patient was continuously infused with a dose of 0.05 mg/kg/hr (total dose 700 mg). No change in the course of the patient's leukemia was observed. However, no significant side effects were noted except for a transient elevation in gamma glutamyl transpeptidase, which returned to normal after discontinuation of the drug. Therefore, pharmacokinetic considerations do not appear to be serious constraints for ODN based therapy. Clearly, more testing needs to be conducted.

Deceptively simple, antisense experiments have been notoriously difficult to carry out and have, at times, also been difficult to reproduce [61]. The reasons why these difficulties arise may be placed into three major categories: Firstly, physical problems relating to differences in ODN synthesis, handling, storage and purification are rarely considered in experimental planning. Differences in experimental conditionsand sensitivity between subtypes of cell lines may also be factors. Secondly, problems related to the physical ODN-target mRNA interactions are also difficult to predict. Cellular mRNAs are thought to have highly ordered secondary and tertiary structures. If the targeted sequence is buried within the core of a structurally complicated mRNA molecule or rendered inaccessible by an associated protein, then hybridization of the ODN and mRNA may not be possible. It is generally difficult to predict the secondary structure of mRNA molecules and accessible regions must be determined empirically by synthesizing ODN multiple sequences targetted at various sites and selecting those which are most effective. Thirdly, an additional important problem relates to non-specific interactions of the ODN with non-targeted mRNAs or proteins [62]. Recent studies indicate that some of the observed activities of ODN, in tissue cultures, may be produced predominantly through non-antisense mechanisms. For example, it has been suggested that exposing cells to an ODN with four consecutive internal guanines (G-4 tract) may result in non-specific inhibition of cell growth [63]. The products of ODN degradation, such as 2'-deoxyadenosine and high concentrations of thymidine, are toxic to cells and inhibit proliferation [64, 65]. It is useful to note that nonspecific interactions may prove to be advantageous if their effects are additive to those obtained by the desired ODN-mRNA interaction.

We acknowledge that the current molecular targets, our groups have concentrated, on may be of lesser utility than others which have yet to be investigated. Optimization of this approach will also require greater knowledge of the AS-ODN mechanism of cell killing. Increasing DNA uptake may increase these compounds' cytostatic and cytocidal efficiency. If uptake is primarily by pinocytosis then increasing extracellular DNA concentration and exposure time may be all that is required. Our results with repeated DNA infusions [66] support this notion and our toxicity data suggest that this will be feasible. Nevertheless, AS-ODN mediated cytotoxicity may also be dependent on the trafficking of the compounds once they enter the cell. Translocating [S-O-P] AS-ODN from pinocytotic vesicles to the endoplasmic reticulum, where they presumably associate with ribosomes may be a crucial problem to solve. Finally, cellular 'defense' mechanisms, such as increasing transcription of the targeted message, may also be factors to consider in planning effective treatment strategies using these agents. Clearly, there are many problems to solve. Nevertheless, for the majority of patients, cancer remains an incurable disease. The antisense approach, with its promise of exquisite molecular specificity and concordant limited toxicity, remains attractive. While this area remains in its scientific infancy, our studies and those of our colleagues [67–69] convince us that modulation of gene expression with AS-ODN is a therapeutic strategy worth pursuing.

1.2   The C-myb Protooncogene

Of the genes that we have targeted for disruption using the AS-ODN strategy [26, 36, 74], one that has been of particular scientific interest in our laboratory, and one where therapeutically motivated disruptions are now in clinical trial, is the c-myb gene [73]. C-myb is the normal cellular homologue of v-myb, the transforming oncogene of the avian myeloblastosis virus (AMV) and avian leukemia virus E26. It is a member of a family, composed of at least two other highly homologous genes designated A-myb and B-myb. The molecular and cell biology of the Myb family genes and their encoded proteins has recently been reviewed in detail by Lyon et al. [75].

Located on chromosome 6q in humans, c-myb's predominant transcript encodes a ~72 kDa protein (Myb) [73, 75]. Myb protein consists of three primary functional regions [76] (Fig. 1-3). At the NH_2 terminus is a DNA binding domain which recognizes the core consensus sequence 5'-pyAAC(G/Py)G-3' [71]. This region consists of three imperfect repeats consisting of 51–52 amino acids which have been designated R1, R2, and R3 respectively. R2 and R3 have been shown to be absolutely essential for DNA binding while the role of R1 is less clear. R1 contribute to the stability of the DNA-protein interaction but it does not appear to be required for binding. Within each repeat are three perfectly conserved tryptophan residues. Together,

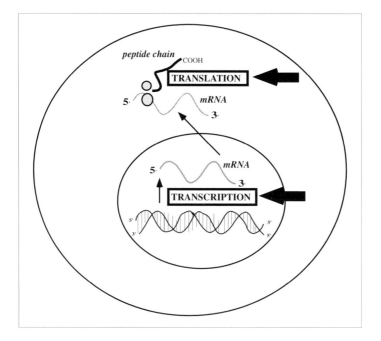

Figure 1-2. Inhibition of gene function may be accomplished by physical destruction of the target gene itself (homologous recombination). It may also be accomplished by interference with expression at the transcriptional level (triplex formation), or at the translational level (antisense strategies). See text for details.

Figure 1-3. Functional map of the c-myb protein indicating the DNA binding, transcriptional activation, and repressor domains. The location of the putative leucine zipper like structure is also indicated.

they form a cluster in the hydrophobic core of the protein which maintains the DNA binding helix-turn-helix structure.

The mid-portion of the Myb protein contains an acidic transcriptional activating domain. In concert with the DNA binding portion of the protein, these two regions are sufficient to transactivate promoters containing Myb binding sites. An adjacent downstream region of the protein also appears to play a role in the activation of some targets, e.g. c-myc [77]. This region, independent of the DNA binding domain, has also been reported capable of transactivating certain targets such as the heat shock protein [78] and DNA polymerase alpha [79]. It has also been reported that some Myb transactivation targets require an interaction between Myb and other transcription factors, such as the nuclear factor-m (NF-M) [80] and the cat enhancer binding protein (C/EBP) [81]. It is therefore likely that some of Myb's transcriptional activating properties may be dependent on the protein partners with which it interacts. This, of course, appears to be true of many transcriptional factors.

Myb also contains a negative regulatory domain which has been localized to the carboxy terminus. Interestingly, the carboxy terminus is deleted in v-myb. This has been thought to contribute to v-myb's transforming ability. Recently, a putative leucine zipper structure was described within the amino terminal portion of Myb's carboxy terminal domain [82]. Leucine zippers, such as those found in the transcription factors Jun, Fos and Myc, are thought to facilitate the protein-protein interactions which permit heterodimerization of DNA binding proteins. As noted above, such dimerization is thought to play a key role in regulating the transcriptional activity of these factors. A Myb dimerizing binding partner has yet to be identified but Myb-Myb homodimerization, which likely occurs through its leucine zipper, does lead to a loss of DNA binding and transactivation ability [83]. One could reasonably postulate that MYB driven gene transactivation might be regulated by the binding of additional protein partners in the leucine zipper domain of the carboxy terminus of the protein [82]. Alternatively, as was true for the transactivating functions of the protein, repressor activity might also require the interaction of Myb with other proteins.

Can one reasonably postulate that aberrant c-myb expression or Myb function might play a role in carcinogenesis? Amplification of c-myb in acute myelogenous leukemia (AML) and c-myb overexpression in 6q⁻ syndrome has been reported [84] suggesting that altered c-myb expression may indeed play a role in leukemogenesis. The exact nature of this role remains to be defined. One potential mechanism is through MYB's ability to regulate hematopoietic cell proliferation. Indeed, regulating progression through the G_1/S gate of the cell cycle seems to be a key function of c-myb [85]. The mechanism whereby c-myb regulates cell proliferation is uncertain but might relate to its role in regulating a number of important cell cycle genes including c-myc [77], PCNA [22] and cdc2 [86]. Myb may also play a role in regulating differentiation [87]. Evidence to support this hypothesis includes the observations that as primitive hematopoietic cells mature, c-myb expression declines [88], and that constitutive expression of c-myb usually [89], but not always [90], inhibits the ability of hematopoietic cells to undergo differentiation. More directly, c-myb has been shown to function as a transcription factor for several cellular genes, including the neutrophil granule protein mim-1 [91], CD4 [92], IGF-1 [93], and CD34 [94], and possibly other growth factors [95] or growth factor receptors such as c-kit [26]. Finally, though once thought to be hematopoietic cell-specific, it has become clear in recent years that c-myb is expressed in non-hematopoietic tumor tissues as well [66]. Such expression is often low level but studies from our laboratory suggest that downregulation of c-myb under such circumstances can slow the growth of some malignant cell types, including malignant melanoma [66]. Accordingly, c-myb may also become a rational target in other malignant diseases, thereby, increasing the scope and relevance of the studies to be proposed below.

1.2.1 Targeting the c-myb gene

Our laboratory has been using AS-ODN since 1988 to understand the role of c-myb and other protooncogenes in normal and malignant human hematopoiesis. These investigations were designed initially to elucidate the role of Myb protein in regulating hematopoietic cell development. The results obtained from these studies had obvious clinical relevance and therefore more translationally-oriented studies were undertaken. These have now culminated in Phase I clinical trials which are presently ongoing at the Hosptial of the University of Pennsylvania. Below is a summary of the c-myb targeted AS-ODN development and a brief allusion to our initial clinical experience with this agent.

1.3 *In vitro* Experience in the Hematopoietic Cell System

1.3.1 Role of c-myb Encoded Protein in Normal Human Hematopoiesis

Attempts to exploit the c-myb gene as a therapeutic target for AS-ODN began as an outgrowth of studies which were seeking to define the role of Myb protein in regulating normal human hematopoiesis. During the course of these studies, it was determined that exposing normal bone marrow mononuclear cells (MNC) to c-myb AS-ODN resulted in a decrease of cloning efficiency and progenitor cell proliferation [96]. The effect was lineage indifferent since c-myb AS-ODN inhibited granulocyte-macrophage colony forming units (CFU-GM), CFU-E (erythroid), and CFU-Meg (megakaryocyte). In contrast, c-myb sense oligomers had no consistent effect on hematopoietic colony formation when compared with growth in control cultures ($p = .778$, $p = .796$, $p = .375$ for CFU-GM, CFU-E, and CFU-Meg respectively). Inhibition of colony formation was also dose related and sequence specific. If progenitor cells were exposed to 0.3, 1.4, 7, and 14 μM of c-myb ODN, the sense-ODN did not significantly affect colony formation while the AS-ODN, in the same concentrations, lead to the formation of 21 ± 3, 17 ± 1,10 ± 2, and 4 ± 2 megakaryocyte colonies respectively [96]. Antisense oligodeoxynucleotides at concentrations of 1.4, 7, and 14 μM led to the formation of 775 ± 127, 515 ± 9, and 244 ± 25 myeloid colonies [96]. The requirement for Myb during normal hematopoiesis has also been confirmed in gene targeting experiments, carried out using the technique of homologous recombination [97]. Other investigations determined that hematopoietic progenitor cells appeared to require Myb protein during specific stages of development [98], in particular, when they were actively cycling.

1.3.2   Myb Protein is also Required for Leukemic Hematopoiesis

Since the c-myb antisense oligomers inhibited normal cell growth, we were also interested in determining their effect on leukemic cell growth. To address this question, we employed a variety of leukemic cell lines, including those of myeloid and lymphoid origin [48, 99]. In addition, we also employed primary patient material [104].

We determined firstly, the effect of myb sense-ODN and AS-ODN on the growth of HL-60, K562, KG-1, and KG-1a cell lines [71]. The antisense oligomer inhibited the proliferation of each leukemia cell line, although, the effect was most pronounced on HL-60 cells. Specificity of this inhibition was demonstrated by the fact that the sense oligomer had no effect on cell proliferation, nor did antisense sequences with 2 or 4 nucleotide mismatches. To determine whether the treatment with c-myb AS-ODN modified cell cycle distribution of HL-60 cells, we measured the DNA content in exponentially growing cells exposed to either myb sense or myb antisense ODN. Control cells and cells treated with c-myb sense oligomers had twice the DNA content of HL-60 cells exposed to the antisense oligomers. The majority of these cells appeared to reside either, in the G1 compartment or were blocked at the G1/S boundary.

To examine the effect of the c-myb oligomers on lymphoid cell growth, we employed a lymphoid leukemia cell line designated CCRF-CEM [99]. As noted earlier in the case of normal lymphocytes [96], the CCRF-CEM cells were extremely sensitive to the anti-proliferative effects of the c-myb AS-ODN. When exposed to the sense oligomers, we found negligible effects on CEM cell growth in short term suspension cultures. In contrast, exposure to c-myb AS-ODN resulted in a daily decline in cell numbers. Compared with untreated controls, antisense DNA inhibited growth ~ 2 logs. Growth reduction was not a cytostatic effect since cell viability was reduced only ~70% after exposure to the antisense oligomers and CEM cell growth did not recover when cells were left in culture for an additional nine days. The c-myb antisense oligomers also appeared effective in inhibiting cell growth of primary patient material, when derived from patients with both acute and chronic myeloid leukemias [104]. Colony inhibition was ~58–95% and, as determined by elimination of bcr-abl expressing CML clones, apparently quite efficient.

1.3.3 Evidence that Normal and Leukemic Progenitor Cells Rely Differentially on c-myb Function

In order to be useful as a therapeutic target, neoplastic cells, leukemic or otherwise, would have to be more dependent on Myb protein than their normal counterparts. To examine this critical issue in hematopoietic cells we incubated phagocyte and T cell depleted normal human MNC, human T lymphocyte leukemia cell line blasts (CCRF-CEM), or 1:1 mixtures of these cells with sense or AS-ODN, to codons 2–7 of human c-myb mRNA (Tab. 1-1). Oligomers were added to liquid suspension cultures at Time 0 and at Time +18 h. Control cultures were untreated. In controls or in cultures to which 'high' doses of sense oligomers were added, CCRF-CEM proliferated rapidly, whereas, MNC numbers and viability decreased < 10%. In contrast, when CCRF-CEM were incubated for four days in c-myb AS-ODN, cultures contained $4.7 \pm 0.8 \times 10^4$ cells/ml (mean ± SD; n = 4) compared with $285 \pm 17 \times 10^4$/ml in controls. At the effective antisense dose, MNC were largely unaffected. After four days in culture, remaining cells were transferred to methyl-cellulose supplemented with recombinant hematopoietic growth factors. Myeloid colonies/clusters were enumerated at day ten of culture inception. Depending on the cell number plated, control MNC formed from 31 ± 4 to 274 ± 18 colonies. In dishes containing equivalent numbers of untreated or sense ODN exposed CCRF-CEM, colonies were too numerous to count (TNTC). When MNC was mixed 1:1 with CCRF-CEM in AS-ODN concentrations ≤ 5 µg/ml, only leukemic colonies could be identified by morphologic, histochemical and immunochemical analysis. However, when antisense oligomer exposure was intensified, normal myeloid colonies could be found in the culture, whereas leukemic colonies could no longer be identified with certainty using the same analytic methods.

Finally, at antisense DNA doses used in the above studies, AML blasts from eighteen of twenty-three patients exhibited ~75% decrease in colony and cluster formation compared with untreated or sense oligomer treated controls. When 1:1 mixing experiments were carried out with primary AML blasts and normal MNC, we were again able to eliminate preferentially, AML blast colony formation, while normal myeloid colonies continued to form [99].

1.3.4 Use of c-myb ODN as Bone Marrow Purging Agents

The experiments described seem to suggest that leukemic cell growth could be inhibited preferentially after exposure to c-myb AS-ODN and application, in the area

Table 1-1. Effect of c-myb oligomer exposure on colony/cluster formation by T cell leukemia and normal bone marrow progenitor cells (MNC).

Cells Plated	No. added	oligomer/amt. added°		Colony/cluster
MNC	5×10^4/ml	NONE		24 ± 4
		MYB S	(20; 5.0)	31 ± 4
		MYB AS	(20; 5.0)	30 ± 6
T LEUKEMIA	5×10^4/ml	NONE		TNTC*
		MYB S	(20; 5.0)	TNTC
		MYB AS	(20; 5.0)	1 ± 1
MNC+LEUKEMIA	5×10^4/ml	NONE		TNTC
	of each	MYB S	(20; 5.0)	TNTC
		MYB AS	(2; 0.5)	TNTC
		MYB AS	(5; 1.0)	TNTC
		MYB AS	(10; 2.5)	41 ± 5
		MYB AS	(20; 5.0)	34 ± 1

Cells were exposed to oligomers for four days in suspension cultures, and then transferred to semi-solid media as described in the methods section. After twelve days in culture, colonies and clusters were counted in paired dishes with an inverted microscope. Colony/cluster counts are presented as mean ± SD.
° g/ml added to the culture medium at time 0, and +18 h respectively
* Too numerous to count (> 1000 colonies)

of bone marrow transplantation, seemed compelling with regard to a clinical use for this observation. Exposure conditions, in this setting, were completely under the control of the investigators and patient exposure to the material was minimal. This made approval by regulatory agencies less difficult. We, therefore, attempted to determine if AS-ODN could be utilized as *ex vivo* bone marrow purging agents [99, 104]. To test this, normal MNC was mixed (1:1) with primary AML or Chronic Myelogenous Leukemia (CML) blast cells and then exposed to the oligomers using a slightly modified protocol designed to test the feasibility of a more intensive antisense exposure. Towards this end, an additional oligomer dose (20 µg/ml) was given, just prior to plating the cells in methylcellulose. In control growth factor-stimulated cultures, leukemic cells formed 25.5 ± 3.5 (mean ± SD) colonies and 157 ± 8.5 clusters (per 2×10^5 cells plated). Exposure to c-myb sense ODN did not significantly alter these numbers (19.5 ± 0.7 colonies and 140.5 ± 7.8 clusters; p >.1). In contrast, equivalent concentrations of antisense oligomers totally inhibited colony and cluster formation by the leukemic blasts. Colony formation was also inhibited in the plates containing normal MNC, but only by ~50% in comparison with untreated, control plates (control colony formation, 296 ± 40 per 2×10^5 cells plated; treated colony formation, 149 ± 15.5 per 2×10^5 cells).

1.4 Evaluation of the Efficiency of an Antisense Purge

In antisense treated co-cultures of normal MNC and primary AML blast cells (1:1 ratio), only normal colonies could be identified. To assess the potential effectiveness of an antisense purge, we conducted similar co-culture studies with CML cells obtained from patients in blast crisis and in chronic phases of their disease. Since these cells express the tumor-specific, and easily detected, bcr-abl fusion gene, it was an easy task to measure residual leukemic cells in oligonucleotide treated cultures [99, 104].

To carry out these studies, RNA was extracted from cells cloned in methylcellulose cultures after exposure to the highest c-myb AS-ODN dose. The RNA was then reverse transcribed and the resulting cDNA amplified. For each patient studied, mRNA was also extracted, from a comparable number of cells, derived from untreated control colonies, using the same technique. Eight cases were evaluated and in each case, bcr-abl expression, as detected by RT-PCR, correlated with colony growth in cell culture. Bcr-abl expression was also either greatly decreased or nondetectable in colonies/or cells which were inhibited by exposure to c-myb AS-ODN [7/11].

These results suggest that bcr-abl expressing CFU might be substantially or entirely eliminated from a population of blood or MNC by exposure to the AS-ODN. To explore this possibility further, re-plating experiments were carried out on samples from two patients. We hypothesized that if CFU, belonging to the malignant clone, were present at the end of the original 12 day culture period but not detectable because of failure to express bcr-abl, then they might re-express the message upon re-growth in fresh cultures. Accordingly, cells from these patients were exposed to ODN and were then plated into methylcellulose cultures, formulated to favor growth of either CFU-GM or CFU-GEMM. As was found with the original specimens, untreated control cells and cells exposed to sense oligomers had RT-PCR detectable, bcr-abl transcripts. Those which were exposed to the c-myb antisense oligomers had none. One of the paired dishes from these cultures was then solubilized with fresh medium and all cells contained therein were washed, disaggregated, and re-plated into fresh methylcellulose cultures without re-exposing the cells to oligomers. After 14 days, CFU-GM and CFU-GEMM colony cells were again probed for bcr-abl expression. Control and sense treated cells had RT-PCR detectable mRNA but none was found in the antisense treated colonies. These results suggest that elimination of bcr-abl expressing cells and CFU was highly efficient and perhaps permanent.

1.5   *In vitro* Experience in Non-Hematopoietic Cell Neoplasms
C-myb is a Target of Unexpected Utility in Maligant Melanoma

As noted above, c-myb is located on chromosome 6q 22-23, in humans. Structural abberations in this chromosomal location have also been linked to some human melanomas. For this reason, we noted, with interest, a report which suggested that altered c-myb expression might play a role in the pathogenesis of maligant melanoma [101]. We targeted the c-myb gene in human melanoma cells with AS-ODN to learn more about the biologic importance of its expression and the therapeutic potential of disrupting its function. To this end, we screened five human melanoma cell lines (Hs294T, SK-MEL-37, A375, C32, WM39) for c-myb mRNA by Northern analysis. Total RNA (20 µg) from each cell line was blotted to nitrocellulose and then probed with a ^{32}P-labeled human c-myb cDNA. None of the lines gave a positive signal. However, when a sensitive RT-PCR was employed, a technique which effectively amplifies small quantities of mRNA, c-myb mRNA was detected unambiguously in all cell lines.

   To determine the biological significance of this low level c-myb expression in the melanoma cells, we targeted c-myb mRNA in SK-MEL-37 and Hs294T cells with unmodified or P-ODN and control DNA sequences. In Hs294T cells, exposure to c-myb control sequences had no statistically significant effect on cell proliferation. In contrast, the c-myb AS-ODN inhibited growth in a dose responsive manner, up to ~60% (p<.001) of control cell values. Growth inhibition was accompanied by loss of RT-PCR detectable c-myb but not β-actin mRNA (the levels of which are constant in all cells), suggesting that growth inhibition was secondary to perturbation of c-myb expression. Visual examination of the cultures revealed some clue regarding the inhibitory mechanism. Hs294T cells appeared to undergo cytolysis after exposure to the c-myb AS-ODN, suggesting that c-myb perturbation could be a lethal event in these cells. This contrasted with the effect observed on SK-MEL-37 cells, which appeared to undergo growth arrest with or without what appeared, morphologically, to represent differentiation towards a more mature phenotype.

1.5.1   Relationship Between DNA Dose, Frequency of Exposure, and Inhibition of Cell Growth

We also examined cell growth inhibition as a function of oligodeoxynucleotide concentration and frequency of exposure. When SK-MEL-37 were exposed to AS-ODN, the most important factor for achieving growth inhibition was initial exposure to high concentrations of material. For example, at concentrations ± 50 µg/ml, no effect on cell growth was observed when the ODN were added to cultures in divided doses of 20 µg/ml/day × 2 days, or 10 µg/ml/day × 5 days. In contrast, when cells were exposed to a single bolus of 50 µg/ml, cell growth was inhibited ~25% compared with untreated controls. This relationship was even more apparent at higher doses. A single total ODN dose of 100 µg/ml inhibited growth much more significantly than the same total dose delivered in divided doses of 20 µg/ml/day × 5 doses. Even at doses up to 250 µg/ml, 50 µg/day for 5 days was not as effective as a total dose of 200 µg given as 100 µg/ml/day × 2 doses (50% vs 70% inhibition respectively).

To determine if these results were influenced by possible degradation of unmodified ODN, we carried out similar experiments with Hs294T cells exposed to the more stable S-O-P AS-ODN. A similar but less strict relationship between extracellular concentration and inhibition of cell growth was again observed. That is, initial high concentrations were more effective than equivalent final concentrations built up by cumulative lower doses. It appears that for either type of compound, sufficient cellular uptake, to inhibit the target gene, is only achieved by initial exposure to some critical 'high' concentration of compound. Whether or not this relationship would hold for leukemic cells has not been examined formally but we have no reason to believe that this would not be true for these cell types as well.

1.6   *In vivo* Treatment Models

The experience gained with *in vitro* culture systems suggested, but did not prove, that *in vivo* activity with the c-myb targeted antisense molecule might be expected as well. To test this, we developed SCID mouse human chimeras bearing either human leukemia [37] or human melanoma [104]. This animal system has the obvious advantage of allowing activity against a human tumor to be determined in a setting simulating actual clinical use.

1.6.1   *In vivo* Treatment of Human Leukemia in a SCID Mouse Model

We established human leukemia-SCID mouse chimeras with K562 cells and treated diseased animals with S-O-P AS-ODN [37]. K562 cells express the c-myb protooncogene, which served as the target for the antisense DNA as well as tumor specific bcr/abl oncogene, which was utilized to track the human cells in the mouse host. Once animals had detectable circulating leukemic blast cells, the mean (± SD) survival of untreated control mice was 6 ± 3 days. The survival of animals, treated for 7 or 14 days with either sense or scrambled sequence c-myb ODN, was not statistically different from the control animals. In distinct contrast, animals treated for similar lengths of time with c-myb AS-ODN survived at least 3.5 times longer than the various control animals (Fig. 1-4). In addition, animals receiving c-myb antisense DNA had significantly less disease at the two sites most frequently manifesting

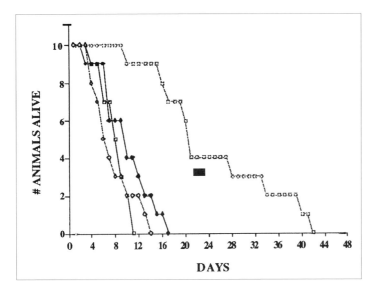

Figure 1-4. Survival curves of SCID-human chimeric animals transplanted with K562 chronic myelogenous leukemia cells. Animals received a 14 day infusion of oligomers at a dose of 100 µg/day.
—□— CONTROL, —◆— SENSE, —□— ANTISENSE,
—◇— SCRAMBLED.
(Adapted from [37])

leukemic cell infiltration: the central nervous system and the ovary (Fig. 1-5). These results suggested that phosphorothioate modified antisense DNA might be efficacious for the treatment of human leukemia *in vivo* and was an important component in justifying the clinical trials which have been activated at our institution.

1.6.2 *In vivo* Treatment of Human Melanoma in a SCID Mouse Model. Utility of c-myb as a Target

Knowing that myb mutations have been found in melanoma, we examined the effect of the c-myb antisense DNA on human melanoma cell growth in a SCID mouse model [66]. In the first of three types of experiments to assess this question, 41 mice were inoculated with Hs294T cells. When tumor nodules became palpable, animals were randomly assigned to receiv either no treatment (13 animals) or 7 day infusions (500 µg/day; 25 µg/g body weight) of c-myb sense (14 animals) or (14 animals) S-O-P AS-ODN. Animals were examined daily for 40 days to determine the effects of the P-ODN on survival and tumor growth. The antisense treatment significantly in-

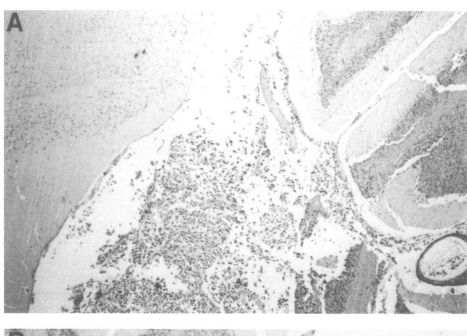

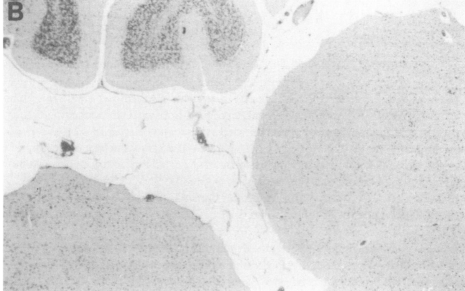

Figure 1-5. Composite photomicrographs (× 400) of mouse brain obtained from chimeric SCID-hu K562 leukemia-bearing mice treated with (top) sense, or (bottom) antisense phosphorothioate c-myb ODN. Note extensive meningeal and sub-arachnoid infiltration with leukemic blasts in (A) and lack of involvement in (B).(From [37].)

hibited local tumor growth in comparison with the control and sense [S-O-P] AS-ODN treated groups. In fact, until ~day 35, calculated tumor weights in the antisense group were ~50% lower than the other groups. After this time, growth in the antisense treated group recovered and essentially, paralleled that in the control and sense treated animals. Nevertheless, when the animals were sacrificed on day 40, tumors removed from the antisense treated animals were significantly smaller (p<.05) than those in the control groups. The mean (± SD) weight of tumors taken from the animals in the control, sense, and antisense groups was 3.5 ± 1.7 g, 3.3 ± 1.2 g and 2.5 ± 0.5 g respectively.

We then examined the growth inhibitory effects of the c-myb [S-O-P] AS-ODN against a sub-clinical tumor burden. In this experiment, mice were inoculated subcutaneously with 2×10^6 Hs294T tumor cells. Three days later, animals were randomized to receive either no treatment (9 mice) or a 7 day infusion (500 µg/day) of either c-myb sense (8 mice) or [S-O-P] AS-ODN (10 mice). Tumor growth was evaluated over a 65 day period. While no control animals were lost during this period, 3 sense and 4 antisense treated animals died of uncertain causes. In the remaining animals, inhibition of tumor growth in the antisense treated group was again noted throughout the observation period and appeared to be greater than that observed in the first experiment. When sacrificed at 60 days after the pumps were implanted, mean (± SD) tumor weights of control, sense and antisense groups were 4.5 ± 1.7 g, 4.0 ± 1.5 g and 2.1 ± 1.2 g respectively. The differences between these groups were statistically significant (p<.05). Figure 1-6 illustrates typical tumors observed in these mice.

Lastly we examined the effect of a repeat infusion on tumor growth. In this experiment, animals (10 per group) were again inoculated with 2×10^6 tumor cells. Three days later they were randomized to receive no treatment (control) or an infusion of sense or [S-O-P] AS-ODN (500 µg/day × 14 days). Sixteen days after the first infusion ended, a repeat infusion of identical dose and duration was begun. In this experiment, 3 control and 1 sense treated animal died tumor-related deaths during the observation period. In the antisense treated animals, tumor growth inhibition was more dramatic than in the previous experiments and persisted throughout the observation period. When animals were sacrificed 85 days after the first pump was implanted, mean ± SD of tumor weights of control (n = 7), sense (n = 9) and antisense (n = 10) groups were 3.0 ± 2.0 g, 1.7 ± 1.5 g and 0.7 ± 0.5 g respectively. The difference in tumor weights between the control and antisense treated groups was highly significant (p<.01) as was the difference between the sense and antisense treated groups (p<.05). Though it appeared that tumor sizes differed between the control and sense treated groups, the mean sizes were not of statistical significance (p>.05). It is important to note that in contrast to the experiments carried out with a lower total dose of [S-O-P] AS-ODN, none of the animals in the high dose, repeat infusion, sense or antisense treated groups died before the experiment was terminat-

CONTROL

SENSE

 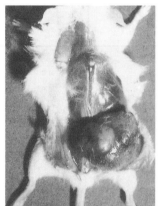

ANTISENSE

Figure 1-6. Representative photomicrographs of human melanomas in situ in animals treated with oligodeoxynucleotides targeted to the c-myb gene. Control and sense treated animals are shown in the top and middle panels, in situ (left) and with the skin reflected to display the tumor (right). Corresponding views of animals treated with the antisense compound are shown in the bottom panel. (From [66].)

ed. These results suggest that toxicity of the [S-O-P] AS-ODN was an unlikely cause of animal deaths.

1.7   Pharmacokinetic/Dynamic Studies in the Melanoma Model

An important issue in all of these studies is whether or not tumor responses can be shown to correlate with changes in the targets gene's (in this case c-myb) expression. We first sought to satisfy ourselves that the oligomers were indeed entering the tumor tissue. To determine ODN uptake in tumor tissue and to correlate effects on c-myb expression with tumor growth, five animals with established tumors (~0.5 g) were infused with c-myb [S-O-P] AS-ODN (500 µg/day) for 7 days. On days 7, 9 and 11 post infusion, the animal was sacrificed and its tumor excised to determine tissue c-myb mRNA levels. As shown in Figure 1-7, c-myb mRNA levels were decreased measurably (normalized to β-actin), but not completely extinguished, in comparison with control expression. This decrement in c-myb expression persisted for approximately two days after the infusion finished but returned towards the baseline thereafter.

Normalization of c-myb expression may be related to [S-O-P] AS-ODN concentration in tissue falling below a critical level. In support of this hypothesis, [S-O-P] AS-ODN levels in the tumor tissue decreased rapidly from levels estimated to be ~500 ng (per 50 µg extracted DNA), during the infusion to a level of <50 ng but > 10 ng (per 50 µg extracted DNA) on day 8, one day after the infusion finished. It should be pointed out that while human c-myb is selectively suppressed by the antisense ODN, the PCR primers employed for detection will amplify both human and murine c-myb mRNA. Since murine blood and stromal elements gradually infiltrate the tumor, some of the c-myb mRNA detected could be contributed from this source. Regardless, it is clear that the downregulation of the target is transient, which is expected. An important goal of future studies will be to determine the length of time target gene suppression is required to produce a useful clinical effect.

1.8   Use of Antisense Oligonucleotides in a Clinical Setting

Based on the type of data presented above, a favorable therapeutic index in toxicology testing and more detailed knowledge of the pharmacokinetics of AS-ODN has begun clinical evaluation. At the University of Pennsylvania we have initiated two such trials: One of these is a pilot study to evaluate the effectiveness of [S-O-P] AS-ODN to c-myb as a marrow purging agent in patients with chronic phase (CP) or accelerated phase (AP) CML. The other is a PhaseI/II study evaluating the toxicity of systemically infused c-myb [S-O-P] AS-ODN in patients with CML in blast crisis, or refractory acute leukemia.

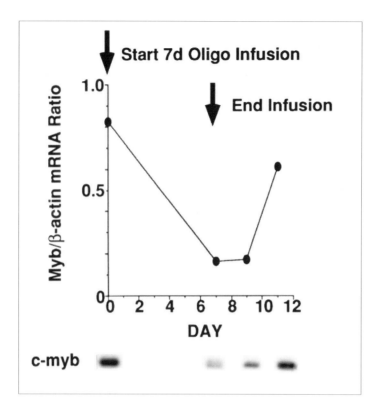

Figure 1-7. Effect of Myb directed P-ODN on c-myb mRNA expression in human melanomas growing in SCID mice. P-ODN were infused into mice (500 µg/day × 7 days) with ~1 gm tumors. On days 7, 9, and 11 tumors were excised and total RNA was extracted for determination of c-myb mRNA levels. The relative amount of Myb mRNA was normalized to b-actin mRNA detected in the same sample. (From [66].)

When designing a clinical protocol it is important that a general format be followed. The background for the study should be clearly and succinctly stated in an introductory section, followed by equally clear objectives for the study. Patient eligibility for treatment on the protocol should be well defined, including specific exclusionary criteria. For example, in the case of patients who are being transplanted with a c-myb purged marrow, we have adopted the following rules: The growth of the patient's marrow cells must be inhibited *in vitro* by exposure to the anitsense c-myb DNA and not by exposure to a control DNA sequence, and/or both c-myb and bcr/abl gene expression must be specifically down-regulated in the patient's marrow cells. Any patient with an allogeneic marrow transplant donor in not eligible for treatment. A detailed protocol outlining the specific treatment then needs to be prepared.

Details of the intercurrent management plans, during the transplant period, should also be provided, along with a detailed plan as to how the effect of the transplant will be measured. Criteria which need to be met in order to decide if the patient has had a response should be listed. Categories for response include complete, partial, stable disease, and no response.

The studies which will be performed on patients treated on the protocol need to be described and it must be stated, explicitly, when the studies need to be carried out. An example of this, adapted from the University of Pennsylvania protocol is given in Table 1-2. The nature of any companion laboratory tests, what they are and when they will be performed should also be described. We, for example, study periodically, bcr/abl gene expression and look for the Philadelphia chromosome in all patients that we have treated. It should be obvious that a mechanism for identifying, following and reporting to the appropriate authorities, potential drug toxicity, must also be provided.

1.8.1   Initial Experience with c-Myb P-ODN as a Marrow Purging Agent

Earlier mention was made of our recently initiated clinical trial to evaluate the effectiveness of phosphorothioate modified c-myb antisense ODN as a marrow purg-

Table 1-2. Study Parameters

	Prior	ABMT-daily	ABMT-weekly	Follow-up monthly
Physical exam	X	X		X
Weight	X	X		X
Hemoglobin, Hematocrit, WBC, Differential, Platelets	X	X		X
Elecrolytes,BUN,Creat.	X	X		X
Bilirubin,SGOT/PT,Uric Acid, LDH, Alk.Phos.	X		X	
Chest X-ray	X		prn	
EKG	X		prn	
MUGA Scan	X			
PFT's/DLCO	X			
Creatinine Clearance	X			
Bone Marrow Asp/Biopsy	X			X[1]
Adverse Reactions		X[2]		X[2]

All scans and x-rays should be done ≤ 6weeks before registration. CBC with differential should be done ≤ 48 h before registration. All chemistries should be done ≤ 2 weeks before registration. If abnormal, they must be repeated within 48 h prior to registration.
[1]Every 2–3 months for 2 years unless evidence for relapse. Detection of minimal residual disease will also be performed. [2]Patients will be monitored daily during therapy and monthly thereafter for any adverse reactions which may be treatment related. Adverse reactions will be reported to the IRB and FDA. *VIRAL TITERS: HEP A, B, & C; HIV; HSV; VZV0

ing agent in Chronic Phase (CP) or Accelerated Phase (AP) CML patients [102]. Five patients have been treated thus far. Four did well with engraftment (MNC>500) occurring between 11–18 days post transplant. One patient failed to engraft with either purged or untreated back-up marrow and died on Day +50 with pulmonary aspergillosis. Clinical and laboratory responses, including cytogenetics and RT-PCR for bcr/abl gene expression, are being evaluated in all patients. Two CP and one AP patient have been more fully evaluated. Both CP patients demonstrated, pre-transplant, *in vitro* antisense ODN sensitivity which was dose and sequence dependent. One CP patient (#1) also demonstrated unequivocal elimination of bcr/abl positive cells in the purged specimen and was reconstituted with bcr/abl negative blood cells. Patient #1 continued to manifest complete disappearance of the Ph+ chromosome and was bcr/abl negative by RT-PCR for several months post-transplant. Both CP patients remain in complete clinical and partial cytogenetic remission ($\sim$15% Ph$^+$ metaphases). The AP patient was unresponsive *in vitro*, although, in a clinical CR, his metaphases were 100% Ph$^+$. Although follow-up was short and patient numbers were small, we suggest that purging with $^{S-O-P}$ AS-ODN is a safe and potentially useful procedure in CML patients without an allogeneic marrow donor.

As noted earlier, we have also employed phosphorothioate myb AS ODN in a Phase I study to examine the toxicity of infused c-myb antisense DNA. We have studied eighteen refractory leukemia patients thus far (2 patients were treated at 2 different dose levels; 13 had AP or BC CML). Myb $^{S-O-P}$ AS ODN was delivered by continuous infusion at dose levels ranging between (0.3 mg/kg/day $\times$ 7 days) and (2.0 mg/kg/day $\times$ 7 days). No recurrent dose related toxicity has been noted, although, idiosyncratic toxicities, not clearly drug related, were observed (1 transient renal insufficiency; 1 pericarditis). One BC patient survived $\sim$14 months with transient restoration of CP disease. These studies show that $^{S-O-P}$ AS ODN may be administered safely to leukemic patients. Whether or not patients treated on either study derived clinical benefit is uncertain, but the results of these studies suggest that ODN may eventually demonstrate therapeutic utility in the treatment of human leukemias.

1.9   Conclusions

The above discussion was meant to recapitulate how the c-myb gene was identified as a potential therapeutic target and how its utility was explored, step-wise, using *in vitro* and *in vivo* models. A frequently asked question is, "why not target the bcr/abl gene with antisense DNA"? This target has the advantage of being both tumor specific and of clear importance in the pathogenesis of this disease. Nevertheless, recent studies suggest that the most primitive stem cells may not express bcr/abl mRNA [103]. Since CML is a stem cell disorder, bcr/abl mRNA may not be an appropriate target for these most primitive cells. Furthermore, it is not certain that once transformed, continuous expression of bcr/abl is required for maintenance of the malignant phenotype. Finally, since bcr/abl likely functions as a signal transducing protein, and the signaling apparatus in cells appear highly redundant, targeting bcr/abl may allow for resistance to emerge. Other antisense targets, for example c-myb, may then turn out to be more effective in the long term.

It is straightforward that much of the translational aspects of this development are dependent on the "antisense approach" to gene therapy. Though simple in theory and execution, antisense experiments may be far more difficult to conduct and interpret than the above discussion would indicate. The reasons for this are multifactorial but relate either to an often frustrating inability to identify a suitable region of an mRNA to target, or, to non-sequence dependent effects on cell function. These problems are likely to be related to mRNA secondary and tertiary structure and the particular chemistry of the oligo employed, respectively. This does not mean, however, that carefully controlled experiments can not be informative. Rather, the caveat here is not to confuse a biologic effect with an "antisense effect", although, non-specific effects could obviously be advantageous if appropriately exploited. Whether c-myb will turn out to be a useful target in hematologic malignancies, or in the less obvious case of solid tumors, remains to be seen but the question will be answered in the near future. If encouraging results are obtained, this work will be all the more exciting because at the same time that c-myb proves itself a useful target it will simultaneously open up a new approach to anticancer therapeutics.

References

1. Slamon, D. J., deKernion, J. B., Verma, I. M. and Cline, M. (1984) Expression of cellular oncogenes in human malignancies. *Science* 224: 256–262.
2. Sawyers, C. L., Denny, C. T. and Witee, O. N. (1991) Leukemia and the disruption of normal hematopoiesis. *Cell* 64: 337–350.
3. Slamon, D. J., Boone, T. C., Murdock, D. C., Keith, D. F., Press, M. F., Larson, R. A. and Souza, L. M. (1986) Studies of the human c-myb gene and its product in human acute leukemias. *Science* 232: 347–350.
4. Burck, K. B., Liu, E. T. and Larrick, J. W. (1988) *In*: Burck, K. B., Liu, E. T. and Larrick, J. W. (eds), *Oncogenes: An introduction to the Concept of Cancer Genes*. Springer-Verlag, New York, pp. 38–66.
5. Cooper, G. M. (1990) *In*: G. M. Cooper (ed.), *Oncogenes*. Johns and Barlett Boston, pp 35–49.
6. Doolite, R. F., Hunkapiller, M. W., Hood, L. E., Deare, S. G., Robbins, K. C., Aaronson, S. A. and Antonaides, H. N. (1983) Simian sarcoma virus oncogene, v-sis, is derived from the gene (or genes) encoding a platelet-derived growth factor. *Science* 221, 275-277.
7. Sherr, C. J., Rettenmier, C. W., Sacca, R., Roussel, M. F., Liook, A. T. and Stanley, R. F. (1985) The c-fms protooncogene product is related to the receptor for the mononuclear

phagocyte growth factor, CSF-1. *Cell* 41: 665–676.

8. Majumder, S., Ray, P. and Besmer, P. (1990) Tyrosine protein kinase activity of the H24-feline sarcoma virus p80gag-kit-transforming protein. *Oncogene Res.* 5: 329–335.

9. Bos, J., L. (1989) Ras oncogenes in human cancer: Review. *Cancer Res.* 49: 4682–4689.

10. Bertani, A., Polentarutti, N., Sica, A., Rambaldi, A., Mantovani, A. and Colotta, F. (1989) Expression of c-jun protooncogene in human myelomonocytic cells. *Blood* 74: 1811–1816.

11. Turner, R. and Tjian, R. (1989) Leucine repeats and an adjacent DNA binding domain mediate the formation of functional cFos-cJun heterodimers. *Science* 243: 1689–1694.

12. Gonda, T. J. (1991) Targets for trans-activation by myb. *Cancer Cells* 3: 22–23.

13. Hunter, T. (1991) Cooperation between oncogenes. *Cell* 64: 249–270.

14. Bishop, J. M. (1991) Molecular themes in oncogenesis. *Cell* 64: 235–248.

15 Sin, E., Muller, W., Pattengate, P., Tepler, I., Wallace, R. and Ledger, P. (1987) Coexpression of MMTV/v-Ha-ras and MMTV/c-myc genes in transgenic mice: synergistic action of oncogenes *in vivo*. *Cell* 49: 465–475.

16. Waldman, A. S. (1992) Targeted homologous recombination in mammalian cells. *Crit. Rev. Oncology/Hematology* 12: 49–64.

17. Cooney, M., Czernuszewicz, E., Postel, E., Flint, S. and Hogan, M. (1988) Site-specific oligonucleotide binding represses transcription of the human c-myc gene *in vitro*. *Science* 241, 456.

18. Carter, G. and Lemoine, N. R. (1993) Antisense technology for cancer therapy: does it make sense. *Fr. J. Cancer* 67: 869–876.

19. Zaug, A. J. and Cech, T. R. (1986) The intervening sequence RNA of tetrahymena is an enzyme. *Science* 231: 470–475.

20. Zamecnik, P. C. and Agrawal, S. (1991) Oligodeoxynucleotide hybridization inhibition of HIV and influenza virus. *Nucleic Acids Symp. Ser.* 127–131.

21. Heikkile, R., Schwab, G., Wickstrom, E., Loke, S. L., Pluznik, D. H., Watt, R. and Neckers, L. M. (1987) A c-myc antisense oligodeoxynucleotide inhibits entry into S phase but not progress from G- to G1. *Nature* 328: 445–449.

22. Jaskulski, D., DeRiel, K., Mercer, W. E., Calabretta, B. and Baserga, R. (1988) Inhibition of cellular proliferation by antisense oligodeoxynucleotides to PCNA cyclin. *Science* 240: 1544–1546.

23. Lallier, T. and Bronner-Fraser, M. (1993) Inhibition of neural crest cell attachment by integrin antisense oligonucleotides. *Science* 259: 692–695.

24. Chiang, M. Y., Chan, H., Zounes, M., Freier, S., Lima, W. and Bennett, C. F. (1991) Antisense oligonucleotides inhibit intercellular adhesion molecule 1 expression by two distinct mechanisms. *J. Biol. Chem.* 266: 18162–18171.

25. Hermine, O., Beru, N., Pech, N. and Goldwasser, E. (1991) An autocrine role for erythropoietic in mouse hematopoietic cell differentiation. *Blood* 9: 2253–2260.

26 Ratajczak, M. Z., Luger, S. M., DeRiel, K., Abrahm, J., Calabretta, B. and Gewirtz, A. M. (1992) role of the KIT protooncogene in normal and malignant human hematopoiesis. *Proc. Natl. Acad. Sci. USA* 89: 1710–1714.

27. Becker, D., Lee, P. L., Rodeck, U. and Herlyn, M. (1992) Inhibition of the fibroblast growth factor receptor 1 (FGFR-1) gene in human melanocytes and malignant melanomas leads to inhibition of proliferation and signs indicative of differentiation. *Oncogene* 7: 2303–2313.

28. Saison-Behmoaras, T., Tocque, B., Rey, I., Chassignol, M., Thuong, N. T. and Helene, C.

(1991) Short Modified antisense oligonucleotides directed against Ha-ras point mutation induce selective cleavage of the mRNA and inhibit T24 cells proliferation. *EMBO J.* 10: 1111–1118.

29. Skorski, T., Szczylik, C., Ratajczak, M. Z., Malaguarnera, L., Gewirtz, A. M. and Calabretta, B. (1992) Growth factor-dependent inhibition of normal hematopoiesis by N-ras antisense oligodeoxynucleotides. *J. Exp. Med.* 175: 743–750.

30. Prochownik, E. V., Kukowska, J. and Rodgers, C. (1988) C-myc antisense transcripts accelerate differentiation and inhibit G1 progression in murine erythroleukemia cells. *Mol. Cell. Biol.* 8: 3683–3695.

31. Leonetti, J. P., Mechti, N., Degols, G., Gagnor, C. and Lebleu, B. (1991) Intracellular distribution of microinjected antisense-oligonucleotides. *Proc. Natl. Acad. Sci. USA* 88: 2702–2706.

32. Leonetti, J. P., Machy, P., Degols, G., Lebleu, B. and Leserman, L. (1990) Antibody targeted liposomes containing oligodeoxyribonucleotides complementary to viral RNA selectively inhibit viral replication. *Proc. Natl. Acad. Sci. USA* 87: 2448–2451.

33. Van der Krol, A. R., Mol JNM and Sruitje, A. R. (1988) Modulation of eukaryotic gene expression by complementary RNA or DNA sequences. *BioTechniques* 6: 958–976.

34. Zamecnik, P. C. and Stephenson, M. L. (1978) Inhibition of rous sarcoma virus replication and cell transformation by a specific oligodeoxynucleotide. *Proc. Natl. Acad. Sci. USA* 75: 280–284.

35. Wickstrom, E., Bacon, T. A., Wickstrom, E. L., Werking, C. M., Palmiter, R. D., Brinster, R. L. and Sandgren, E. P. (1991) Antisense oligodeoxynucleotide methylphosphonate inhibition of mouse c-myc p65 protein expression in E mu-c-myc transgenic mice. *Nucleic Acids Symp. Ser.* 151–154.

36. Gewirtz, A. M. and Calabretta, B. (1988) A c-myb antisense oligodeoxynucleotide inhibits normal human hematopoiesis *in vitro*. *Science* 242: 1303–1306.

37. Ratajczak, M. Z., Kant, J. A., Luger, S. M., Hijiya, N., Zhang, J., Zon, G. and Gewirtz, A. M. (1992) *In vivo* treatment of human leukemia in a scid mouse model with c-myb antisense oligodeoxynucleotides. *Proc. Natl. Acad. Sci. USA* 89: 11823–11827.

38. Kitajuma, I., Shinohara, T., Bilakovics, J., Brown, D. A., Su, S. and Nerenberg, M. (1992) Ablation of transplanted HTLV-1 Tax transformed tumors in mice by antisense inhibition of NF-kB. *Science* 258: 1792–1795.

39. Trojan, J., Johnson, T. R., Rudin, S., Ilan J. and Tykicinski, M. L. (1993) Treatment and provention of rat glioblastoma by immunogenic C6 cells expressing antisense insulin-like growth factor I RNA. *Science* 259: 94–96.

40. Whitesell, L., Rosolen, A. and Neckers, L. M. (1991) *In vivo* modulation of N-myc expression by continuous perfusion with an antisense oligonucleotide. *Antisense Res. Dev.* 1: 343–350.

41. Stein, C. A. (1992) Anti-sense oligodeoxynucleotides-promises and pitfalls. *Leukemia* 6: 967–974.

42. Yakubov, L. A., Deeva, D. A., Zarytova, V. F., Ivanova, D. M., Ryte, A. S., Yurchenki, L. V. and Vlassov, V. V. (1989) Mechanism of oligonucleotide uptake by cells: Involvement of specific receptors. *Proc. Natl. Acad. Sci. USA* 86: 6454–6458.

43. Loke, S. L., Stein, C., Zhang, X., Avigan, M., Cohen, J. and Neckers, L. M. (1988) Delevery of c-myc antisense phosphorothioate oligodeoxynucleotides to hematopoietic cells in culture by liposome fusion: specific reduction in c-myc protein expression correlates with inhibition of cell growth and DNA synthesis. *Curr. Topics Microbiol.*

Immunol. 141: 282–289.

44. Gao, W. -Y., Storm, C., Egan, W. and Cheng, Y. -C. (1993) Cellular pharmacology of phosphorothioate homooligodeoxynucleotides in human cells. Mol. Pharm. 43: 45–50.

45. Loke, S. L., Stein, C. A., Zhang, X. H., Mori, K., Nakanishi, M., Subansinghe, C. and Cohen, J. S. (1989) Characterization of oligonucleotide transport into living cells. *Proc. Natl. Acad. Sci. USA* 86: 3474–3478.

46. Stein, C. A., Neckers, L., Nair, B., Mumbauer, S., Hoke, G. and Pal, R. (1991) Phosphorothioate oligodeoxycytidine interferes with binding of HIV-1 gp 120 to CD4. *J. Acquired Immune Defin. Syndrome* 4: 686–693.

47. Miller, A. D. (1992) *Hum. Gene Ther.* comes of age. *Nature* 357: 455–460.

48. Ratajczak, M. Z., Hijiya, N., Catani, L., DeRiel, K., Luger, S. M., MacGrave, P. and Gewirtz, A. M. (1992) Acute-and chronic myelogenous leukemia colony-forming units are highly sensitive to the Grwoth Inhibitory Effects of C-myb antisense oligodeoxynucleotides. *Blood* 79: 1956–1961.

49. Eder, P. S., DeVine, R. J., Dagle, J. M. and Walder, J. A. (1991) Substrate specificity and kinetics of degradation of antisense oligonucleotides by a 3' exonuclease in plasma. *Antisense Res. Dev.* 1: 141–151.

50. Howe, K. M., Reakes, C. F. and Watson, R. J. (1990) Characterization of the sequence-specific interaction of mouse c-myb protein with DNA. *EMBO J.* 9: 161–169.

51. Miller, P. (1989) in Oligodeoxynucleotides: antisense inhibitors of gene expression. (Cohen J ed.): pp 79–95, Macmillan, London.

52. Stein, C. and Cohen, J. (1989) in Oligodeoxynucleotides: antisense inhibitors of gene expression, pp 97–117, Macmillan, London.

53. Helene, C. and Toulme, J. (1990) Specific regulation of gene expression by antisense, sense and antigene nucleic acids. *Biochim. Biophys. Acta* 1049: 99–125.

54. Mukhopadhyay, T., Tainsky, M., Cavender, A. C. and Roth, J. A. (1991) Specific inhibition of K-ras expression and tumorigenicity of lung cancer cells by antisense R N A . *Cancer Res.* 51: 1744–1748.

55. Watson, P. H., Pon, R. T. and Shiu, P. C. (1991) Inhibition of c-myc expression by phosphorothioate antisense oligonucleotide identifies a critical role for c-myc in the role of human breast cancer. *Cancer Res.* 51: 3996–4000.

56. Collins, J. F., Herman, P., Schuch, C. and Bagby, G. C. (1992) C-myc antisense oligonucleotides inhibit the colony-forming capacity of colo320 carcinoma cells. *J. Clin. Invest.* 89: 1523–1527.

57. Agarawal, S., Temsamani, J. and Tang, J. Y. (1991) Pharmacokinetics, biodistribution and stability of oligodeoxynucleotide phosphorothioates in mice. *Proc. Natl. Acad. Sci. USA* 88: 7595–7599.

58. Iversen P. (1993) in *In vivo* studies with phosphorothioate oligonucleotides: rationale for systemic therapy. Crooke, S. T. and Lebleu, B. (eds), *Antisense Research and Application.* CRC Press, Ann Arbor, pp. 461–469.

59. Goodchild, J., Kim, B. and Zamecnik, P. C. (1991) The clearance and degradation of oligodeoxynucleotides following intravenous injection into rabbits. *Antisense Res. Dev.* 1: 153–160.

60. Bayever, E., Iversen, P., Smith, L., Spinolo, J. and Zon, G. (1992) Guest editorial: Systemic human antisense therapy begins. *Antisense Res. Dev.* 2: 109–110.

61. Tidd DM. (1990) A potential role for antisense oligodeoxynucleotide analogues in the development of oncogene targeted cancer chemotherapy. *Anticancer Res.* 10, 1169.

62. Stein, C. A. and Cohen, J. S. (1989) Phosphorotioate oligodeoxynucleotide analogues. *In*: J. S. Cohen (ed.), *Oligodeoxynucleotides. Antisense Inhibitors of Gene Expression.* Basingstoke, The Macmillian Press Ltd. pp 97–117.

63. Yaswen, P., Stampfer, M. R., Gosh, K. et al. (1993) Effects of sequence of thioated oligonucleotides on cultured human mammary epithelial cells. *Antisense Res. Dev.* 3: 67.

64. Doida, Y. and Okada, S. (1967) Synchronization of L5178Y cells by successive treatment with excess thymidine and colcemid. *Exp. Cell. Res.* 48: 540.

65. Scharenberg, J. G. M., Rijkers, G. T., Toebes, E. A. H., Stall, G. E. J., Zegers, B. J. M. (1988) Expression of deoxyadenosine and deoxyguanosine toxicity at different stages of lymphocyte activation. *Scand. J. Immunol.* 28: 87.

66. Hijiya, N., Zhang, J., Ratajczak, M. Z. et al. (1994) The biologic and therapeutic significance of c-myb expression in human melanoma. *Proc. Natl. Acad. Sci. USA*

67. Skorski, T., Nieborowska-Skorska M., Barietta, C., Malaguarnera, L., Szczylik, C., Chen, S. T., Lange, B., Calabretta, B. (1993) Highly efficient elimination of Philadelphia leukemic cells by exposure to bcr/abl antisense oligodeoxynucleotides combined with mafosfamide. *J. Clin. Invest.* 92: 194.

68. Kitajima, I., Shinohara, T., Bilakovics, J., Brown, D. A., Xu, X., Nerenberg, M. (1992) Ablation of transplanted HTLV-I Tax-transformed tumors in mice by antisense inhibition of NF-kB. *Science* 258: 1792.

69. Wickstrom, E., Bacon, T. A., Wickstrom, E. L. (1992) Down-regulation of c-Myc antigen expression in lymphocytes of Em-c-myc transgenic mice treated with anti-c-myc DNA methylophosphonates. *Cancer Res.* 52: 6741.

70. Ransone, L. J., Verma, I. M. (1990) Nuclear proto-oncogenes fos and jun. *Annu. Rev. Cell Biol.* 6: 539–557.

71. Biedenkapp, H., Borgmeyer, U., Sippel, A. E. and Klempnauer, K. H. (1988) Viral myb oncogene encodes a sequence-specific DNA-binding activity. *Nature* 335: 835–837.

72. Cerutti, P.; Hussain, P.; Pourzand, C.; Aguilar, F. (1994) Mutagenesis of the H-ras protooncogene and the p53 tumor suppressor gene. *Cancer Res.* 54 (7 Suppl): 1934s–1938s.

73. Lueschger, B. and Eisenman, R. N. (1990) New light on Myc and Myb. Part II. Myb. *Gene. Develop.* 4: 2235–2242.

74. Small, D., Levenstein, M., Kim, E., Carow, C., Amin, S., Rockwell, P., Witte, L., Burrow, C., Ratajczak, M. Z., Gewirtz, A. M. and Civin, C. I. (1994) STK-1, the human homologue of Flk-2/Flt-3, is selectively expressed in CE34+ human bone marrow cells and is involved in the proliferation of early progenitor/stem cells. *Proc. Natl. Acad. Sci. USA* 91: 459–463.

75. Lyon, J., Robinson, C. and Watson, R. (1994) The role of the Myb proteins in normal and neoplastic cell proliferation. *Crit. Rev. Oncogen.* 5: 373–388.

76. Sakura, H., Kanei-Ishii, C., Nagase, T., Nakagoshi, H., Gonda, T. J. and Ishii, S. (1989) Delineation of three functional domains of the transcription activator encoded by the c-myb protoonocogene. *Proc. Natl. Acad. Sci. USA* 86: 5758–5762.

77. Cogswell, J. P., Cogswell, P. C., Kuehl, M., Cuddihy, A. M., Bender, T. M., Engelke, U., Marcu, K. B. and Ting, J. P. -Y. (1993) Mechanism of c-myc regulation by c-myb in different cell lineages. *Mol. Cell. Biol.* 13: 2858–2869.

78. Foos, G., Natour, S. and Lempnauer, K. H. (1993) TATA-box dependent transactivation of the human HSP70 promoter by Myb proteins. *Oncogene* 8: 1775.

79. Watson, R. J., Robinson, C. and Lam EWF (1993) Transcription regulation by murine B-myb is distinct from that by c-myb. *Nucl. Acid. Res.* 21: 267-.

80. Ness, S. A., Kowenz-Leutz, E., Casini, T. and Graf, T. (1993) Myb and NF-M: combinatorial activators of myeloid genes in heterologous cell types. *Gene. Dev.* 7: 749.

81. Burk, O., Mink, S., Ringwald, M. and L. Klempnauer, K. H. (1993) Synergistic activation of the chicken mim-1 gene by v-myb and C/EBP transcription factors. *EMBO J.* 12: 2027–2038.

82. Kanei-Ishii, C., MacMillan, E. M., Nomura, T., Sarai, A., Ramsay, R. G., Aimoto, S., Ishii, S. and Gonda, T. J. (1992) Transactivation and transformation of MYB are negatively regulated by a leucine-zipper structure. *Proc. Natl. Acad. Sci. USA* 89: 3088–3092.

83. Nomura, T., Sakai, N., Sarai, A., Sudo, T., Kanei-Ishii, C., Ramsay, R. G., Favier, D., Gonda, T. J. and Ishii, S. (1993) Negative autoregulation of c-myb activity by homodimer formation through the leucine zipper. *J. Biol. Chem.* 268: 21914–21923.

84. Barletta, C., Pelicci, P. G., Kenyon, L. C., Smith, S. D., Dalla-Favera, R. (1987) Relationship between the c-myb locus and the 6q- chromosomal aberration in leukemias and lumphomas. *Science* 235: 1064–1067.

85. Gewirtz, A. M., Anfossi, G., Venturelli, D., Valpreda, S., Sims, R., Calabretta, B. (1989) G_1/S transition in normal human T-lymphocytes requires the nuclear protein encoded by c-myb. *Science* 245: 180–183.

86. Ku DH, Wen SC, Engelhard, A., Nicolaides NC, Lipson, K., E., Marino TA, and Calabretta, B. (1993) C-myb transactivates cdc2 expression bia MYB binding sites in the 5'-flanking region of the human cdc2 gene. *J. Biol. Chem.* 268: 2255–2259.

87. Weber, B. L., Westin, E. H. and Clarke, M. F. (1990) Differentiation of mouse erythroleukemia cells enhanced by alternatively spliced c-myb mRNA. *Science* 249: 1291–1293.

88. Todokoro, K., Watson, R. J., Higo, H., Amanuma, H., Kuramochi, S., Yanagisawa, H. and Ikawa, Y. (1988) Downregulation of c-myb gene expression is a prerequisite for erythropoietin-induced erythroid differentiation. *Proc. Natl. Acad. Sci. USA* 85: 8900-.

89. Clarke, M. F., Kukowska, L. J., Westin, E., Smith M, and Prochownik, E. V. (1988) Constitutive expression of a c-myb cDNA blocks Friend murine erythroleukemia cell differentiation. *Mol. Cell. Biol.* 8: 884-.

90. Rosson, D. and O'Brien, T. G. (1995) Constitutive c-myb expression in K562 cells inhibits induced erythroid differentiation but not tetradecanoyl phorbol acetate-induced megakaryocytic differentiation. *Mol. Cell. Biol.* 15: 772–9.

91. Ness, S. A., Marknell, A. and Graf, T. (1989) The v-myb oncogene product binds to and activates the promyelocyte-specific mim-1 gene. *Cell* 59: 1115-.

92. Nakayama, K., Yamamoto, R., Ishii, S. and Nakauchi, H. (1993) Binding of c-Myb to the core sequence of the CD4 promoter. *Int. Immunol.* 5: 817–824.

93. Reiss, K., Ferber, A., Travali, S., Porcu, P., Phillips, P. D. and Baserga, R. (1991) The protooncogene c-myb increases the expression of insulin-like growth factor 1 and insulin growth factor 1 receptor messenger RNAs by a transcriptional mechanism. *Cancer Res.* 51: 5997.

94. Melotti, P., Ku DH, and Calabretta, B. (1994) Regulation of the expression of the hematopoietic stem cell antigen CD34: Role of c-myb. *J. Exp. Med.*

95. Szczylik, C., Skorski, T., Ku, D. -H., Nicolaides, N. C., Wen, S. -C., Rudnicke, L., Bonati, A., Malaguarnera, L. and Calabretta, B. (1993) Regulation of proliferation and cytokine expression of bone-marrow fibroblasts: Role of c-myb. *J. Exp. Med.* 178: 997–1005.

96. Gewirtz, A. M. and, B. Calabretta (1988) A c-myb antisense oligodeoxynucleotide inhibits normal human hematopoiesis *in vitro*. *Science* 242: 1303–1306.

97. Mucenski, M. L., McLain, K., Kier, A. B., Swerdlow, S. H., Schreiner, M., Miller, T. A., Pietryga, D. W., Scott, W. J. and Potter, S. S. (1991) A functional c-myb gene is required for normal murine fetal hepatic hematopoiesis. *Cell* 65: 677-.
98. Caracciolo, D., Venturelli, D., Valiteri, M., Peschle, C., Gewirtz, A. M. and Calabretta, B. (1990) Stage-related proliferative activity determines c-myb functional requirements during normal human hematopoiesis. *J. Clin. Invest.* 85: 55–61.
99. Calabretta, B., Sims, R. B., Valiteri, M., Caracciolo, D., Szczylik, C., Venturelli, D., Ratajczak, M., Beran, M. and Gewirtz, A. M. (1991) Normal and leukemic hematopoietic cells manifest differential sernsitivity to inhibitory effects of c-myb antisense oligodeoxynucleotides: An *in vitro* study relevant to bone marrow purging. *Proc. Natl. Acad. Sci. USA* 88: 2351–2355.
100. Anfossi, G., Gewirtz, A. M. and Calabretta, B. (1989) An oligomer complementary to c-myb encoded mRNA inhibits proliferation of human myeloid leukemia cell lines. *Proc. Natl. Acad. Sci. USA* 86: 3379–3383.
101. Linnenback, A. J., Huebner, K., Prddy, E. P., Herlyn, M., Parmiter, A. H., Nowell, P. C. and Koprowski, H. (1988) Structural alteration in the MYB protooncogene and deletion within the gene encoding a-type protein kinase C in human melanoma cell lines. *Proc. Natl. Acad. Sci. USA* 85: 74–78.
102. Luger, S. M., Ratajczak, M. Z., Stadtmauer, E. S., Mangan, P., Magee, D., Silberstein, L., Mdelstein, M., Nowell, P. and Gewirtz, A. M. (1994) Autografting for chronic myelogenous leukemia (CML) with c-myb antisense purged bone marrow: a preliminary report. *Blood* 84 (Suppl 1): 151a.
103. Bedi, A., Zehnbauer, B. A., Collector, M. I. et al. (1993) BCR-ABL gene rearrangement and expression of primitive hematopoietic progenitors in chronic myeloid leukemia. *Blood* 81: 2898-.

2 Thymidine Kinases

R. G. Vile

2.1   Introduction

The principal objective of gene transfer therapy in cancer patients, as it is with chemo- or radiotherapy, is to kill the target (tumor) cells. This contrasts with the gene therapy of most other diseases in which the aim is both to preserve the target cells and to correct the underlying genetic defects which are responsible for the relevant pathology [1]. Although efforts directed at classical genetic correction of cancer cells, using tumor suppressor gene replacement or antisense strategies, have shown some encouraging results in animal models [2, 3], even these ultimately seek to promote cell death, for example by the induction of apoptosis [4].

Tumor cell killing, using gene transfer, can be achieved in two ways [5]: In the first, genes are transferred to the tumor cells to stimulate an effective anti-tumor immune response such that the majority of tumor cells are killed by the immune system and not directly by the transferred gene(s) (immunotherapy) [6]. In the second, conditionally cytotoxic genes are delivered and their expression can lead directly to cell death (cytotoxic gene therapy) [7].

Since surgery can often be used to remove primary tumor deposits, the induction of anti-tumor immunity using immunotherapy is most attractive for the treatment of patients with disseminated disease. However, there are also many clinical situations in which surgery is not possible and gene transfer therapy leading to localised tumor cell killing, would be of significant clinical benefit. It is for these situations that cytotoxic gene therapy has been proposed and has now reached clinical trials in the United States. Although a variety of genes are now being developed for such approaches [7], the gene for which most preclinical and clinical data is available is the Herpes Simplex Virus thymidine kinase gene (HSVtk). In this chapter we will discuss the properties of viral tk genes which have suggested that they may be useful in clinical treatment of cancer. In addition, we will discuss recent results which suggest that genes, such as HSVtk, may be useful not only in treatment of localised, inoperable tumor deposits but may also have a wider role in protocols aimed at enhancing the generation of anti-tumor immunity and/or as a safety feature of gene transfer vectors.

T. Blankenstein (ed.) Gene Therapy
©1999, Birkhäuser Verlag Basel

2.2 Viral Thymidine Kinase Genes

The HSVtk gene [8] was discovered as a novel thymidine kinase activity (thymidine 5' phosphotransferase) in cells infected with HSV type 1. HSVtk was subsequently shown to be an early viral gene which is turned off late in infection as a result of expression of other viral gene products acting at a transcriptional level. Although there are several different strains of HSV-1, the sequence of the tk gene is very similar between them and only minor differences exist. Other herpes viruses also possess thymidine kinase genes, including HSV-2 and Varicella-zoster virus (VZV), although these tks have been far less studied than that of HSV-1.

In the pharmaceutical effort to combat infection with Herpes viruses, the viral-specific tk activity was identified as a potential intervention point. Screening of a variety of nucleoside analogues of thymidine revealed compounds which are preferential substrates of the viral enzyme but which eventually form dead-end derivatives which block DNA synthesis in tk-expressing cells. Two such compounds, acyclovir (AC) [9] and ganciclovir (GC) [10] have proved to be very effective in treating her-

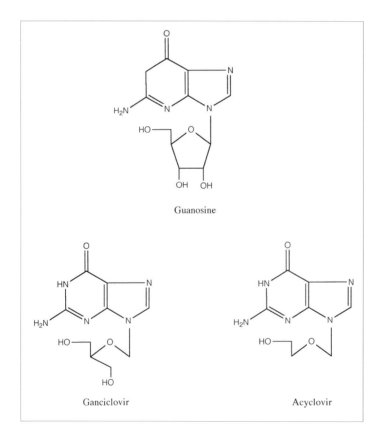

Guanosine

Ganciclovir

Acyclovir

Figure 2-1. The structure of Ganciclovir and Acyclovir which can be phosphorylated into triphosphate inhibitors of DNA synthesis].

pes and cytomegalovirus infections (Fig. 2-1). Other thymidine analogues have also been identified, including bromovinyldeoxyuridine [11] and the tk gene of VZV has been shown to mediate sensitivity to the nucleoside analogue 6-methoxypurine arabinonucleoside [12].

In infected cells, GC or AC are phosphorylated by the HSVtk gene into monophosphorylated compounds which form the substrates for cellular kinases. Following the conversion of the monophosphorylated form to triphosphate derivatives, these compete with the normal nucleotides which are incorporated into replicating DNA by cellular DNA polymerase. When a growing strand of DNA incorporates the AC or GC derivative, chain termination results leading to cell death (Fig. 2-2) [13]. Pharmaceutically, these drugs have proved to be very effective since they will only be converted to toxic triphosphate inhibitors in virally infected (tk-expressing) cells. It is also significant that the mechanism of action of these drugs depends upon cell division, since only actively dividing cells can incorporate the toxic derivatives into newly synthesised DNA.

Just how the cell dies following this inhibition of DNA synthesis has recently become the focus of considerable interest. Several studies have reported that cells dying under the influence of the HSVtk/GC system do so by the induction of apoptosis [14, 15] (so-called programmed cell death [16, 17]). However, other investiga-

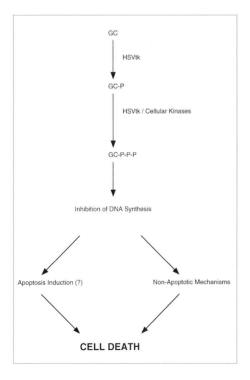

Figure 2-2. The metabolic mechanism of direct cell killing by the HSVtk-Ganciclovir (GC) system.

tors have not been able to detect the hallmarks of apoptotic cell death following HSVtk/GC-mediated cell killing [18] (Fig. 2-2). It seems likely that different cell lines might die by different mechanisms, perhaps in part dependent upon their genetic abilities to undergo apoptosis efficiently [19, 20]. The importance of the distinction between apoptotic and necrotic cell death induced by HSVtk is discussed in more detail below.

2.3   HSVtk in Cancer Gene Therapy

The cloning of the viral tk gene has made it possible to contemplate its introduction and expression in cell types of choice, both *in vitro* and *in vivo*. As such, HSVtk has been used as a positive selection marker to select cells deficient in cellular thymidine kinases [21]. However, the potential of HSVtk for cancer treatment became clear when it was shown that tumor cells, transduced with the gene *in vitro* became sensitive to treatment with GC [22]. Moreover, cells stably expressing HSVtk, implanted *in vivo* as actively growing tumors, could be completely eradicated by systemic treatment with the drug [23]. Therefore, Moolten and others suggested that transfer of HSVtk to tumor cells *in vivo* might be used to treat otherwise inoperable tumor deposits by subsequent treatment with GC or AC. In essence, accurate gene delivery followed by administration of the prodrug could be used to induce the transduced cells to commit 'suicide'. The attractions of this type of therapy are:
1) Its tumor selectivity – provided that the gene can be delivered both *efficiently* and *specifically*;
2) The ability to use a drug which is already in widespread clinical use and
3) The dependence of the treatment upon the need for cell division which should add to the tumor selectivity.

2.3.1   The Problems of Efficiency of Gene Delivery

In the original papers proposing the use of HSVtk in suicide gene therapy for cancer, Moolten suggested that the gene could be delivered to tumor cells using retroviral vectors [22, 23]. Subsequently, many groups have reported on different methods of delivering recombinant vectors to tumor cells *in vivo*, including injection of plasmid DNA, retroviral and adenoviral stocks either directly into tumors or systemically into the bloodstream supplying areas of tumor deposition [24, 25]. However, then, as now, the inability of current vectors, viral or otherwise, to deliver even a single gene copy to every tumor cell in even moderately sized tumors means that it would be highly improbable that sufficient genetic material could be delivered to ensure transduction of every cell in the targeted deposit [5]. Added to this is the reality that not every tumor cell is dividing at any given point so that the

HSVtk/GC combination would not target all the cells of the tumor population. Finally, despite efforts to target gene therapy vectors for tumor specific transduction and/or expression [26, 27], for anything other than completely localised injection of HSVtk vector directly into tumors with minimal leakage away from the tumor cells to surrounding cells, no delivery vehicle is currently sufficiently targetable so that it only transduces tumor cells and not any other cell type.

Therefore, the concept of HSVtk-mediated suicide gene therapy for cancer, as initially proposed, suffers from the Achilles Heel of nearly all gene therapy protocols – that of inefficient and poorly targeted delivery. Although *in vivo* delivery of HSVtk genes using plasmid [28], retroviral [29–31] and adenoviral vectors [32, 33] have all shown significant efficacy in reducing and/or eradicating experimantal tumors in animal models, the clinical problems in humans are likely to be of a greater order of magnitude [25, 34] and gene delivery remains the central hindrance to effective suicide gene therapy.

2.3.2 The Bystander Effect

The problem of inefficient levels of gene delivery are not unique to suicide gene therapy for cancer and plague efforts to achieve effective gene therapy of most diseases. However, HSVtk-mediated gene therapy has achieved a notable boost from findings which show that gene delivery probably does not have to achieve the 'one hit-one kill' levels of tumor cell transduction as might initially be thought. Several groups have shown, using HSVtk-transduced cells *in vitro*, that when populations of expressing cells are admixed with populations of parental, non-transduced cells, drug treatment can eradicate a much greater proportion of cells than are known to be expressing the gene [14, 22, 28, 30, 35]. In some cell lines, total cell eradication can be achieved in populations in which as few as 10% of the cells harbour the gene [14, 22, 35]. This phenomenon has been called the 'bystander effect' and has important implications for gene therapy of cancer (and other diseases as well), since it removes the burden of the need for delivery of the gene to 100% of the target cell population [5].

The mechanism of the bystander killing of non-expressing cells is currently unclear and may differ between cell lines. One explanation postulates that the toxic derivatives of HSVtk-mediated phosphorylation of GC are passed between directly adjoining cells, possibly via intercellular communication channels such as gap junctions, leading to death of these cells [35] (Fig. 2-3). In support of this 'metabolic cooperation' [36] mechanism is the finding that bystander killing is only effective in cultures in which the cells are in close juxtaposition rather than sparsely plated so that they do not physically communicate. In addition, passage of labelled GC analogue compounds can be followed between cells which show bystander killing and typically pass only a few cell layers away from the expressing cells [35, 37]. Finally, several groups have shown that cell lines, which normally show only a low efficiency of bystander killing *in vitro*, can be converted to a high efficiency bystander phenotype by the transfection of the cDNA encoding the connexin protein, which is a central component of the cellular gap junction [38]. Another hypothesis has proposed that killing is achieved via uptake of apoptotic vesicles released from dying, HSVtk-expressing cells [14] (Fig. 2-3). These vesicles may contain active drug, toxic metabolites or even apoptosis-inducing signals which induce cell death in the recipient cell. These two explanations of the bystander effect do not necessarily have to be mutually exclusive and it seems probable that the same cell line might employ both under suitable conditions. Similarly, cells lacking functional gap junctions or with a low intrinsic ability to undergo apoptosis may use only one of the two available alternatives. It may not be possible to make general statements about the mechanisms involved in the local bystander effect and detailed studies on each separate cell line and *in vivo* model are likely to be required.

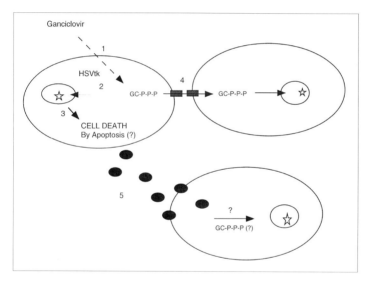

Figure 2-3. Bystander killing of non transduced cells by HSVtk-expressing tumor cells. In the presence of ganciclovir, a cell expressing HSVtk generates a triphosphate derivative (GC-P-P-P) (1) which is toxic to DNA synthesis (2) and leads to death of the cell (3), possibly by the induction of apoptosis. Cells closely juxtaposed to the HSVtk+ cell can be killed either by the transfer of any of the GC-P, GC-P-P or GC-P-P-P metabolites via cell-cell communication channels such as gap junctions (4) or by the uptake of apoptotic vesicles released from the dying cell (5). These vesicles might contain the toxic metabolites themselves or may transmit apoptotic signals directly to the recipient cells.

Whatever the mechanism, the ability of expressing cells to lead to the killing of non-expressing cells may well explain, at least in part, the success of the *in vivo* delivery experiments which have successfully eradicated growing tumors despite the improbability of having delivered the HSVtk gene to every tumor cell [39].

2.4 Clinical Trials of HSVtk-mediated Suicide Gene Therapy

On the basis of these findings and the currently available systems of delivering genes to tumors *in vivo*, suicide gene therapy appears to be well suited to treatment of clinical situations in which an accessible tumor poses a threat to the life of a patient but cannot be surgically removed. The most obvious candidate situations for such a therapy are brain tumors (gliomas or metastases to the brain from tumors of other histological types), mesothelioma [40] or, for instance, metastases in a localised organ (such as the liver) [12].

Several groups have used the glioma model in animal systems to demonstrate the potential of HSVtk gene transfer therapy [30–32, 41]. Predominant amongst these were the studies of Culver et al. who showed that delivery of retroviruses encoding the HSVtk gene to growing gliomas in rats could be used to eradicate even large tumors [30, 41]. This group used stereotactic injection of (allogeneic) retroviral producer cells, directly into the tumor deposit followed by administration of ganciclovir. The use of producer cells rather than viral supernatant was designed to increase the efficiency of viral transfer to tumor cells since the producer cells provide a continual source of virus in close aposition to the tumor cells for several days until they are rejected by the immune system. Retroviral delivery is cited as an extra safety measure since retroviral vectors will only productively infect actively dividing cells [34] (i.e. the tumor cells and the expanding endothelial cells of the tumor vasculature) but should not infect the quiescent surrounding neural tissue [41, 42].

Results of these, and similar studies in animal model systems have been extremely encouraging (with some exceptions [43]) and extensive tumor eradication has been possible. Tumor cell killing has been attributed to gene transfer and the accompanying bystander killing effects [44]. In addition, other mechanisms of tumor clearance may have been important, including the possible destruction of the (replicating) tumor vasculature [41] and immune reactivity induced by the presence of allogeneic producer cells as well as other possible immune effects (see below).

The results of preclincal models have been sufficiently impressive that clinical trials have been initiated and are now well progressed in the United States for patients with advanced gliomas and brain metastases [45]. Although the results have not been completely reported, it is clear that the results of animal models have not translated into the human situation as well as might have been hoped [46]. Although tumor volumes have been decreased by the implantation of producer cells, *in vivo* transduction efficiencies have been low (<0.17%) and it is not clear whether the effects which have been seen are due to gene transfer or as an adjunct to inflammation resulting from the implantation of murine cells in the tumors. However, the transition from animal to human trials would not be expected to be trouble free and several mitigating factors make it likely that improvements will still be possible. For example, human tumors are much larger than the glioma counterparts which were treated so successfully in rats and the rate at which the human tumor cells divide *in vivo* is much slower than that of the experimental cells. Since both the HSVtk/GC combination and retroviral infection depend upon the presence of cycling cells, such factors will produce significant reductions in efficacy.

Therefore, modifications to the clinical trials may offer improved prospects. These have involved initially, the use of higher titre retroviral producer cells and multiple injections into the same lesion. For the future, it may be that the higher titres of adenoviral transfer systems will allow transduction efficencies to be increased leading to better proportions of tumor cell killing [24, 32].

The disappointing results of these clinical trials in treatment of highly localised, accessible tumors does not bode well for more problematic clinical situations, such as the treatment of dispersed tumor deposits, even in a localised region, such as colorectal metastases in the liver [12]. Once again, the major obstacle to effective suicide gene therapy in such situations remains the efficiency and selectivity of the delivery vehicles. Thus, although systemic delivery of HSVtk [29, 47] and other genes [48] in some animal models have produced notable effects on tumor burdens, the transition to the human situation may well be even more difficult than in the case of the brain tumor trials; for instance, the retroviral vectors currently used for *in vivo* delivery studies in animal models are rapidly inactivated by human complement and so will have extremely short half-lives when injected into patients [34].

A second series of patient trials based on HSVtk-mediated killing, but which does not depend upon *in vivo* gene delivery using viral vectors, relies even more heavily upon the local bystander killing effect. The studies of Freeman and colleagues have demonstrated that when HSVtk-modified ovarian tumor cells are injected intraperitoneally, they appear to localise to sites of pre-existing tumor growth [14, 49, 50]. It is hoped that the modified cells will traffic to tumor deposits so that, when ganciclovir is administered, sufficient tumor cells will be located within range of the modified cells so that the bystander effect from their killing will generate clinically beneficial effects. These trials have the advantage of not relying upon direct gene transfer to the tumor cells but suffer from the drawback that the bystander effect may not have sufficient range to eradicate all the tumor cells *in situ*, especially if the delivery of the HSVTK-modified cells is not uniformly effective. However, the effects of this sort of therapy may also be boosted by a contribution from immune stimulation against the tumor cells as a result of an inflammatory response against the allogeneic modified HSVtk+ cells [50, 51] and also, possibly, as a response to dying tumor cells *in vivo* (see below).

2.5 HSVtk-mediated Anti Tumor Immunopotentiation in the Treatment of Metastatic Disease?

The treatment modalities envisaged for HSVtk so far have relied upon the ability to deliver the gene into target cells, or to very close neighbours. In contrast, the treatment of untransduced cells by immune clearance, at distant sites to that where gene transfer occurs, has focused upon expression of immunomodulatory genes such as cytokines and co-stimulatory molecules in tumor cells, transduced either *ex vivo* [6] or directly *in vivo* [52]. Results from several groups are now emerging and suggest that delivery of HSVtk to tumor cells may have a role in treatment of disseminated disease which was previously underestimated. Different groups using HSVtk gene transfer therapy have shown that the *in vivo* effects of HSVtk/GC are severely diminished in animals which lack an intact immune system, particularly T cells [29, 50, 53, 54]. In one report, the impressive effects of *in vivo* delivery of retroviruses encoding HSVtk were largely abolished in nude mice compared with immunocompetent animals [29]. Therefore, it seems that tumor cell killing by HSVtk gene expression may be attributable to several separate, but probably interdependent, effects (Fig. 2-4), the principal of which are:

1) Direct killing of cells transduced with, and expressing, HSVtk;
2) Local bystander killing of neighbouring, but non-transduced, cells and
3) (T cell) immune-mediated killing of untransduced tumor cells, both locally and at a distance from the site of gene transfer, activated by *in vivo* killing of tumor cells.

Hence, the HSVtk/GC system has two potential levels at which the efficiency of the initial delivery step can be amplified – a feature which would support the development of such treatments for the future in the current absence of truly targeted and efficient delivery vehicles (Fig. 2-4).

The mechanisms involved in stimulation of anti-tumor immunity by the *in vivo* killing of tumor cells by HSVtk/GC are currently unclear but are likely to involve both specific and non-specific mechanisms of tumor immunity (Fig. 2-4) [55]:

a) Non-specific inflammatory reactions may be responsible for the additional clearance of local tumor deposits in response to, for example, the implantation of allogeneic retroviral producer cells [30, 41] or HSVtk-modified cells [49–51] used to deliver the HSVtk gene to tumor deposits *in vivo*. In addition, in systems in which no evidence of apoptotic cell death can be detected [18], or even in those in which some death occurs by apoptosis, any necrosis at the site of the dying tumor cells might be expected to induce a brisk inflammatory reaction which would attract immune effector cells to the tumor site. This reaction would contribute to the creation of an immunostimulatory environment for non-specific lysis of tumor cells [50]. Apoptosis, on the other hand, is not associated with inflammation *in vivo* [16] and so the precise ways in which HSVtk mediated cell killing occurs *in vivo* may be important in improving the generation of the im-

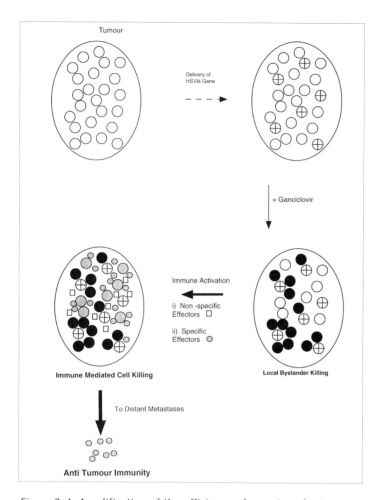

Figure 2-4. Amplification of the efficiency of gene transfer in HSVtk-mediated killing by both the local, and the immune, by-stander effects. HSVtk gene transfer to tumor cells (open circles) in vivo is usually a relatively inefficient process leading to only a fraction of the tumor cells being trasnduced (crossed circles). Administration of ganciclovir leads to direct cell killing of the transduced cells and killing of close neighbours by the local by-stander effect (black circles). Cell killing in vivo leads to an activation of an immune response at the tumor site which will have both non-specific (local inflammatory response) and, hopefully, specific anti-tumor components,both of which contribute further to local killing of tumor cells. Finally, any immune effector cells with specific anti-tumor activity (e.g. T cells activated against shed tumor antigens) can leave the primary tumor deposit and travel through the body to attack metastatic deposits.

munopotentiating effects of this system.

b) However, tumor specific immunity has also been reported following tumor cell killing by suicide gene systems such as HSVtk [29, 50, 53, 54, 56, 57]. It seems likely that this phenomenon may be related to the fact that, as the tumor cells die, they increase the availability to the immune system of putative tumor antigens [58–63] (Fig. 2-5). It is well known that tumor cells evade the immune system by a variety of mechanisms, including deficiencies in their ability to present tumor antigens efficiently [55] – perhaps by the lack of MHC or co-stimulatory [64] molecules on the tumor cell surface and/or by the lack of efficient antigen

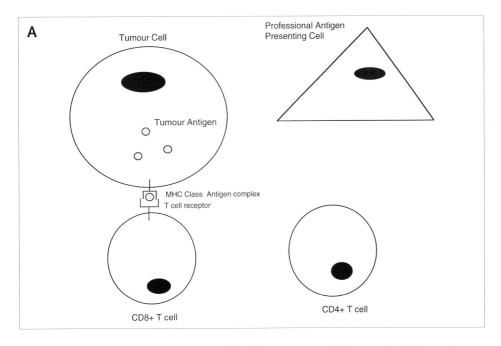

Figure 2-5. A possible mechanism to explain the generation of improved anti-tumor immunity following in vivo tumor cell killing by HSVtk. A. Tumor cells (top left) possess tumor antigens but these are not sufficiently available to the professional antigen presenting cells (APC), such as dendritic cells or macrophages (top right). Therefore, CD4+ T cell (bottom right) help is not generated and anti-tumor CD8+ cytotoxic T cells are not activated (bottom left). B. In the presence of HSVtk/GC the tumor cell is killed and releases high concentrations of tumor cell debris and antigens locally at the tumor site. This allows putative rejection antigens to be taken up and efficiently presented by APCs to CD4+ T cells. Activation of tumor specific helper cells leads to cytokine secretion and activation of tumor specific CD8+ CTL. C. This in vivo activation of anti tumor immunity might be enhanced by co-delivery of HSVtk (to kill tumor cells) along with appropriate cytokines such as GM-CSF, for example, to augment the recruitment of APCs to the tumor site to present the tumor antigens or IL-2 to promote the activation of CD8+ CTL.

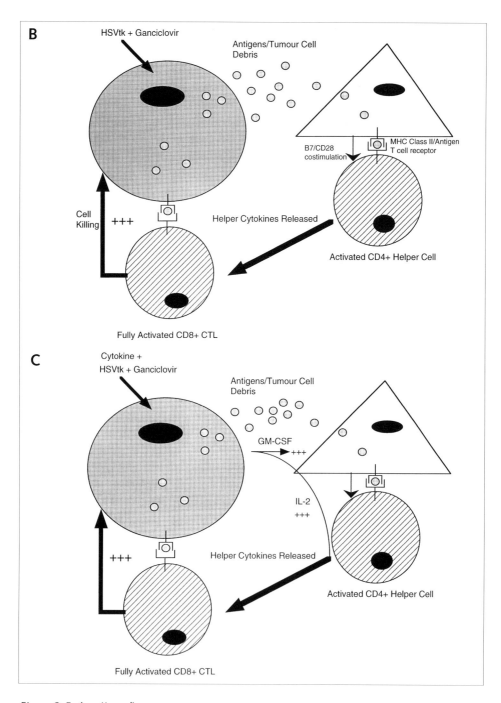

Figure 2-5. (continued)

presenting cell-mediated generation of CD4+ T cell help [6, 65] (Fig. 2-5a). It may be that HSVtk-mediated cell killing liberates high concentrations of tumor cell debris and antigens to appropriate cells of the immune system – dendritic cells and macrophages – which are highly specialised for the efficient presentation of shed antigens [66] (Fig. 2-5b). Without this mechanism of localised release, those antigens may remain immunologically hidden. In support of this theory are the findings of groups, working on other suicide genes such as cytosine deaminase, who have shown similar immunopotentiating effects of suicide gene killing, indicating that these effect are probably a feature of specific types of cell killing *in vivo* rather than being unique to the HSVtk/GC system. However, the outgrowth of cells that are resistant to GC treatment in treated tumor deposits (possibly due to loss, or low expression of the gene) [23, 67–69] shows that immune activation and clearance of untransduced cells is not necessarily all pervasive and different tumor types are likely to have different inherent abilities to activate anti-tumor immunity when killed *in vivo*.

The relative contribution of the local bystander killing (Fig. 2-3) and these immune mediated killing effects (Fig. 2-4) to clearance of tumor cells *in vivo* is also unclear. For example, Velu and colleagues have shown that the immune mediated component is probably almost entirely responsible for *in vivo* therapy of a glioma model, when implanted subcutaneously, even though the glioma cells in culture show pronounced local bystander killing [70]. Once again, it seems likely that different experimental systems will rely, to differing extents, upon local and immune-mediated cell killing, and generalised predictions should be avoided.

These results are exciting for the future development of suicide gene therapies for cancer. It would be extremely attractive to be able to treat local tumors and simultaneously activate anti-tumor immunity, which would be effective against occult metastases posing a long term threat to the life of the patient. In this respect, we, and others, are developing vectors for the co-delivery of HSVtk along with other immunostimulatory genes, which can be expressed at the same time as the cells are killed and can augment the generation of anti-tumor immune responses [47, 71, 72] (Fig. 2-5c).

2.6 Other Roles for tk

The ability of tk to be transferred to specific target cells and then conditionally to kill those cells, on administration of GC, makes it an attractive gene for a variety of uses other than those described above. For example, HSVtk has been proposed as a means to kill virally infected cells (such as HIV-infected T cells) [73] and to kill inappropriately dividing cells in diseases such as restenosis [74, 75].

In addition, it has been proposed that HSVtk could be incorporated into genetic vectors, for both cancer and other therapies, as a 'failsafe' feature, in the event that a genetic therapy were to require termination in a treated patient. Hence, vectors designed to express therapeutic genes, other than sucide genes such as immunomodulatory, corrective or chemoprotective genes, can now be constructed to incorporate the HSVtk gene. If non target cells were to become inadvertently transduced with the vector, or if the transduced cells were begin to behave in an inappropriate manner (for example, become transformed), treatment of the patient with GC would rapidly eradicate the rogue cells and terminate the treatment before long term damage could be done [7, 76–78].

Finally, it has also been proposed that genes such as HSVtk could be inserted into normal cell progenitor populations, such that when the tumors do develop, treatment with GC would be effective in curing the patient [7, 79]. Such 'preemptive' insertion of suicide genes would be particularly suited to individuals with a high risk of developing cancer later in life, such as those identified by genetic screening to carry defective tumor suppressor genes, such as p53. Enough cells should be left untransduced that the tissue can be effectively regenerated following ablation of the aberrant population or a mosaicism could be created by multiple transductions such that almost all the cells in a given tissue are potentially sensitised to at least one drug [7]. Although this is an attractive theory for a cancer-protective form of gene therapy, it still remains somewhat futuristic, owing in large part to the ethical problems associated with the risks of gene delivery to healthy individuals who have not yet developed disease [78] and to practical problems associated with achieving efficient and completely targeted delivery/expression of the protective gene to the appropriate target cell population.

2.7 Future Developments for Thymidine Kinase Therapies in Cancer

Gene transfer therapies using HSVtk have proved remarkably effective in certain situations in preclinical models of cancer. They have, however, as yet, failed to make an effective transition to the clinic. The reasons for this have been discussed above and there are various improvements which will be incorporated in the protocols of the future.

Despite the multiple levels at which HSVtk action might potentially act to help to clear tumor cells (Fig. 2-6), the overriding limitation remains that of inefficient (and imprecise) delivery of the gene to tumor cells. Even if a single copy of HSVtk could be expressed in every tumor cell, whilst being transcriptionally silent in any other cells in the body, it is still possible that the cancer may not be curable due to low levels of expression in some cells [68]. The reality is that no delivery system can yet provide the specificity or the efficiency required [5]. Therefore, improvements have to be made to the existing systems to improve:
1). The levels of gene transfer to the tumor cells and
2). The efficacy of expression following transfer.

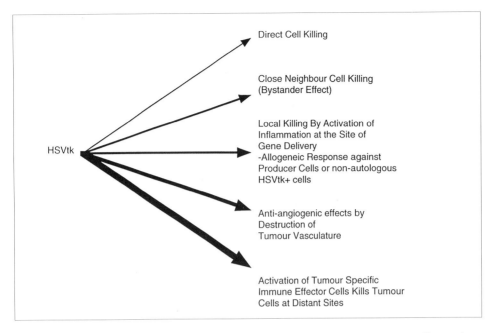

Direct Cell Killing

Close Neighbour Cell Killing
(Bystander Effect)

Local Killing By Activation of
Inflammation at the Site of
Gene Delivery
-Allogeneic Response against
Producer Cells or non-autologous
HSVtk+ cells

Anti-angiogenic effects by
Destruction of
Tumour Vasculature

Activation of Tumour Specific
Immune Effector Cells Kills Tumour
Cells at Distant Sites

HSVtk

Figure 2-6. Summary of the multiple elements of HSVtk-mediated anti-tumor effects. The potential relative importance, in terms of numbers of tumor cells affected by each mechanism, is indicated; however, different tumor systems are likely to be affected to differing extents by each process and the rank order shown is predictive rather than absolute.

Obvious ways of increasing the levels of gene transfer are new improved vectors with higher titres. Adenoviral vectors have higher titres than retroviruses and may well achieve better levels of transduction in the patient. Newer vectors, perhaps even with restricted capabilities to replicate within tumors [34, 80], would also improve this stage. Regarding systemic delivery of the gene to disseminated tumor deposits, the development of tumor-selective vectors of high enough titre is well underway but remains a prospect for the future.

Improvements in efficacy of expression of the tk gene may be more possible in the short term future. Stronger promoters, which stay active in tumor cells for longer, might increase the levels of tk protein available for reaction with GC. Alternatively, mutation of the HSVtk gene has already yielded so-called 'super-suicide' genes which have a higher enzymatic activity and so can generate more toxic derivatives from the same amount of GC than can the standard tk gene [81, 82]. Similarly, cloning of different tk genes from other herpes viruses may yield more potent enzymes. On the other hand, improved nucleoside analogues might be developed which are either better substrates for tk and/or lead to enhanced killing of expressing cells and/or generate toxic metabolites which can be passed more efficiently to neighbouring cells to increase the potency of the bystander killing effect.

As mentioned above, development of new vectors which co-express immunostimulatory genes may improve the efficacy of tk expression by recruiting the appropriate immune cells to augment cytotoxic cell killing with immune-mediated cell killing. This may have benefits for killing tumor cells both at the local tumor site but also, hopefully, also at distant sites of disease as well. However, whether or not improvements can be made in the potency of the gene or the prodrug, or in the nature of genes co-expressed with tk, the most significant boost for the prospects of tk-mediated suicide gene therapy will surely come with improvements in the targeted efficiency of the gene delivery systems of the future.

References

1. Anderson, W. F. (1992) *Hum. Gene Ther. Science* 256: 808–813.
2. Fujiwara, T., Grimm, E. A. and Roth, J. A. (1994) Gene Therapeutics and Gene Therapy for Cancer. *Curr. Opin. Oncol.* 6: 96–105.
3. Mercola, D. and Cohen, J. S. (1995) Antisense approaches to cancer gene therapy. *Cancer Gene Therapy* 2: 47–59.
4. Liu, T. -J. et al. (1995) Apoptosis Induction Mediated by Wild-Type p53 Adenoviral Gene Transfer in Squamous Cell Carcinoma of the Head and Neck[1]. *Cancer Res.* 55: 3117–3122.
5. Vile, R. G. and Russell, S. J. (1994) Gene transfer technologies for the gene therapy of cancer. *Gene Therapy* 1: 88–98.
6. Pardoll, D. M. (1995) Paracrine Cytokine Adjuvants in Cancer Immunotherapy. *Annu. Rev. Immunol.* 13: 399–415.
7. Moolten, F. L. (1994) Drug sensitivity ("suicide") genes for selective cancer chemotherapy. *Cancer Gene Ther.* 1: 279–287.
8. Wagner, M. J., Sharp, J. A. and Summers, W. C. (1981) Nucleotide sequence of the thymidine kinase gene of herpes simplex virus type 1. *Proc. Natl. Acad. Sci. USA* 78: 1441–1445.
9. Elion, G. B. The biochemistry and mechanism of action of acyclovir. *Antimicrob. Chemother.* 12: 9–17 (1983)
10. Nishiyama, Y. and Rapp, F. (1979) Anticellular efects of 9-(2-hydroxyethoxymethl) gguanine against herpes simplex virus-transformed cells. *Gen. Virol.* 45: 227–230.
11. Balzarini, J., De Clercq, E., Verbruggen, A., Ayusawa, D. and Seno, T. (1985) Highly selective cytostatic activity of (E)-5-(2-bromovinyl)-2'-deoxxyuridine derivatives for murine mammary carcinoma (FM3A) cells transformed with the herpes simplex virus type 1 thymidine kinase gene. *Mol. Pharmacol.* 28: 581–587.
12. Huber, B. E., Richards, C. A. and Krenitsky, T. A. (1991) Retroviral-mediated *gene therapy* for the treatment of hepatocellular carcinoma: An innovative approach for cancer therapy. *Proc. Natl. Acad. Sci. USA* 88: 8039–8043.
13. Reid, R., Eng-Chung, M., Eng-Shang, H. and Topal, M. D. (1988) Insertion and extension of acyclic, dideoxy, and ara nucleotides by herpesvirdae, human alpha and human beta polymerases. *J. Biol. Chem.* 263: 3898–3904.
14. Freeman, S. M. et al. (1993) The "bystander effect": tumor regression when a fraction of

the tumor mass is genetically modified. *Cancer Res.* 53: 5274–5283.

15. Samejima, Y. and Meruelo, D. (1995) 'Bystander killing' induces apoptosis and is inhibited by forskolin. *Gene Ther.* 2: 50–58.
16. Wyllie, A. H. (1993) Apoptosis (The 1992 Frank Rose Memorial Lecture). *Brit. J. Cancer* 67: 205–208.
17. Stewart, B. W. (1994) Mechanisms of Apoptosis: Integration of Genetic, Biochemical, and Cellular Indicators. *J. Nat. Cancer Inst.* 86: 1287–1293.
18. Kaneko, Y. and A., T. (1995) Gene therapy of hepatoma: bystander effects and non-apoptotic cell death induced by thymidine kinase and ganciclovir. *Cancer Lett.* 96: 105–110.
19. Lowe, S. W., Schmitt, E. M., Smith, S. W., Osborne, B. A. and Jacks, T. (1993) p53 is required for radiation-induced apoptosis in mouse thymocytes. *Nature* 362: 847–849.
20. Fisher, D. E. (1994) Apoptosis in Cancer Therapy: Crossing the Threshold. *Cell* 78: 539–542.
21. Wigler, E. A. (1977) Transfer of purified herpes virus thymidine kinase gene to cultured mouse cells. *Cell* 11: 223–232.
22. Moolten, F. L. (1986) Tumor chemosensitivity conferred by inserted Herpes thymidine kinase genes: Paradigm for a prospective cancer control strategy. *Cancer Res.* 46, 5276.
23. Moolten, F. L. and Wells, J. M. (1990) Curability of tumors bearing Herpes Thymidine Kinase genes transferred by retroviral vectors. *J. Nat. Cancer Inst.* 82: 297–300.
24. Jolly, D. (1994) Viral vector systems for gene therapy. *Cancer Gene Ther.* 1: 51–64.
25. Crystal, R. G. (1995) Transfer of genes to humans: early lessons and obstacles to success. *Science* 270: 404–410.
26. Miller, N. and Vile, R. G. (1995) Targeted vectors for gene therapy. *FASEB J.* 9: 190–199.
27. Vile, R. G. (1994) Tumor specific gene expression. *Seminars in Cancer Biology* 5: 429–436.
28. Vile, R. G. and Hart, I. R. (1993) Use of tissue-specific expression of the Herpes Simplex Virus thymidine kinase gene to inhibit growth of established murine melanomas following direct intratumoral injection of DNA. *Cancer Res.* 53: 3860–3864.
29. Vile, R. G., Nelson, J. A., Castleden, S. C., Chong, H. and Hart, I. R. (1994) Systemic gene therapy of murine melanoma using tissue specific expression of the HSVtk gene involves an immune component. *Cancer Res.* 54: 6228–6234.
30. Culver, K. W. et al. (1992) *In vivo* gene transfer with retroviral vector-producer cells for treatment of experimental brain tumors. *Science* 256: 1550–1552.
31. Short, M. P. et al. (1990) Gene delivery to glioma cells in rat brain by grafting of a retrovirus packaging cell line. *J. Neurosci. Res.* 27: 427–439.
32. Chen, S. -H., Shine, H. D., Goodman, J. C., Grosman, R. G. (1994) Gene therapy for brain tumors: Regression of experimental gliomas by adenovirus-mediated gene transfer *in vivo*. *Proc. Natl. Acad. Sci. USA* 91: 3054–3057.
33. Roy Smythe, W. et al. (1994) Use of recombinant adenovirus to transfer the Herpes Simplex Virus thymidine kinase (HSVtk) gene to thoracic neoplasms: an effective *in vitro* drug sensitisation system. *Cancer Res.* 54: 2055–2059.
34. Vile, R. G. and Russell, S. J. (1995) Retroviruses as vectors. *Brit. Med. Bull.* 51: 12–30.
35. Bi, W. L., Parysek, L. M., Warnick, R. and Stambrook, P. J. (1993) *In vitro* evidence that metabolic cooperation is responsible for the bystander effect observed with HSV tk retroviral gene therapy. *Hum. Gene Ther.* 4: 725–731.
36. Hooper, M. L. and Subak-Sharpe, J. H. (1981) Metabolic cooperation between cells. *Int.*

Rev. Cytol. 69: 45–104.

37. Pitts, J. D. (1994) Cancer gene therapy: a bystander effect using the gap junctional pathway. *Molec. Carcinogen.* 11: 127–130.

38. Colombo, M. B. et al. (1995) Retroviral transduction of the connexin 43 gene increases the efficacy of 'suicide' gene transfer in malignant gliomas. *Gene Ther.* 2 S1, 80.

39. Marini III, F. C., Nelson, J. A. and Lapeyre, J. -N. (1995) Assessment of bystander effect potency produced by intratumoral implantation of HSVtk-expressing cells using surrogate marker secretion to monitor tumor growth kinetics. *Gene Ther.* 2: 655–659.

40. Smythe, W. R., Hwang, H. C. and Amin, K. J. et al. (1994) Use of recombinant adenovirus to transfer the herpes simplex virus thymidine kinase (HSV-tk) gene to thoracic neoplasms: an effective *in vitro* drug sensitization system. *Cancer Res.* 54: 2055–2059.

41. Ram, Z., Culver, K. W., Walbridge, S., Blaese, R. M. and Oldfield, E. H. (1993) *In situ* retroviral mediated gene transfer for the treatment of brain tumors in rats. *Cancer Res.* 53: 83–88.

42. Ram, Z. et al. (1993) Toxicity studies of retroviral-mediated gene transfer for the treatment of brain tumors. *J. Neurosurg.* 79: 400–407.

43. Tapscott, S. J., Miller, A. D., Olson, J. M. and Berger, M. S. (1994) Gene therapy of rat 9L gliosarcoma tumors by transduction with slectable genes does not require drug selection. *Proc. Natl. Acad. Sci. USA* 91: 8185–8189.

44. Culver, K. W. and Blaese, R. M. (1994) Gene therapy for cancer. *Trends Genet.* 10: 174–178.

45. Oldfield, E. H. et al. (1993) Clinical Protocol: Gene therapy for the treatment of brain tumors using intra-tumoral transduction with the thymidine kinase gene and intravenous ganciclovir. *Hum. Gene Ther.* 4: 39–69.

46. Ram, Z. e. a. (1995) Summary of results and conclusions of the *Gene Ther.* apy of malignant brain tumors: clincal study. *J. Neurosurg* 82, 343A.

47. Chen, S. -H. et al. (1995) Combination gene therapy for liver metastasis of colon carcinoma *in vivo. Proc. Natl. Acad. Sci. USA* 92: 2577–2581.

48. Hurford, J. R. K., Dranoff, G., Mulligan, R. C. and Tepper, R. I. (1995) Gene therapy of metastatic cancer by *in vivo* retroviral gene targeting. *Nat. Genet.* 10: 430–435.

49. Freeman, S. M., McCune, C., Angel, C., Abraham, G. N. and Abboud, C. N. (1992) Treatment of ovarian cancer using HSV-TK gene modified vaccine-regulatory issues. *Hum. Gene Ther.* 3: 342–349.

50. Whartenby, K. A., Abboud, C. N., Marrogi, A. J., Ramesh, R. and Freeman, S. M. The biology of cancer gene therapy. *Lab. Investig.* 72, 131–145 (1995)

51. Freeman, S. M., Ramesh, R., Marrogi, A. J., Jensen, A. and Abboud, C. N. (1994) *In vivo* studies on the mechanism of the bystander effect. *Cancer Gene Ther.* 1, 326.

52. Nabel, G. J. et al. (1992) Clinical Protocol: Immunotherapy of malignancy by *in vivo* gene transfer into tumors. *Hum. Gene Ther.* 3: 399–410.

53. Caruso, M. et al. (1993) Regression of established macroscopic liver metastases after *in situ* transduction of a suicide gene. *Proc. Natl. Acad. Sci. USA* 90: 7024–7028.

54. Barba, D., Hardin, J., Sadelain, M. and Gage, F. H. (1994) Development of anti tumor immunity following thymidine kinase-mediated killing of experimental brain tumors. *Proc. Natl. Acad. Sci. USA* 91: 4348–4352.

55. Vile, R. G., Chong, H. C. and Dorudi, S. (1996) Immunosurveillance of cancer: specific and non-specific mechanisms. *In*: Dalgleish, A. G., Browning, M. J. (eds), *Tumor Immunology.* Cambridge University Press, pp. 7–38; *in press.*

56. Mullen, C. A., Coale, M. M., Lowe, R. and Blaese, R. M. (1994) Tumors expressing the cytosine deaminase suicide gene can be eliminated *in vivo* with 5-fluorocytosine and induce protective immunity to wild type tumor. *Cancer Res.* 54: 1503–1506.

57. Consalvo, M. et al. (1995) 5-fluorocytosine-induced eradication of murine adenocarcinomas engineered to express the cytosine deaminase suicide gene requires host immune competence and leaves an efficient memory. *J. Immunol.* 154: 5302–5312.

58. Cavallo, F. et al. (1992) Role of neutrophils and CD4+ T lymphocytes in the primary and memory response to non immunogenic murine mammary adenocarcinoma made immunogenic by IL-2 gene transfer. *J. Immunol.* 149: 3627–3635.

59. Colombo, M. P. and Forni, G. (1994) Cytokine gene transfer in tumor inhibition and tumor therapy: where are we now? *Immunol. Today* 15: 48–51.

60. Dranoff, G. et al. (1993) Vaccination with irradiated tumor cells engineered to secrete murine granulocyte macrophage colony stimulating factor stimulates potent, specific, and long lasting anti-tumor immunity. *Proc. Natl. Acad. Sci. USA* 90: 3539–3543.

61. Huang, A. Y. C. et al. (1994) Role of bone marrow derived cells in presenting MHC Class I-restricted tumor antigens. *Science* 264: 961–965.

62. Raychaudhuri, S. and Morrow, W. J. W. (1993) Can soluble antigens induce CD8+ cytotoxic T cell responses? A paradox revisited. *Immunol. Today* 14: 344–348.

63. Udono, H., Levey, D. L. and Srivastava, P. K. (1994) Cellular requirements for tumor-specific immunity elicited by heat shock proteins: tumor rejection antigen gp96 primes CD8+ T cells *in vivo*. *Proc. Natl. Acad. Sci. USA* 91: 3077–3081.

64. Hellstrom, K. E., Hellstrom, I. and Chen, L. (1995) Can co-stimulated tumor immunity be therapeutically efficacious? *Immunol. Rev.* 145: 123–145.

65. Pardoll, D. M. Cancer vaccines. *Immunol. Today* 14, 310–316. (1993)

66. Grabbe, S., Beissert, S., Schwarz, T. and Granstein, R. D. (1995) Dendritic cells as initiators of tumor immune responses: a possible strategy for tumor immunotherapy? *Immunol. Today* 16: 117–121.

67. Golumbek, P. T. et al. (1992) Herpes simplex-1 virus thyumidine kinase gene is unable to completely eliminate live, nonimmunogenic tumor cell vaccines. *J. Immunother.* 12: 224–230.

68. Moolten, F., Wells, J. M. and P. J., M. (1992) Multiple transuction as a means of preserving ganciclovir chemosensitivity in sarcoma cells carrying retrovirally transduced herpes thymidine kinase genes. *Cancer Lett.* 64: 257–263.

69. Moolten, F. L., Wells, J. M., Heyman, R. A. and Evans, R. M. (1990) Lymphoma regression induced by ganciclovir in mice bearing a herpes thymidine kinase transgene. *Hum. Gene Ther.* 1: 125–134.

70. Cool, V. et al. (1995) The bystander effect associated to HSVtk gene transfer is absent in intracerebral tumors and seems to be mainly immune mediated in subcutaneous tumors. *Gene Ther.* 2 S1, 84.

71. Chong, H. c., Castleden, S., Diaz, R. M., Hart, I. R. and Vile, R. G. (1995) Combination Gene Therapies for the immunotherapy of cancer. *Gene Ther.* 2 S1, 80.

72. Ram, Z. et al. (1994) *In vivo* transfer of the human interleukin-2 gene: negative tumoricidal results in experimental brain tumors. *J. Neurosurg.* 80: 535–540.

73. Dropulic, B. and Jeang, K. T. (1994) Gene therapy for human immunodeficiency virus infection: genetic antiviral strategies and targets for intervention. *Hum. Gene Ther.* 5: 927–939.

74. Guzman, R. J., Hirschowitz, E. A., Brody, S. L., Crystal, R. G., Epstein, S. E. and Finkel, T. (1994) *In vivo* suppression of injury-induced vascular smooth muscle cell accumulation using adenovirus-mediated transfer of the herpes simplex virus thymidine kinase gene. *Proc. Natl. Acad. Sci. USA* 91: 10732–10736.

75. Ohno, T., Gordon, D., San H., Pompili, V. J., Imperiale, M. J., Nabel, G. J. and Nabel, E. G. (1994) Gene Therapy for Vascular Smooth Muscle Cell proliferation After Arterial Injury. *Science* 265: 781–784.

76. Tiberghien, P., Reynolds, C. W. and Keller, J. et al. (1994) Ganciclovir tratment of herpes simplex thymidine kinase-transduced primary T lymphocytes: an approach for specific *in vivo* donor T-cell depletion after bone marrow transplantation. *Blood* 84: 1333–1341.

77. Lupton, S. D., Bruton, L. L., Kalberg, V. A. and Overell, R. W. (1991) Dominant positive and negative selection using a hygromcyin phosphotransferase-thymidine kinase fusion gene. *Mol. Cell Biol.* 11: 3374–3378.

78. Moolten, F. L. and Cupples, L. A. (1992) A model for predicting the risk of cancer consequent to retroviral gene therapy. *Hum. Gene Ther.* 3: 479–486.

79. Moolten, F. L. (1990) Mosaicism produced by gene insertion as a means of improving chemotherapeutic selectivity. *Crit. Rev. Immunol.* 10: 203–233.

80. Russell, S. J. (1994) Replicating vectors for cancer therapy: a question of strategy. *Sem. Cancer Biol.* 5: 437–443.

81. Black, M. E. and Loeb, L. A. (1993) Creation of HSV-1 thymidine kinase mutants for gene therapy. *Cancer Gene Ther.* 1 (Suppl. 1): 2.

82. Black, M. E. and Loeb, L. A. (1993) Identification of important residues within the putative nucleoside binding site of HSV-1 thymidine kinase by random sequence selection: analysis of selected mutants *in vitro*. *Biochemistry* 32: 11618–11626.

3 Strategies for Enhancing Tumor Immunogenicity (or how to transform a tumor cell in a Frankenstenian APC)

F. Cavallo, P. Nanni, P. Dellabona, P. L. Lollini, G. Casorati, G. Forni

3.1 The Rationale of the Problem

The immunological approach to cancer therapy springs from the observation that the onset and growth of a tumor are events so strictly connected with immune reactivity as to be markedly influenced by it. Significant tumor impairment should, thus, be achievable if changes can be induced in a few immune features of this relationship. The recent characterization of many tumor associated antigens (TAA) on human tumor cells [1] has provided concrete support for the concept of antitumor immunization. In addition, the frequent, persistent evidence of minimal residual disease after current treatments makes the immune approach highly relevant, since an immune memory may be able to suppress recurrences originating from the few neoplastic cells that remain after surgery or chemotherapy [2].

3.2 The Ambiguity of Tumor Immunogenicity

If immunogenicity is defined as the ability to modify the activation state of the immune system, then tumors are immunogenic. The series of antigenic and regulatory signals they deliver in the course of their growth, however, do not lead to an efficient antitumor response.

Either enhanced or diminished expression of some normal membrane structures, the new expression of altered or normally repressed molecules and the presence of new peptides associated with the major histocompatibility complex (MHC) or heat-shock glycoproteins may act as TAA on the tumor cell surface [1, 3]. A few of these alterations can be recognized by granulocytes, macrophages, NK cells and T- and B-lymphocytes with different degrees of selectivity and specificity [4]. In addition, tumor cells themselves modulate the immune response. They secrete characteristic repertoires of biological response modifier molecules [5], use them as paracrine factors to recruit and suppress reactive leukocytes and modulate the activity of endothelial and stromal cells [6]. Tumor cell expansion can, thus, be seen as the proliferation of many potentially immunogenic minipumps. The factors secreted increase in amount as the tumor expands. Initially confined to the local microenvi-

T. Blankenstein (ed.) Gene Therapy
©1999, Birkhäuser Verlag Basel

ronment, they progressively subvert systemic reactivity and alter brain and metabolic functions, resulting in anorexia and cachexia [7].

The dynamic evolution of the tumor-host interactions increases the intrinsic ambiguity of this relationship, since tumor-activated reactions that may hamper the tumor in a given stage may subsequently become ineffective or even be instrumental in its expansion [8]. This ambiguity is already apparent in the preclinical stages, when immune reactivity can lead to either surveillance against initial tumors or their promotion [9]. Lastly, all tumors that reach a clinical status must have been (and still be) able to sneak through immune controls [10].

3.3   Checkpoints of Tumor Immunogenicity

This ability to elude an immune response, proceed without eliciting any response at all or even exploit a response to achieve better growth is the hallmark of transformed cells that become tumorigenic [10]. Any antigenic signal they may present to the immune system is both ignored by natural immune mechanisms and not sufficient to bring about those of specific and adaptive immunity.

What makes tumors defective immunogens? This is a provocative question and the several answers put forward have themselves moulded more than one immunotherapy strategy, each based on the implicit assumption that the immune mechanisms have already been unraveled to a sufficient degree. The past, however, is littered with the failures of awkward strategies inspired by naive immunological concepts. It must be conceded that the emotional impact associated with cancer justifies the impatience of these attempts, while their very lack of success has served to put fancy experimental data in a more realistic perspective.

Even if tumors are evasive immunogens, once a firm reaction is mounted their ability to soak up and paralyse the immune cells is no longer sufficient to prevent their inhibition. Most tumors, suppressive and lethal in syngeneic mice, are in fact, recognized and rejected when injected in allogeneic hosts, even if their foreignness only takes the form of a few minor antigens [11]. Work dating back to the beginning of the century, on the rejection of allogeneic tumors, provides one of the oldest and more convincing proofs that the immune system can eradicate established tumors [12]. Furthermore, it is possible to preimmunize syngeneic mice towards apparently nonimmunogenic tumors and bring about their rapid rejection [13].

Promotion of effective tumor immune recognition, therefore, is a major but tenable challenge. Identification of the checkpoints, set up during the establishment of an immune response, provides a clue to the potential intervention sites. There are several ways of making tumors more immunogenic. The most rational, at present, is gene engineering, since this flexible technology allows the stable insertion of a functioning gene into the genome of a host cell, thus, endowing it with new capabilities. It is also possible to make a tumor cell express, following their characterization, the molecules on which immune recognition depends.

The poor immunogenicity of TAA stems from several tumor features: Absence of adhesion molecules and other costimulatory signals, inefficient, indirect presentation by bone-marrow-derived professional antigen presenting cells (APC), absence of the cluster of cytokines required for T-lymphocyte activation and limping presentation of TAA peptides by the few MHC glycoproteins expressed on the tumor cell surface seem to be the most important. All these obstacles can be overcome by genetic engineering. Substance is, thus, being given to the long-cherished immunological dream of enhancing a tumor's immunogenicity and building up an efficient antitumor vaccine. Little can so far be claimed with respect to the clinical effectiveness of this approach. The fact that it rests on a technological basis, however, enables one to analyse the reasons for both its achievements and its failures, with a view to improving its efficiency [14].

Which gene should be transferred and with what rationale, and what kind of strategies should be designed for enhancing tumor immunogenicity are challenging questions. Four approaches can readily be imagined, namely insertion in the tumor cells of genes for:

a) cytokines,
b) suicide genes,
c) costimulatory molecules and
d) MHC glycoproteins.

Paradoxically, the concepts behind these approaches are not new and their implementation with more primitive technologies has so far proved of little clinical efficacy.

3.4 Immunogenicity of Tumor Cells Transduced with Cytokine Genes

Cytokines are a set of proteins distinguished by their interactivity and their formation of a communication code between immune cells. Therefore, the possibility that they could be used to alter the scenario naturally elicited by tumor growth was promptly explored [15]. Firstly, several experimental and clinical data obtained with low doses of recombinant cytokines, injected at the tumor growth site, showed that local cytokines may evoke a strong antitumor reactivity and lead to the establishment of an immune memory [15, 16]. Later, it was found that similar reactions are activated by repeated local injections of cytokines, by those released by cytokine gene-engineered (cytokine-engineered, for short) tumor cells, and those released by other cytokine-engineered cells, even those not antigenically related to the tumor itself [17, 18]. However, cytokine-engineered tumor cells appear to elicit more effective reactions than exogenous cytokines. The growth of most cytokine-engineered tumors *in vivo*, in fact, is impaired by a quick reaction, mounted a few hours after their introduction [19] and unrelated neighboring tumors are occasionally inhibited as well. These nonspecific reactions are governed in a feedback fashion by their own intensity, since the quantity of cytokine increases at first, as the cytokine-engineered cells expand, and then decreases as a function of the efficacy of the reaction being activated [6]. The repertoire of cells involved in these reactions was, in many cases, unexpected in that granulocytes and macrophages massively infiltrated the tumor mass and killed the cytokine-engineered cells. T-lymphocytes appear to guide their lytic activity by both establishing characteristic membrane-membrane contacts and releasing proinflammatory cytokines. The observation that the antitumor reaction of inflammatory cells becomes marginal in mice, selectively deprived of either T-lymphocytes or granulocytes, underlines the importance of this crosstalk between specific and nonspecific immune cells [16, 19]. Surprisingly, tumor cell destruction is not associated with damage to adjacent normal tissues. Guidance by T-lymphocytes, preferential recognition of certain aspects of the neoplastic cell membrane and the concentration gradient of the secreted cytokine may play a role in this restriction of killing to the tumor cells [16].

In the late stages of the cytokine-activated reaction, the dialogue between the leukocytes and T-lymphocytes flows in the opposite direction. Leukocytes appear to be instrumental in the indirect presentation of TAA to CD4+ and CD8+ memory lymphocytes. This new crosstalk, plus the release of many secondary cytokines, builds up a particularly immunogenic environment that allows the generation of an immune memory even against poorly or non-immunogenic tumors [6]. Moreover, secretion of a given cytokine by the engineered cells can lead to the selective promotion of particular memory mechanisms [17, 20, 21].

3.5 Immunogenicity of Tumor Cells Transduced with Suicide Genes

The results obtained with cytokine-engineered cells suggest that a tumor becomes more immunogenic when its immune recognition takes place in a microenvironment rich in a given cytokine. When cytokine-engineered cells are used as a vaccine, live cells appear to elicit a stronger immune response than irradiated or mitomycin-C treated cells [22, 23]. Because the use of cytokine-engineered cells, to induce an efficient immune response against a small tumor burden, is now an attractive clinical possibility [24], these findings open important issues: It is not clear how much a cytokine-engineered cell's stronger immunogenicity is directly dependent on the signals delivered by the cytokine it secretes on the memory lymphocyte, or on the cytokine ability to attract leukocytes and induce the secretion of proinflammatory cytokines at the tumor site, thus, a particularly immunogenic microenvironment is built up, allowing efficient TAA recognition. Alternatively, it may simply stem from the fact that they give rise to an initial tumor that regresses when the reaction triggered by the cytokine they secrete is enough to block its proliferation.

Since it is difficult to evaluate the weight of each of these three events in enhancing tumor immunogenicity, the lack of appropriate controls has led to the impression that live cytokine-engineered cells are also more immunogenic than parental cells [25]. The use of suicide gene-engineered tumors may allow some independent evaluation of the significance of these immunogenic signals.

As specifically indicated in another chapter of this book, suicide genes code for enzymes which metabolize a given nontoxic compound, generating lethal metabolites. Their insertion into the cells of a species, normally devoid of such activity, makes them selectively sensitive to the compound metabolized because of the inserted gene. In the absence of this compound, these engineered cells live normally. In its presence, they only generate toxic metabolites and die.

Insertion of the cytosine-deaminase (CD) gene into the cells of a poorly immunogenic, spontaneous mammary adenocarcinoma of BALB/c mice (TSA) exposes them to poisoning by the nontoxic prodrug 5-fluorocytosine (5-FC). After an initial growth, CD-gene transfected TSA (TSA-CD) cells are totally rejected when 800 mg/kg body weight of 5-FC are injected daily i.p. for 30 days, starting 1, 3 and 7 days after challenge [26]. Morphological observations showed that during this treatment, many cells display abnormal mitoses, atypical chromatin aggregation and over-dilated nuclear pores or absence of nuclear membrane. The slack areas caused by necrosis were infiltrated by lymphocytes, mostly CD8+, and granulocytes, which eventually surrounded the remaining tumor cells. Unexpectedly, 5-FC no longer fully protected mice that had been depleted of T-lymphocyte and granulocytes against TSA-CD, showing that the host immune reactivity plays an important role in their total elimination. The systemic immune memory that follows 5-FC-induced regression of TSA-CD cells, on the other hand, ensures specific and complete protection. Mice, in which TSA-CD tumors were no longer palpable in the left flank, rechallenged with wild-type TSA cells in the right flank, show that the slow expansion and regression of TSA-CD cells provide an efficient immunizing signal [26]. The combination of CD engineering and 5-FC, indeed, can be regarded as an elegant equivalent of the early-20th-century practice of ligating a tumor's blood vessels to provoke its necrosis and induce immunity.

The fact that regression is itself a major immunogenic stimulus sets the data obtained with cytokine-engineered cells in a different perspective. Indeed, one of the major factors determining the higher immunogenicity of cytokine-engineered cells rests on the ability of the cytokine they release to make the tumor regress.

But how important are the regulatory signals delivered by these cytokines? Do they elicit tumor regression only? Apparently, the local cytokine and the inflammatory leukocytes it recruits, influence the immune response in a much more subtle way. In the TSA adenocarcinoma system, for example, when regression takes place in the presence of a given cytokine, the selective promotion of particular memory mechanisms is exploited by the encouragement (or inhibition) of certain types of reactivity to obtain a more effective response. A vigorous CTL-mediated memory is

271

acquired in the presence of the Th-1 cytokines IL-2, IL-7 and IFN-γ, whereas the development of CTL is inhibited and the memory is mediated by non-cytotoxic CD8$^+$ lymphocytes and antibodies in the presence of the Th-2 cytokine IL-4.

Deflection of memory mechanisms towards a Th-1 or Th-2 response may result from direct influence of the cytokine on memory T-lymphocytes. It can also depend on the cytokine's own ability to recruit distinct leukocyte populations and prompt TAA presentation by different sets of APCs, as well as its ability to induce the release of characteristic repertoires of secondary cytokines. Inhibition of a subsequent challenge is conferred by both Th-1 and Th-2 reactivity. The regression of already established tumors, on the other hand, requires the reactivity induced by Th-1 cytokines [23].

3.6 Immunogenicity of Tumor Cells Transduced with the Genes for Costimulatory Factors

Initiation of a primary T-lymphocyte response takes place in the secondary lymphatic organ, where it is accounted for by professional APC [27]. These cells capture antigens, process them into peptides and present them at high density, in the context of membrane MHC glycoproteins, to specific T-lymphocytes. In addition to this extraordinary ability to generate and display the antigenic signal for the responding T-lymphocytes, professional APCs are unique in providing T-lymphocytes with a second set of activatory signals, globally defined as costimulation. Activation of resting primary T-lymphocytes and reactivation of resting memory T-lymphocytes, in fact, require both antigen and costimulatory signals [28, 29].

Thus, no cell, other than a bone marrow derived APC, can initiate an immune response. This key mechanism for maintaining peripheral tolerance [30] explains why tumors are evasive immunogens. They are potentially antigenic [31, 32] but lack crucial costimulatory signals. Their transformation into quasi-professional APC, by engineering them to express costimulatory molecules, is currently sought in many laboratories.

A prerequisite to this goal is the definition of the minimal co-stimulatory signal. Much experimental evidence indicates that B7-CD28 interaction plays a major costimulatory role. Both B7-1 and B7-2 are expressed on professional APC, and both interact with the CD28 and CTLA4 receptors expressed on all CD4$^+$ and on the majority of CD8+ lymphocytes [33–38]. Indeed, transfection of B7-1 genes into mouse experimental tumors has turned on specific protective antitumor immunity. However, it soon became clear that B7-1 costimulatory effects are only evident in immunogenic tumors, namely those which naturally express a strongly antigenic TAA together with a set of accessory molecules, such as ICAM-1, that potentiate B7-1 activity [39–44]. B7-1 alone is thus unable to provide the minimal costimulatory signal for the elicitation of antitumor T-lymphocytes.

It is still debated whether or not the costimulatory effects of B7-2, are confined to immunogenic tumors as well. All the comparative data on the costimulatory activity of B7-1 and B7-2 shows that they are equally efficient [45, 46]. However, all these data were obtained with immunogenic tumors. In contrast, our findings show that poorly immunogenic TSA adenocarcinoma cells, engineered to express B7-2, are superior to those engineered to express B7-1 in the induction of both protective and curative immunity. Thus, in the presence of a limiting tumor immunogenicity, B7-2 can still costimulate T-lymphocytes, whereas B7-1 cannot. Since we have also observed similar protective effects with the poorly immunogenic B16.F1 melanoma, we can rule out the possibility that this greater efficiency is peculiar to a single type of tumor. Instead, a wider application of B7-2 over B7-1 in the costimulation of T-lymphocytes reacting against poorly immunogenic tumors can be envisaged [47].

What are the bases for the induction of a strong antitumor immunity by B7 molecules? Costimulation of T-lymphocytes, via CD28, allows the sustained secretion of wide range of lymphokines, which act in an autocrine as well as a paracrine manner on both tumor-specific and bystander lymphocytes and leukocytes. Indeed, irrespective of their costimulatory potency, tumor cells expressing B7-1 or B7-2 share the ability to trigger a pro-inflammatory response at the site of their inoculum. Both lymphocytes and granulocytes are rapidly recruited within the tumor cells [44, 47]. Furthermore, *in vivo* depletion experiments have confirmed that CD8[+] T-lymphocytes, NK cells and granulocytes are activated by the B7-expressing vaccines and are critically involved in tumor rejection [44, 47]. The conspicuous inflammatory reaction that follows the injection of B7-1 or B7-2 expressing tumor cells opens up the possibility that professional APC are recruited at the tumor site, allowing the indirect presentation of TAA to T-lymphocytes. This potent cross-priming for TAA could enable the use of tumor cell vaccines expressing B7-1 or B7-2 to be extended to MHC-mismatched individuals, provided the vaccine and the endogenous tumor share a common TAA.

Although the inflammatory responses triggered by tumor cells, engineered to express B7-1 or B7-2, appear macroscopically similar, they may be induced by a qualitatively or quantitatively different set of cytokines. It has been reported recently, that B7-1 or B7-2 costimulation can drive the differentiation of CD4[+] T-lymphocytes towards a polarized Th-1 or Th-2 phenotype [48, 49], and the possibility that this is equally true of antitumor CD8[+] effector T-lymphocytes is an intriguing prospect. Most probably, both the density of accessory molecules and the expression of accessory molecules, such as ICAM-1, serve to determine the outcome of the costimulatory signal delivered by B7-1 and B7-2. When expressed on poorly immunogenic tumors, both B7-1 and B7-2 trigger secretion of inflammatory cytokines, but only those induced by B7-2-mediated costimulation allow the maturation of an efficient antitumor immunity. Whether the reduced effect observed for B7-1 is due to its lower costimulatory capacity or to active homoeostatic control of the ensuing CD8[+] T-lymphocyte response remains to be determined.

3.7 Immunogenicity of Tumor Cells Transduced with the Genes for Allogeneic MHC Glycoproteins

Vaccination with tumor cells, engineered to express allogeneic MHC membrane gly-coproteins (allo-MHC), allows the associated recognition of TAA and allo-MHC on the same cell membrane. The simultaneous expression of target antigen and allo-MHC has investigated over the years with conventional F1 hybrid cells [50], so-matic cell hybrids [51], and finally genetic engineering. Even so, it is still not clear whether their associated recognition is a prerequisite for enhancing antigen im-munogenicity. If it is not, then a much simpler vaccine, made with a mixture of tu-mor and normal allogeneic cells, would lead to similar results.

The immunogenic weight of TAA and allo-MHC recognition on the same or sep-arate cells, can be inferred by examining the reaction mechanisms activated in the light of current immunological wisdom. The first point to consider is that effector T-lymphocytes, activated following associated or separate recognition, will react against wild-type tumors expressing TAA and syngeneic-MHC only.

By itself, the recognition of allo-MHC triggers numerous allo-reactive T-lym-phocytes to act as the "initiator" lymphocytes described by Cohen and Livnat many years ago [52]. Their timely release of a cocktail of appropriate cytokines recruits several leukocyte populations to the tumor site, boosts NK activity, activates macrophages, attracts professional APCs and induces the expansion of helper or cy-tolytic T-lymphocytes, endowed with anti-tumor activity [53]. Both TAA-specific poised lymphocytes and autoreactive lymphocytes directed to the TAA also physio-logically expressed by a few normal cells, such as MAGEs or tyrosinase, might thus, be rescued from anergy [54, 55]. Since allo-MHC acts just like a non-specific poly-clonal stimulus, initiation of an anti-TAA response can be adequately provoked by allo-MHC expressed on leukocytes [53]. On the other hand, closer cell-cell interac-tions, due to TAA and allo MHC associated recognition, can induce a more effec-tive response. Studies with somatic cell hybrids [51] or tumor cells transduced with genes coding for multiple antigens [55] suggest that coexpression of allo-MHC and target antigen forms a stronger immunogenic signal than their expression on sepa-rate cells, a question that can only be settled through the use of induction models, in which expression of allo-MHC and a molecularly defined TAA on both separate cells and on the same cell is compared.

Coexpression by the same cell, on the other hand, may play a central role if the allo-MHC are instrumental for a TAA peptide presentation, or if TAA peptides must be presented along with costimulatory molecules. In the case of peptide presenta-tion, the type of antigen that could be then recognized by effector T-lymphocytes on wild-type tumor cells remains an open question.

Two points deserve further discussion in view of the tumoral nature of the engi-neered cells: Many tumors have a defective expression of MHC glycoproteins [56]. Strong promoters in the vector carrying allo MHC genes bypass or overcome autologous gene alterations, such as deletions, but not those of peptide processing nor those connected with post-translational steps. Many tumor cells transduced with allo-MHC may thus continue to display low levels of membrane MHC glycoproteins. We have shown recently, that TSA adenocarcinoma cells, transfected with an allo-MHC gene plus the IFN-γ gene, display a high MHC expression, unlike cells which do not secrete IFN-γ [57]. Moreover, allo-MHC expression by neoplastic cells frequently leads to tumor rejection, an event which is in itself immunogenic, as made clear by the results obtained with suicide genes. Appropriate controls will therefore, be needed to determine whether allo-MHC genes are just another technological way of inducing tumor regression, or whether they afford qualitatively or quantitatively superior tumor immunogenicity compared with thymidine kinase or cytosine deaminase suicide genes.

The use of genes from the highly polymorphic and polygenic MHC also poses the obvious problem of choosing the best locus and allele. The few comparative data reported so far have shown that some gene products induce stronger antitumor responses than others (e.g. H-2K^b versus H-2D^b) [58, 59]. Even less is known about the immunogenicity of a given MHC glycoprotein transduced in tumor cells of different haplotype [60].

Can we draw any firm conclusions concerning the capabilities of allo-MHC in tumor immunotherapy? Some studies indicate that transfected allo-MHC genes offer no particular advantages [59, 60], whereas, others have shown that they can enhance the immune response against parental cells and even bring about their rejection [60, 61]. Comparisons, however, are rendered almost meaningless by the fact that earlier vectors comprised genomic MHC fragments driven by natural promoters, whereas, they are now constructed with cDNAs and highly efficient viral promoters, and the expression levels and experimental frameworks reported rarely coincide [62].

Allogeneic HLA genes have been chosen by Nabel and coworkers for direct gene transfer into tumors [63]. Their pioneering clinical trials have demonstrated clearly, that this approach is feasible and devoid of unwanted side-effects. Additionally, several cytokine-based therapeutic protocols use vaccines made from semi-syngeneic tumor cell lines (e.g. HLA-A2) [64]. The ensuing exposure of patients to allogeneic antigens could turn out to be beneficial. The absence of controls, however, usually prevents discrimination between the effects of the cytokine and those of the allo-MHC.

Transduction of allo-MHC genes and cytokines results in a tumor-specific immune response [18, 57, 65]. TSA cells, transduced with the IFN-γ gene, elicit a primary response based mainly on macrophages [19, 66], followed by a T-lymphocyte-mediated response that cures about one-third of mice bearing parental TSA cell lung micrometastases [67]. To obtain a faster and more effective T-lymphocyte activation, IFN-γ transfectants were re-transduced with allogeneic (H-2^b) genes. The high MHC expression induced by the autocrine activity of IFN-γ caused a strong antitu-

275

mor response, sustained by both macrophages and T-lymphocytes [57]. IFN-γ single transfectants are tumorigenic and therefore, require mitomycin C pretreatment before *in vivo* administration. The IFN-γ + H-2D^b transfectant constitutes a live vaccine that cures about 80% of mice with TSA lung micrometastases, compared with only 0–30% when a single transfectant is used.

3.8   Immunogenicity of Tumor Cells Transduced with the Genes for Syngeneic MHC Glycoproteins

Transduction of tumor cells with syngeneic MHC class I genes has been chiefly used to investigate the consequences of the MHC class I glycoprotein rarefaction, typical of many tumors, and only occasionally to enhance tumor immunogenicity or cure wild-type tumors [61]. The subject of these studies is significantly different from the one we are considering. Even so, they have led to important results that are usually ignored by those using allo-MHC genes. Transduction with MHC genes can profoundly alter several intrinsic (i.e. independent from the host immune system) properties of the recipient tumor cells, including proliferation [68], adhesion [69], expression of melanoma-associated antigens [70] and release of superoxide radicals [71]. It has recently been shown that similar alterations are induced by allo-MHC genes [72], and that this could influence in the transduced cells some intrinsic properties that are directly related to tumor malignancy, for example Matrigel invasion and the expression of collagenase inhibitors (TIMP).

Class II MHC antigens have been transduced in tumor cells in an attempt to transmute them into APCs. However, the APC function probably entails a complex balance of MHC class II and costimulatory signal expression, the release of particular cytokines and selective homing ability. Ostrand-Rosenberg et al. [58] have shown that tumor cells, simultaneously transduced with MHC class II and B7-1 genes, are better stimulators of antitumor immunity than cells carrying either gene alone or mixtures of single gene transduced cells [41]. In principle, transduction of the class II transactivator gene CIITA should be a simple way to induce class II expression in tumor cells. In practice, however, it has been shown that CIITA transfectants are less immunogenic than cells transduced with class II α and β genes (73). The lower immunogenicity obtained with CIITA was attributed to the simultaneous up-regulation of class II-associated invariant chain (Ii) and DM, which probably prevent the loading of endogenous peptides. Thus the "physiologic" antigen presentation enabled by CIITA transduction in tumor cells is actually less effective than the artificial expression of class II αβ genes (and B7) in the absence of associated chains.

3.9 Immunogenicity of cells transduced with multiple genes

Immunological gene therapy is now following the route traced by cancer chemotherapy, with a shift from single agents to multiple agents. A number of recent studies combined multiple genes in the same cell to obtain vaccines with a higher degree of immunogenicity and of therapeutic activity. In a recent survey of the literature [74] we counted more than 30 therapeutic studies using more than one gene. In most cases the use of multiple genes yielded a significant improvement over single genes.

The most complicated issue is now the choice of which genes should be combined to maximize immungenicity and antitumor immune responses. A daunting perspective, given the huge number of possible combinations. In the long run, however, it might be simpler to administer true professional APCs after *ex vivo* pulsing with tumor antigens rather than trying to create Frankensteinian monsters by cumulation of APC genes into cancer cells.

3.10 Provisional Conclusion

Enhancement of immunogenicity by gene engineering is an attractive prospect. The most effective ways of making a tumor cell more immunogenic and transforming it into a Frankensteinian professional APC must be determined by evaluating the weight of several variables in dissimilar tumor systems and in different laboratories. The questions to which future use of these engineered tumor vaccines should give a consensus answer are:
a. Which gene or gene set is the most effective in enhancing tumor immunogeneity?
b. Do engineered tumors possess effective curative capability?
c. Under what clinical conditions can they be employed?

These questions are not marginal. Neat answers will provide a rationale for the clinical use of gene engineered tumor vaccines. When compared with conventional forms of management, therapeutical vaccination is a "soft", noninvasive treatment free from particular distress and side-effects. Any clinical efficacy that a gene-engineered tumor may display could be viewed as a major biological and clinical achievement.

Acknowledgments

We thank Dr. John Iliffe for careful review of the manuscript. This work was supported by grants from the Italian Association for Cancer Research (AIRC), the National Research Council (CNR PF-ACRO), and Italian Ministry of University and Scientific and Technological Research (MURST 40% and 60%).

References

1. Boon, T., Gajewski, T. F., Coulie, P. G. (1995) From defined tumor antigens to effective immunization? *Immunol. Today* 16: 334–36.
2. Longo, D. L. (1994) New clinical prospects. *In*: Forni, G., Foa' R, Santoni, A., Frati, L. (eds), *Cytokine-induced tumor immunogenicity*. London: Academic Press, pp. 469–81.
3. Pardoll, D. M. (1994) Tumor antigens. A new look for the 1990s. *Nature* 369: 357–59.
4. Forni, G., Santoni, A. (1984) Immunogenicity of non-immunogenic tumors. *J. Biol. Resp. Modif.* 3: 128–31.
5. Pekarek, L. A., Weichselbaum, R. R., Beckett, M. A. et al. (1993) Footprinting of individual tumours and their variants by constitutive cytokine expression patterns. *Cancer Res.* 53: 1978–81.
6. Colombo, M. P., Modesti, A., Parmiani, B. et al. (1992) Perspectives in cancer research: local cytokine availability elicits tumor rejection and systemic immunity through granulocyte-T-lymphocyte cross-talk. *Cancer Res.* 52: 1–5.
7. Plescia, O. J., Smith, A. H., Grinwich, K. (1975) Subversion of immune system by tumor cells and role of prostaglandins. *Proc. Natl. Acad. Sci. USA* 72: 1848.
8. Mantovani, A. (1990) Tumor associated macrophages. *Curr. Opin. Immunol.* 2: 689–95.
9. Prehn, R. T. (1972) The immune response as stimulator of tumor growth. *Science* 176: 170–73.
10. Lewis, A. M. Jr., Cook, J. L. (1985) A new role for DNA virus early proteins in viral carcinogenensis. *Science* 227: 15–20.
11. Forni, G., Landolfo, S., Giovarelli, M. et al. (1982) Immune recognition of tumor cells *in vivo*. I. Role of H-2 gene products in T lymphocyte activation against minor histocompatibility antigens displayed by adenocarcinoma cells. *Eur. J. Immunol.* 12: 664–70.
12. Klein, J. (1986) Natural history of the Major Histocompatibility Complex. *New York: J. Wiley and Sons*.
13. Boon, T. (1985) Tumor variants: Immunogenic variants obtained by mutagen treatment of tumor cells. *Immunol. Today* 6: 307–10.
14. Forni, G., Parmiani, G., Guarini, A. et al. (1994) Gene transfer in tumor therapy. *Ann. Oncol.* 5: 789–94.
15. Bubenik, J. (1994) Utilization of IL-2 and Il-2 gene transfer for regional immunotherapy. *In*: Forni, G., Foa' R, Santoni, A., Frati, L. (eds), *Cytokine-induced tumor immunogenicity*. London: Academic Press, pp. 113–131.
16. Forni, G., Giovarelli, M., Cavallo, F. et al. (1993) Cytokine-induced tumor immunogenicity: from exogenous cytokines to gene therapy. *J. Immunother.* 14: 253–57.
17. Cavallo, F., Giovarelli, M., Gulino, A. et al. (1992) Role of neutrophils and CD4⁺ T lymphocytes in the primary and memory response to nonimmunogenic murine mammary adenocarcinoma made immunogenic by IL-2 gene transfection. *J. Immunol.* 149: 3627–35.
18. Roth, C., Delassus, S., Even, J. et al. (1994) Inhibition of tumor growth by histoincompatible cells expressing IL-2. *In*: Forni, G., Foa' R, Santoni, A., Frati, L. (eds), *Cytokine-induced tumor immunogenicity*. London: Academic Press, pp. 163–181.
19. Musiani, P., Allione, A., Modica, A. et al. (1995) Role of neutrophils and lymphocytes in inhibition of a mouse mammary adenocarcinoma engineered to release IL-2, IL-4, IL-7, IL-10, IFN-alpha, IFN-gamma, and TNF-alpha. *Lab. Invest.* 74: 146–157.
20. Pericle, F., Giovarelli, M., Colombo, M. P. et al. (1994) An efficient Th-2-type memory

follows CD8[+] lymphocyte driven and eosinophil mediated rejection of a spontaneous mouse mammary adenocarcinoma engineered to release IL-4. *J. Immunol.* 153: 5659–73.

21. Giovarelli, M., Musiani, P., Modesti, A. et al. (1995) The local relelase of IL-10 by transfected mouse mammary adenocarcinoma cells does not suppress but enhances antitumor reaction and elicits a strong cytotoxic lymphocyte and antibody dependent immune memory. *J. Immunol.* 155: 3112–23.

22. Hock, H., Dorsch, M., Kunzendorf, U. et al. (1993) Vaccination with tumor cells genetically engineered to produce different citokines: effectivity not superior to a classical adjuvant. *Cancer Res.* 53: 714–16.

23. Allione, A., Consalvo, M., Nanni, P. et al. (1994) Immunizing and curative potential of replicating and nonreplicating murine mammary adenocarcinoma cells engineered with IL2, IL4, IL6, IL7, IL10, TNFα, GM-CSF, and IFNγ gene or admixed with conventional adjuvants. *Cancer Res.* 54: 6022–26.

24. Colombo, M. P., Forni, G. (1994) Cytokine gene transfer in tumor inhibition and tentative tumor therapy: Where are we now? *Immunol. Today* 15: 48–51.

25. Tepper, R. I., Mule', J. J. (1994) Experimental and clinical studies of cytokine gene-modified tumor cells. *Hum. Gene Ther.* 5: 153–64.

26. Consalvo, M., Mullen, C. A., Modesti, A. et al. (1995) 5-Fluorocytosine-induced eradication of murine adenocarcinomas engineered to express the cytosine deaminase suicide gene requires host immune competence and leaves an efficient memory. *J. Immunol.* 154: 5302–12.

27. Matzinger, P. (1994) Tolerance, danger, and the extended family. *Annu. Rev. Immunol.* 12: 991–1045.

28. Lafferty, K. J., Prowse, S. J., Simeonovic, C. J. et al. (1983) Immunobiology of tissue transplantation: a return to the passenger leukocyte concept. *Annu. Rev. Immunol.* 1: 143–59.

29. Mueller, D. L., Jenkins, M. K., Schwartz, R. H. (1989) Clonal expansion versus clonal inactivation: a costimulatory signalling pathway determines the outcome of T-lymphocyte antigen receptor occupancy. *Annu. Rev. Immunol.* 7: 445–61.

30. Bretscher, P., Cohn, M. (1970) A theory of self-nonself discrimination. Paralysis and induction involve the recognition of one and two determinants on an antigen respectively. *Science* 169: 1042–49.

31. Talmage, D. W., Woolnough, J. A., Hemmingsen, H. et al. (1977) Activation of cytotoxic T-lymphocytes by nonstimulating tumor cells and spleen factor(s). *Proc. Natl. Acad. Sci. USA* 90: 5687–92.

32. Chen, L., Linsley, P. S., Hellstrom, K. E. (1993) Costimulation of T-lymphocytes for tumor immunity. *Immunol. Today* 14: 483–86.

33. Freeman, G., Freedman, A. S., Segil, J. M. et al. (1989) B7, a new member of the Ig superfamily with unique expression on activated and neoplastic B cells. *J. Immunol.* 143: 2714–22.

34. Freeman, G., Gray, G. S., Gimmi, C. D. et al. (1991) Structure, expression, and T-lymphocyte costimulatory activity of the murine homologue of the human B lymphocyte activation antigen B7. *J. Exp. Med.* 174: 625–31.

35. Vandergerghe, P., Delabie, J. De Boer, M. et al. (1993) *In situ* expression of B7/BB1 on antigen presenting cells and activated B cells: an immunohistochemical study. *Int. Immunol.* 3: 229–36.

36. Linsley, P. S., Brady, W., Grosmarie, L. et al. (1991) Binding of the B cell activation B7 to

CD28 costimulates T-lymphocyte proliferation and interleukin 2 mRNA accumulation *J. Exp. Med.* 173: 721–28.

37. Koulova, L. K., Clark, E. A., Shu, G. et al. (1991) The CD28 ligand B7/BB1 provides the costimulatory signal for alloactivation of CD4⁺ T-lymphocytes. *J. Exp. Med.* 173: 759–64.

38. Azuma, M., Cayabyab, M., Phillips, J. H. et al. (1993) Requirements for CD28-dependent T-lymphocyte-mediated cytotoxicity. *J. Immunol.* 150: 2091–97.

39. Townsend Sa, Allison JA (1993) Tumor rejection after direct costimulation of CD8⁺ T-lymphocytes by B7-transfected melanoma cells. *Science (Wash. DC)* 259: 368–71.

40. Chen, L., Ashe, S., Brady, W. A. et al. (1992) Costimulation of antitumor immunity by the B7 counterreceptor for T lymphocyte molecules CD28 and CTLA4. *Cell* 71: 1093–104.

41. Baskar, S., Ostrand-Rosenberg, S., Nabavi, N. et al. (1993) Constitutive expressin of B7 restores immunogenicity of tumor cells expressing truncated major histocompatibility complex class II molecules. *Proc. Natl. Acad. Sci. USA* 90: 5687–95.

42. Ramarathinam, L., Caste, M., Wu, Y. et al. (1994) T-lymphocyte costimulation by B7/BB1 induces CD8 T-lymphocyte dependent tumor rejection: an important role of B7/BB1 in the induction, recruitment, and effector function of antitumor R cells. *J. Exp. Med.* 179: 1205–14.

43. Chen, L., McGowan, P., Ashe, S. et al. (1994) Tumor immunogenicity determines the effects of B7 costimulation on T-lymphocyte-mediated tumor immunity. *J. Exp. Med.* 179: 523–32.

44. Cavallo, F., Martin-Fontecha, A., Bellone, M. et al. (1995) Coexpression of B7-1 and ICAM-1 on tumors is required for rejection and establishment of a memory response. *Eur. J. Immunol.* 25: 1154–62.

45. Hodge, J. W., Abrams, S., Schlom, J. et al. (1994) Induction of antitumor immunity by recombinant vaccinia viruses expressing B7-1 or B7-2 costimulatory molecules. *Cancer Res.* 54: 5552–55.

46. Yang, G., Hellstrom, K. E., Hellstrom, I. et al. (1995) Antitumor immunity elicited by tumror cells transfected with B7-2, a second ligand for CD28/CTLA4 costimulatory molecule. *J. Immunol.* 154: 2794–800.

47. Martin-Fontecha, A., Cavallo, F., Bellone, M. et al. (1996) Heterogenous effects of B7-1 and B7-2 in the induction of both protective and therapeutic anti tumor immunity against different mouse tumors. *Eur. J. Immunol.* 26: 1851–1859.

48. Bluestone, A. J. (1995) New perspective of CD28-B7-mediated T-lymphocyte costimulation. *Immunity* 2: 555–59.

49. Thompson, C. B. (1995) Distinct role for the costimulatory ligands B7-1 and B7-2 in T helper differentiation? *Cell* 81: 979–82.

50. Di Marco AT, Franceschi, C., Prodi, G. (1972) Helper activity of histocompatibility antigens on cell-mediated immunity. *Eur. J. Immunol.* 2: 240–45.

51. Barbanti-Brodano, G., Di Marco AT, Franceschi, C. et al. (1974) Increased immunogenicity of TSTA on heterokaryocytes of syngeneic tumoral and allogeneic normal cells. *Experientia* 30: 947–51.

52. Cohen, I. R., Livnat, S. (1976) The cell-mediate immune response: Interactions of initiator and recruited T lymphocytes. *Transplantat. Rev.* 29: 24–35.

53. Giovarelli, M., Santoni, A. and G. Forni (1985) Alloantigen-activated lymphocytes from mice bearing a spontaneous "non-immunogenic" adenocarcinoma inhibit its grown by

recruiting host immunoreactivity. *J. Immunol.* 133: 3596–3603.

54. Nanda, K. K., Sercarz, E. E. (1995) Induction of anti-self-immunity to cure cancer. *Cell* 82: 13–7.

55. Baskar, S., Glimcher, L., Nabavi, N. et al. (1995) Major Histocompatibility Complex class II⁺ B7-1 tumor cells are potent vaccines for stimulating tumor rejection in tumor-bearing mice. *J. Exp. Med.* 181: 619–29.

56. Roth, C., Rochlitz, C., Kourilsky, P. (1994) Immune response against tumors. *Adv. Immunol.* 37: 281–351.

57. Lollini, P. L., De Giovanni, C., Landuzzi, L. et al. (1995) Transduction of genes coding for a histocompatibility (MHC) antigen and for its physiological inducer gamma-interferon in the same cell. Efficient MHC expression and inhibition of tumor and metastasis growth. *Hum Gene Ther.* 6: 743–52.

58. Ostrand-Rosenberg, S., Garcia, E. P., Roby, C. A. et al. (1991) Influence of major histocompatibility complex class I, class II and TLA genes on tumor rejection. *Semin Cancer Biol* 2: 311–19.

59. Plautz, G. E., Nabel, G. J. (1994) Direct gene transfer for immunotherapy of cancer. *In*: Forni, G., Foa' R, Santoni, A., Frati, L. (eds), *Cytokine-induced tumor immunogenicity*. London: Academic Press, pp. 345–364.

60. Itaya, T., Yamagiwa, S., Okada, F. et al. (1987) Xenogenization of a mouse lung carcinoma (3LL) by transfection with an allogeneic class I major histocompatibility complex gene (H2-Ld). *Cancer Res.* 47: 3136–40.

61. Feldman, M., Eisenbach, L. (1991) MHC class I genes controlling the metastatic phenotype of tumor cells. *Semin. Cancer Biol.* 2: 337–46.

62. Lollini, P. L., Nanni, P. (1995) Minimal requirements for characterization of cytokine gene-transduced tumor cells: A proposal. *J. Nat. Cancer Inst.* 87: 1718–1718.

63. Nabel, G. J., Chang, A. E., Nabel, E. G. et al. (1994) Clinical protocol: Immunotherapy for cancer by direct gene transder into tumors. *Hum. Gen. Ther.* 5: 57–77.

64. Cascinelli, N., Foa', R., Parmiani, G. (1994) Clinical protocol: Active immunization of metastatic melanoma patients with interleukin-4 transduced, allogeneic melanoma cells. A phase I-II study. *Hum. Gene Ther.* 5: 1059–64.

65. Kim, T. S., Russel, S. J., Collins, M. K. et al. (1993) Immunization with interleukin-2-secreting allogeneic mouse fibroblasts expressing melanoma-associated antigens prolongs the survival of mice with melanoma. *Int. J. Cancer* 55: 865–72.

66. Lollini, P. L., Bosco, M. C., Cavallo, F. et al. (1993) Inhibition of tumor growth and enhancement of metastasis after transfection of the gamma-interferon gene. *Int. J. Cancer* 55: 320–29.

67. Lollini, P. L., Bosco, M. C., De Giovanni, C. et al. (1994) Reduced oncogenicity and enhanced metastatic spread of IFN-gamma transfected tumor cells: therapeutic implications. *In*: Forni, G., Foa' R, Santoni, A., Frati, L. (eds), *Cytokine-induced tumor immunogenicity*. London: Academic Press, pp. 345–64.

68. Sunday, M. E., Isselbacher, K. J., Gattoni-Celli, S. et al. (1989) Altered growth of a human neuroendocrine carcinoma line after transfection of a major histocompatibility complex class I gene *Proc. Natl. Acad. Sci. USA* 86: 4700–704.

69. De Giovanni, C., Nicoletti, G., Sensi, M. et al. (1994) H-2Kb and H-2Db gene transfections in B16 melanoma differently affect non-immunological properties relevant to the metastatic process. Involvement of integrin molecules. *Int. J. Cancer* 59: 269–74.

70. Gorelik, E., Kim, M., Duty, L. et al. (1993) Control of metastatic properties of BL6

melanoma cells by H-2Kb gene: immunological and nonimmunological mechanisms. *Clin. Exp. Metastasis* 11: 439–52.

71. Chia, K. Y., Lim, S. P., Oei, A. A. et al. (1994) Acquisition of immunogenicity by AKR leukemic cells following DNA-mediated gene transfer is associated with the reduction of constitutive reactive superoxide radicals. *Int. J. Cancer* 57: 216–23.

72. Xu, F., Carlos, T., Li, M. et al. (1998) Inhibition of VLA-4 and up-regulation of TIMP-1 expression in B16BL6 melanoma cells transfected with MHC class I genes. *Clin. Exp. Metastasis* 16: 358–370.

73. Armstrong, T. D., Clements, V. K., Martin, B. K. et al. (1997) Major histocompatibility complex class II-transfected tumor cells present endogenous antigen and are potent inducers of tumor-specific immunity. *Proc. Natl. Acad. Sci. USA* 94: 6886–6891.

74. Nanni, P., Forni, G., Lollini, P. L. (1998) Cytokine gene therapy: hopes and pitfalls. *Ann. Oncol.; in press.*

4 Vaccines using Gene-Modified Tumor Cells

S. Cayeux, Z. Qin, B. Dörken, T. Blankenstein

4.1 Introduction

To date, many cytokines and costimulatory molecules have been transfected in a number of different rodent tumors and in several cases, these gene modified tumor cells, when used as vaccines, induced tumor immunity leading to rejection of a challenge with the parental tumor (Blankenstein et al., 1996). Despite the limited success obtained previously in clinical trials involving tumor cell vaccines (Oettgen and Old, 1991), the recent molecular characterization of tumor associated antigens, the availability of many immunostimulatory molecules (i.e. cytokines) and improved methods for the transfer of genes in mammalian cells have revived the long-standing interest in immunotherapy.

The knowledge that almost all cells are able to process endogeneous proteins and present the derived peptides to T cells in association with MHC class I molecules has made it possible to raise cytotoxic T lymphocytes (CTL) directed against potential antigens found in tumor cells. Using CTL recognizing tumor derived peptides, the first genes encoding tumor associated antigens were cloned, e.g. MAGE-1, MAGE-3, MART-1 and tyrosinase expressed by melanomas (Boon et al., 1994), which are normal cellular genes expressed, not exclusively, by tumor cells. At present, it is widely believed that most, if not all, tumors possess antigens against which an immune response can be triggered, although, these antigens must not necessarily be tumor specific.

However in the majority of tumors, tumor associated antigens have not yet been characterized leaving the use of tumor cells as vaccines as the only source of putative tumor specific or associated antigens. In order to generate an immune response by apparently non-immunogenic tumor cells, a variety of genes encoding immunostimulatory activity, most often cytokines, have been used to modify the tumor cells genetically in order to increase their immunogenicity and induce an anti-tumor immune response, both locally and systemically. Cytokines play a central role in immunoregulation and their anti-tumor activity is usually not mediated by direct cytotoxic or cytostatic effects on tumor cells but results from the triggering of an anti-tumor immune response, which is believed to act primarily through short range communication between different immune and non-immune cells. This finding led

T. Blankenstein (ed.) Gene Therapy
©1999, Birkhäuser Verlag Basel

to the application of cytokines by continuous local release using cytokine-transfected tumor cells. Increased local levels of cytokines can achieve an extremely high therapeutic index and minimise systemic toxic side effects. With regard to cancer vaccines, cytokines locally secreted by the transfected tumor are expected to trigger a systemic immunostimulatory cascade thereby imitating more closely a physiological immune response. Tumor cells engineered to secrete one additional cytokine can be completely rejected in a tumor-specific and T cell dependent fashion (Blankenstein et al., 1996). However, depending on the cytokine used, the amount of secreted cytokine or the tumor model itself, tumor suppression can be either complete, partial or not seen at all (Blankenstein et al., 1996).

A similar approach for vaccine construction was derived from the knowledge that when antigen presenting cells present antigen derived peptides to T cells, in association with MHC class I molecules, a second signal is required for the activation of T cells, namely the interaction of costimulatory molecules, such as B7 (so far two related genes B7.1 and B7.2 have been cloned) with its ligands (CD28 and CTLA-4) on T cells. In the presence of this second signal, activation, proliferation and expansion of anti-tumor T cells might occur. However, the majority of tumor cells lack the B7 molecules and when antigens are presented by tumor cells, the absence of this second signal can generate T cell anergy, tolerance or apoptosis (Linsley and Lettbetter, 1993; Chen et al., 1993; June et al., 1994). In several tumor models it was shown that B7.1 gene transfected tumors, but not the unmodified tumors, were rejected in syngeneic animals and induced tumor immunity (Chen et al., 1992; Townsend and Allison, 1993; Baskar et al., 1993). However, it was also reported that B7.1 expression increased immunogenicity only of such tumors which were, *per se*, to some degree immunogenic and expressed MHC class I molecules (Chen et al., 1994).

Since vaccine cells, transfected to express B7.1, were not effective when tumors were of low immunogenicity (Chen et al., 1994) and since the efficiency of cytokine gene modified tumor cells as vaccines must not necessarily be better than a vaccine consisting of tumor cells to which an adjuvant (e.g. Corynebacterium parvum) was mixed (Hock et al., 1993b), the necessity for improving genetically engineered tumor vaccines became apparent. One strategy was the cotransfection of the costimulatory molecule B7.1 and cytokines in tumor cells (Cayeux et al., 1995, 1996). It was possible to demonstrate that vaccination with IL4/B7.1 or IL7/B7.1 cotransfected tumor cells was superior to a vaccine with only one of these genes or a tumor cell/adjuvant vaccine, thereby, demonstrating that tumor cell vaccines can be further improved. In this chapter, we will discuss some of the findings regarding growth suppression/rejection of genetically engineered tumor cells and the underlying immunological mechanisms, followed by a discussion of experiments addressing the question of how to use such cells as vaccines.

4.2 Cytokine Gene Modified Tumor Cells

The list of molecularly defined cytokines grows continually. These immunomodulatory molecules are involved in a wide range of regulatory functions *in vivo* including regulation of immune response, inflammation and hematopoesis. Cytokines are expressed by various cell types and often have similar activities with respect to each other. Their effects vary depending on the cell type on which they act (differentiation, proliferation, activation, suppression). Since they have a short half live in serum, they seem to provide, primarily, short range signalling between immune and/or non-immune cells. When used in the treatment of tumors, cytokines are often toxic when systemically applied in unphysiologically large amounts but have been shown to provide some therapeutic benefit in selected situations (e.g. IL-2 in some melanoma or renal cell carcinoma patients) when given to cancer patients (Lotze and Rosenberg, 1991).

The local secretion of cytokines by cytokine gene modified cells circumvents the problem of toxicity and increases the concentration of cytokine at the site of tumor growth. The efficacy of cytokine gene modified tumor cells has been investigated in rodent models in two ways (Fig. 4-1): 1) By analysis of the growth characteristics (extent of tumor growth inhibition caused by the tranfected cytokine) of the gene transfected tumor cells themselves (Fig. 4-1A) and 2) by testing the potential of gene modified cells to act as vaccine. The analysis of how efficient a cytokine modified tumor cell is as vaccine consisted, in most cases, in pre-immunization of mice with the cytokine gene modified tumor cells and subsequent contralateral challenge with the parental unmodified tumor cells (Fig. 4-1B). Vaccine efficiency is correlated to the percentage of mice that remained tumor-free (rejected a challenge with a parental tumor). Alternatively, in a therapeutic model, the parental unmodified tumor cells were injected and, shortly afterwards, treatment of the mice with the gene modified cells was started (Connor et al., 1993; Porgador et al., 1992, 1993a, 1993b). The survival rate of animals without tumor or the inhibition of metastasis was monitored during a certain period after immunization. In most cases, but not always (e.g. GM-CSF), growth inhibition of the transfected tumor cells and their vaccine effects seem to correlate.

4.2.1 Many Cytokines when Expressed by Transfected Tumor Cells Reduce Tumorigenicity

A large number of cytokines have been expressed in many different rodent tumor cells. Among at least 16 cytokines, which have been analyzed in gene transfection/tumorigenicity studies (Blankenstein et al., 1996), some selected examples are given which demonstrate the different biological effects observed with different cytokines. The examples include different tumor cell lines transfected to produce any

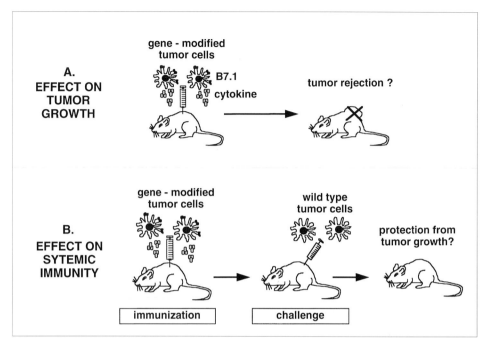

Figure 4-1. Mouse model to evaluate efficiency of gene modified vaccine cells. (A) The extent of tumor growth inhibition caused by genetically modifying tumor cells is performed by injecting mice with gene modified tumor cells and comparing their growth characteristics to parental unmodified tumor cells. (B) The analysis of how efficient are gene modified tumor cells as vaccines consists in most cases in immunization of mice with the gene modified tumor cells and subsequent contralateral challenge with the parental unmodified tumor cells. Vaccine efficiency is correlated to the percentage of mice that remain tumor-free (reject a challenge with a parental tumor).

of the cytokines IL-2, IL-7, TNF or IL-5 (Tab. 4-1). The classification of *decreased* or *unchanged* tumorigenicity of the transfected tumor cells in Table 1 is somewhat arbitrary because of variability in experimental designs, differences in the quantity of cytokine secreted and individual criteria for judgement of changed immunogenicity/tumorigenicity in a particular study. *Decreased* tumorigenicity is defined here by a strongly reduced growth of the cytokine gene transfected tumor cells, e.g. a high percentage of the mice completely rejected an otherwise lethal tumor cell challenge. *Unchanged* tumorigenicity means that expression of the transfected cytokine in tumor cells only slightly reduced tumor growth or not at all.

Quite reliably, IL-2 and IL-7 expression by tumor cells led to their rejection in (syngeneic) mice. The amount of cytokine produced by the tumor cells often correlated with the decrease of tumorigenicity. Some cytokines such as TNF suppressed

Table 1. Cytokine gene transfected rodent tumors

Gene transfected	tumorigenicity of the indicated cell lines		References
	decreased	unchanged	
IL-2	Rat-1		Bubenik et al., 1988
	X63		Bubenik et al., 1990
	CT-26, B16-F10		Fearon et al., 1990
	CMS-5		Gansbacher et al., 1990a
	HSLNV		Russell et al., 1991
	P815		Ley et al., 1991
	TS/A		Cavallo et al., 1992
	J558L		Hock et al., 1993a
	MCA-102		Karp et al., 1993
	3LL		Porgador et al., 1993b
			Ohe et al., 1993
	MBT-2		Connor et al., 1993
	4T07		Tsai et al., 1993
	B16-F10		Dranoff et al., 1993
	EL4		Visseren et al., 1994
	neuro-2a		Katsanis et al., 1994
IL-7	J558L, TS/A		Hock et al., 1991, 1993a
	FSA		McBride et al., 1992
	203-glioma		Aoki et al., 1992
TNF	CHO		Oliff et al., 1987
			Qin et al., 1995a
	J558L		Blankenstein et al., 1991
			Hock et al., 1993a
	1591-RE		Teng et al., 1991
	MCA-205		Asher et al., 1991
	EB	ESB[§]	Qin et al., 1993
		MCA-102	Karp et al., 1993
		B16-F10	Dranoff et al., 1993
	TS/A		Allione et al., 1994
IL-5		TS/A, J558L	Krüger-Krasagakes et al., 1993
		B16-F10	Dranoff et al., 1993

The classification 'decreased' or 'unchanged' tumorigenicity had to be somewhat arbitrary because different criteria were chosen by different investigators. 'Unchanged' tumorigenicity means that the transfected cytokine only slightly or not at all decreased tumorigenicity of the transfected tumor cells eventhough it does not take into account that higher expression levels of the cytokine or variation in number of injected tumor cells could obtain different results. 'Decreased' tumorigenicity means that the tumorigenicity of the transfected cells was strongly diminished, e.g. at least part of the mice completely rejected the tumor cells.
[§]Metastatic potential of the cell was increased.

tumor growth in some, but completely failed or only marginaly reduced tumorigenicity in other tumor models (Qin et al., 1993; Karp et al., 1993; Dranoff et al., 1993). In one tumor model (ESB), it has even been shown that expression of TNF increased the metastatic potential of the cells (Qin et al., 1993). This demonstrates that the particular tumor cell phenotype can drastically influence the biological effect of the transfected cytokine. Other cytokines e.g. IL-5 have been expressed in several tumor cell lines which showed that the tumorigenicity was not changed, eventhough the transfected cytokine induced an immune reaction. IL-5 transfected tumors were heavily infiltrated by eosinophilic granulocytes (Krüger-Krasagakes et al., 1993). Together, various factors seem to influence whether a transfected cytokine can suppress tumor growth by a local anti-tumor reaction: 1) the cytokine itself, 2) the tumor under investigation, 3) the amount of cytokine produced by the tumor and possibly other factors such as mouse strain or site of injection.

4.2.2   Immunological Mechanisms of Tumor Rejection

A large number of different immune cells have been detected infiltrating the various cytokine producing tumors. A frequent observation was the requirement for non-T cells and T cells for tumor rejection. Different cytokines seem to trigger partially similar, partially individual effector mechanisms. Inducer and effector cells are often difficult to distinguish from each other (Blankenstein et al., 1991; Colombo et al., 1992; Pardoll, 1993; Blankenstein et al., 1996). In the various cases where injection of cytokine gene modified tumor cells resulted in decrease of tumorigenicity, the immunological mechanisms leading to tumor rejection have been analyzed in several ways:

1) Comparison of the growth kinetics of gene transfected cells in syngeneic immunocompetent and immunodeficient mice (e.g. T cell deficient nude or T and B cell deficient SCID mice). This has shown that for a variety of cytokines, both T and non-T cells contributed to the tumor suppressive effect. In the absence of T cells, tumor growth was effectively suppressed by a number of cytokines such as IL-2, IL-4, IFNγ and TNF. However, this effect was often transient and most of the mice developed a tumor after a long latency period (Hock et al., 1993a) and often when tumors developed in nude mice, reduced expression of the transfected cytokine in the tumor cells could be found. This demonstrates that one function of the T cells is to prevent the outgrowth of cytokine loss variants. In these experiments the T cell subsets required for complete tumor rejection cannot be characterized.

2) Characterization of the type of immune effector cells infiltrating the tumor tissue using immunohistochemistry or electron microscopy. The detection of certain cell types in the cytokine producing but not parental or mock-transfected tumors suggested that these cells participated in the anti-tumor response. Indeed, sufficient examples, derived from studies where certain immune cell subsets were depleted, ex-

ist which confirmed the anti-tumor activity of immune cells present in the tumor. As example, CD4$^+$ and CD8$^+$ T cells specifically infiltrated J558-IL7 tumors and both cell types were mandatory for complete tumor rejection (Hock et al., 1991, 1993a). On the other hand, immune cells can infiltrate the tumor without being involved in the anti-tumor response (e.g. CD4$^+$ cells in J558-TNF tumors (Hock et al., 1993a). Moreover, despite being heavily infiltrated by immune cells, cytokine producing tumors can progressively grow. This, for example, has been shown for macrophages in M-CSF and eosinophils in IL-5 producing tumors (Dorsch et al., 1993; Krüger-Krasagakes et al., 1993).

3) Selective elimination of defined cell populations by *in vivo* application of appropriate antibodies. By observing the growth kinetics of gene modified tumor cells in these cases, the role of particular cell types in the generation of an anti-tumor response can be clarified. Such *in vivo* depletion studies have been done for the T cell subsets CD4$^+$ and CD8$^+$, Mac1$^+$ cells (predominantly macrophages), granulocytes (neutrophils or eosiniphils) and asialo GM1$^+$ cells (NK cells). However, the question regarding whether or not the secreted cytokine acts directly on the selected effector cell population remains open. We have previously found that, regardless which of the cytokines IL-2, IL-4, TNF and IFNγ was transfected in one tumor model, CD8$^+$ T cells were always needed for long-term tumor elimination eventhough tumor growth was quite effectively suppressed for a certain period in CD8$^+$ T cell depleted mice. Therefore, we have argued that the non-specific inflammatory cells prevent rapid tumor burden without being able to completely eliminate the tumor cells. In most cases, these cells enable CD8$^+$ T cells to develop cytotoxic activity which must not necessarily be caused in all cases directly by the transfected cytokine (Hock et al., 1993a).

4) Analysis of cytotoxic T lymphocyte (CTL) activity of cells derived from spleens of immunized mice. This assay indirectly measures systemic tumor immunity. In several cases, CTL activity correlated with rejection of the cytokine gene transfected tumor (Fearon et al., 1990; Gansbacher et al., 1990). However, *in vitro* cytolytic activity and *in vivo* anti-tumor activity do not always correlate (Barth et al., 1991; Pericle et al., 1994).

The loss of tumorigenicity of cytokine gene transfected tumor cells results from an increased immunogenicity. The secreted cytokines can induce an inflammatory reaction whereby, depending on the cytokine (and the investigated tumor), different immune cells are attracted, activated and eventually destroy the tumor cells. In general, the contribution of a certain cell type to tumor suppression must not always result from a direct effect of the transfected cytokine on these cells but rather can be due to an interaction between different immune cells. As an example, the presence of macrophages and eosinophilic granulocytes infiltrating IL-7 producing tumors was dependent on the presence of CD4$^+$ T cells. When CD4$^+$ T cells were depleted, tumor growth was restored and there was a noticeable absence of macrophages and granulocytes within the tumor (Hock et al., 1991, 1993a).

4.3 Expression of T cell Costimulatory Molecules to Increase Tumor Immunogenicity

It is generally accepted that during antigen presentation the first signal is derived from the interaction of the T cell receptor with MHC class I molecules binding peptides derived from a processed antigen on antigen presenting cells. However, this signal alone is not sufficient to activate T cells. A second signal is necessary for T cell activation, namely the interaction of costimulatory molecules with ligands on T cells. In the presence of this second signal, activation, proliferation and expansion of anti-tumor T cells may occur. One well characterized type of ligand-receptor molecule is the B7 family comprising B7.1 (CD80) and B7.2 (CD86) present on APC. Their counter-receptors are CD28 and CTLA-4 on T cells. Stimulation of T cells by ligation of CD28 leads to the production of IL-2 and other cytokines. The large majority of human and mouse tumors lack costimulatory molecules and in order to improve or to enable these tumor cells to become antigen presenting cells, several tumor cells were subjected to B7.1 gene transfer and tumor rejection as well as systemic immunity were observed (Chen et al., 1992; Townsend and Allison, 1993; Baskar et al., 1993). Depending on whether the tumor cells expressed MHC class I or class II molecules, this effect was mediated by CD8$^+$ or CD4$^+$ T cells. B7.1 expressed by tumor cells seems to act on T cells at the induction and effector phase (Ramarathinam et al., 1994). Similar to cytokines, the level of B7.1 expression appears to play a critical role. B16-F10 tumor cells were only rejected when they expressed high but not low amounts of B7.1 (Wu et al., 1995). In some tumor models where the tumors were known to be non-immunogenic, B7.1 failed to enhance immunogenicity (Chen et al., 1994). In another tumor model, the presence of ICAM-1$^+$ together with B7.1 was found to be necessary to induce an efficient tumor specific immune response (Cavallo et al., 1995).

4.4 Tumor Immunity Induced by Genetically Engineered Tumor Cells

4.4.1 Single Gene Modification of Tumor Cells

The important issues as far as clinical relevance is considered are whether or not genetically engineered tumor cells can act as protective therapeutic vaccines, how strong the vaccine effect is, how reproducible it is in different tumor models, which genetic modification provides the most effective vaccine and which are the immunological mechanisms underlying tumor immunity.

Can cytokine gene modified tumor cells act as vaccine? In order to answer this question, it is important to distinguish whether the vaccine effect results from the transfected cytokine or is simply the effect of tumor cells themselves. Different tumor cell lines used for vaccine experiments can considerably vary in inherent immunogenicity and immunization with irradiated parental tumor cells, occasionally, may be sufficient to protect the mice against a challenge with a tumorigenic dose of the same cells (Dranoff et al., 1993). However, for several cytokines it has been shown that the respectively transfected cells, used for immunization, protected more mice from a tumorigenic parental tumor cell challenge compared with irradiated or Mitomycin C treated parental tumor cells. This has been shown for IL-2, IL-4, IL-6, IL-7, TNF, IFNγ and GM-CSF (Cavallo et al., 1992; Porgador et al., 1992; Hock et al., 1993b; Dranoff et al., 1993; Porgador et al., 1993a, b; Connor et al., 1993; Pericle et al., 1994). The generated immunity was usually tumor specific. The transfected cytokine appeared to actively contribute to the induction of tumor immunity. However, in one tumor model repeated immunization with the irradiated parental cells could completely compensate for a single injection of cytokine (IL-2) secreting cells (Cavallo et al., 1993).

How strong is the vaccine effect? The efficacy of gene modified tumor cells as vaccines has usually been tested by immunization of mice and subsequent (between 1–4 weeks later) challenge with a tumorigenic dose of the parental tumor cells at a distant site, or in a more rigorous experimental setting, mice first received the parental tumor cells and, shortly thereafter, were treated with the gene modified cells. In the case of pre-immunization, the question of vaccine efficacy can be addressed by either increasing the challenge dose while in the therapeutical setting, one can ask how long after injection of the parental tumor cells can the vaccine cells still induce tumor elimination. We have shown that by increasing the challenge dose the vaccine effect of several cytokine (IL-2, IL-4, IL-7, TNF, IFNγ) producing J558L cells became less obvious (Hock et al., 1993b). When the vaccine effect of cytokine gene transfected cells was compared in a prophylactic and a therapeutic setting, it was uniformly observed that the vaccine potency diminished in the latter case, eventhough, vaccination was started within the first 10 days after tumor cell challenge (Golumbek et al., 1991; Dranoff et al., 1993; Connor et al., 1993; Cavallo et al., 1993). The reason for that is not known but most likely is related to tumor burden, tumor induced immune suppression, time needed for tumor immunity (e.g. CTL activity) to develop, or any combination of these factors. Certainly, a therapeutic model is impeded by the extremely rapid growth of most mouse tumors which does not reflect the situation in most cancer patients. If, however, the decrease or loss of vaccine effect in the therapeutic, compared with the prophylactic, model results from the difference in inducing an anti-tumor response in naive animals and breaking tolerance (which seems to be dependent on the presence of the tumor (Schreiber, 1993)) in tumor bearing animals, then, further treatment modalities which abolish tumor induced immune suppression or reduce tumor load have to precede vaccination.

How reproducible is the vaccine effect of cytokine gene modified tumor cells? The answer here is, unfortunately, that the vaccine effect is poorly reproducible when the same cytokine is analysed in different tumor models as shown for the cytokines IL-2, IL-4, IL-6, IFNγ or GM-CSF (Tab. 4-2). This is not surprising if one considers the variability in phenotypes between individual tumors. At the present time, for all cytokines currently employed in clinical trials, there are examples of negative vaccination results in rodent tumor models. There is no reason to assume that similar discrepancies will not also be found in cancer patients. It is clearly of great importance to elucidate the cellular and molecular mechanisms which are responsible for these different results.

Table 2. Poor reproducibility of vaccine effect with gene modified tumor cells

Cytokine gene	Vaccine effect	No vaccine effect
IL-2	TS/A (Cavallo et al., 1992) MBT-2 (Connor et al., 1993)	MCA-102 (Karp et al., 1993) B16-F10 (Dranoff et al., 1993)
IL-4	Renca (Golumbeck et al., 1991) TS/A (Pericle et al., 1994)	J558L (Hock et al., 1993b) B16-F10 (Dranoff et al., 1993)
IL-6	3LL (Porgador et al., 1992)	B16-F10 (Dranoff et al., 1993) J558L (unpublished)
IFN-γ	3LL (Porgador et al., 1993a)	J558L (Hock et al., 1993b)
GM-CSF	B16-F10 (Dranoff et al., 1993) MBT-2 (Saito et al., 1994)	TS/A (Allione et al., 1994) J558L (Qin et al., 1997)

By which mechanisms do gene modified vaccines induce tumor immunity? Tumor immunity induced by gene modified tumor vaccines was consistently shown to be dependent on T cells. Two models have been proposed by which T cells are induced. In the first model, the tumor cells directly present tumor antigens to T cells, which are activated in the presence of tumor cell derived T cell stimulatory gene products. In support of this assumption are experiments with IL-2 or B7.1 gene transfected MHC class I+ class II− tumors, which showed that tumor immunity was exclusively dependent on CD8+ T cells (Fearon et al., 1990; Chen et al., 1992). However, again no conclusive results have been obtained in different models. Tumor immunity induced by another IL-2 gene transfected MHC class I+ class II− tumor required both, CD4+ and CD8+ T cells although CD4+ T cells were not needed for rejection of the transfected cells (Cavallo et al., 1992). This could indicate that, in addition to a direct recognition of the tumor cells by T cells, tumor antigens are taken up by host APC, processed and presented to CD4+ T cells which, in turn, may help CD8+ T cells to sustain immunity. The second model is based on experiments with GM-CSF transduced tumor cells (Dranoff et al., 1993). GM-CSF induces differentiation/activation of dendritic cells (Steinmann, 1991) and the vaccine effect, mediated by GM-CSF producing tumor cells, has been explained by indirect antigen presentation through APC, possibly dendritic cells, which are induced by GM-CSF to engulf tumor antigen and, after homing into secondary lymphoid organs, e.g. lymph nodes, present the antigens to T cells via both MHC class I and class II molecules. Consistently, CD4$^+$ and CD8$^+$ T cells were required for rejection of the challenge tumor. Indeed, the GM-CSF mediated vaccine effect requires bone marrow derived cells for presentation of even MHC class I antigens, as demonstrated by the use of bone marrow chimeras in a tumor model with a surrogate tumor antigen (Levitsky et al., 1994). Furthermore, GM-CSF induced the accumulation of dendritic cells at the vaccine site of respectively transfected tumor cells (Qin et al., 1997). In the majority of cases, however, it is not known whether T cells are activated directly by the tumor cells, indirectly by host APC or both pathways and which consequences these two alternatives have on generation of tumor immunity.

4.4.2 Comparison of Vaccine Effects of Gene Modified Tumor Cells with Traditional Tumor Vaccines

Tumor cell vaccines, consisting of a mixture of tumor cells and adjuvants, have been widely tested clinically with little reported success (Oettgen and Old, 1991). Thus, it is important to compare the new gene modified tumor cell vaccines to this long-standing established approach. One advantage of using cytokines instead of adjuvants is standardization, which was not possible with adjuvants since efficacy was variable from batch to batch. We have demonstrated previously, that J558L mouse plasmacytoma tumor cells transfected to produce IL-2, IL-4 IL-7, TNF or IFNγ were not qualitatively different at generating systemic immunity and protecting mice against a wildtype tumor cell challenge when compared with a classical vaccine consisting of tumor cells admixed with the adjuvant C. parvum (Hock et al., 1993b; Cayeux et al., 1996). In essence, similar results were obtained in another tumor model (Cayeux et al., 1995) and Allione et al. (1994) did not find a marked difference in immunogenicity between cytokine engineered tumor cells and a tumor cell/adjuvant mixture. However, although it is not surprising that adjuvants, which consist of a mixture of pathogens, generate an immune response, it is not clear whether or not an immune response directed against adjuvants is the optimal response to generate against tumor cells. Furthermore, as the mechanisms of action of cytokines as immunostimulators become more clearly defined, the optimal type of immune response required for the destruction of tumor cells will eventually be characterized.

4.4.3   Improvement of Gene Modified Tumor Cell Vaccines by Combined Effects

Considering the results obtained with single gene modified tumor cell vaccines, it is clear that there is room for improvement. While in some experimental models single gene modification of tumor cells with cytokines or with costimulatory molecules led to improved protection against challenge with wildtype tumors, in other models modification with one gene was not sufficient to induce either an efficient or reproducible anti-tumor response (Hock et al., 1993b). Therefore, various combined strategies have been tested. We have previously found in the J558 tumor model that injection of any possible combination of tumor cells producing either of the cytokines IL-2, IL-4, IL-7, TNF or IFNγ did not augment tumor immunity in comparison with that obtained by one cytokine alone (Hock et al., 1993b). Similarly, coexpression of IL-2 with any of nine different genes encoding immunostimulatory activity did not reveal any synergistic protective effect (Dranoff et al., 1993).

Based on previous observations that rejection or lack of growth inhibition or vaccine potency of cytokine gene modified tumors did not always correlate with the extent of T cell infiltration in the tumor, eventhough T cells were required for rejection (Hock et al., 1993a; 1993b), we speculated that T cells in cytokine (IL-2, IL-4, IL-7, TNF, IFNγ) transfected tumors were not appropriately activated. Therefore recently, we have analysed the vaccine efficacy of tumor cells cotransfected with genes for B7.1 and cytokines, e.g. IL-4, IL-7. We found that in two tumor models, J558L and TS/A, coexpression of IL-4/B7.1 and IL-7/B7.1 had an impact on immunogenicity (Cayeux et al., 1995, 1996). In both tumor models expression of cytokine or B7.1 alone did not induce reliable rejection of the transfected cells, whereas mice injected with cytokine/B7.1 coexpressing cells always rejected the tumor cells. This effect was mediated by T cells. Moreover, vaccines using TS/A-IL7/B7.1 or J558-IL4/B7.1 cells were more effective than adjuvant/tumor cell mixture or single gene-transfected cells. For IL-7 and B7.1 cotransfected cells, this observation is compatible with *in vitro* data showing a synergistic effect of IL-7 and anti-CD28 antibodies on T cell proliferation, cytokine secretion and IL-2 receptor expression (Costello et al., 1993). Interestingly, the cytokines induced in this *in vitro* system included IL-4, TNF and GM-CSF which have all been implicated to either enhance immunogenicity of the respective gene transfected tumor cells or to augment tumor antigen presentation by professional antigen presenting cells, such as dendritic cells. When the tumor infiltrating cells were analysed in single or double gene transfected tumors, we found that in B7.1⁺ tumors, a high percentage of T cells were CD28⁺ but few T cells were CD25⁺. In IL-7 tumors, CD28⁺ T cells were virtually absent and a high proportion was CD25⁺. Both activation markers were found on the majority of T cells in IL-7/B7.1 cotransfected tumors. Although the precise mechanism of action of the combined use of IL-7 and B7.1 is not yet known, one can speculate that IL-7 and B7.1 act on tumor infiltrating T cells by inducing a cytokine environment which contributes to an efficient host immune response. As our knowledge improves regarding the precise way to induce, most potently and specifically, an anti-tumor immune response (T cell mediated), more sophisticated and precise strategies to construct potent tumor vaccines can be elaborated.

4.4.4 Viable Versus Non-Proliferating Tumor Cell Vaccines

We have reported, in two different tumor models, that irradiation of tumor cell vaccines decreased the vaccine efficiency. When the potency of irradiated versus live IL7/B7.1 or IL4/B7.1 tumor cell vaccines to protect against tumor cell challenge was compared, we found that live cells generated a better systemic tumor immunity than irradiated cells (Cayeux et al., 1995, 1996). This was not caused by the loss of expression of the transfected genes; both cytokine and B7.1 were still expressed *in vitro* 10 days after irradiation. Similar results were previously obtained with tumor cell/adjuvant as vaccine (Hock et al., 1993b) or with cytokine or B7 transfected tumor cells in other models (Allione et al., 1994; Townsend et al., 1994). In contrast, GM-CSF transduced B16-F10 cells (Dranoff et al., 1993) were equally effective regardless of whether the vaccine cells were irradiated or viable. This comparison was possible because GM-CSF producing B16-F10 cells were rejected when, additionally, IL-2 was expressed by the tumor cells. No clear explanation for these different findings is yet available.

4.5   Conclusion

As our understanding of an anti-tumor response increases, it is clear that improved immuntherapeutic approaches will be developed. In mouse tumor models, it has been shown that gene modified tumor cell vaccines can be more effective than traditional tumor vaccines which they are going to replace in clinical trials. Additionally, the effectiveness of expression of costimulatory molecules and cytokines by tumor cell vaccines has been demonstrated in the studies mentioned above. However, from these experimental tumor models we have learned that problems, resulting from tumor phenotypical diversity and the persistent difficulty of successfully treating tumor bearing animals with immunotherapy, will continually be present. Established therapies for the treatment of cancer remain surgery, radio- and chemotherapy but lack of long term cure as well as the development of metastases well justify the addition of new therapeutic approaches such as immuno-gene therapy as a supplement.

Acknowledgments

This work was supported by grants from the US-Englische Krebshilfe, Mildred Scheel Stiftung, e. V. and the B.M.B.F.

References

1. Allione, A., Consalvo, M., Nanni, P., Lollini, P. L., Cavallo, F., Giovarelli, M., Forni, M., Gulino, A., Colombo, M. P., Dellabona, P., Hock, H., Blankenstein, Th., Rosenthal, F. M., Gansbacher, B., Bosco, M. C, Musso, T., Gusella, L., Forni, G. (1994) Immunizing and curative potential of replicating and nonreplicating murine mammary adenocarcinoma cells engineered with interleukin (IL)-2, IL-4, IL-6, IL-7, IL-10, tumor necrosis factor α, granulocyte-macrophage colony-stimulating factor, and γ-Interferon gene or admixed with conventional adjuvants. *Cancer Res.* 54: 6022–6026.
2. Barth, R. J., Mulé, J. J., Spiess, P. J., Rosenberg, S. A. (1991) Interferon γ and tumor necrosis factor have a role in tumor regression mediated by murine CD8+ tumor-infiltrating lymphocytes. *J. Exp. Med.* 173: 647–658.
3. Baskar, S., Ostrand-Rosenberg, S., Nabavi, N., Nadler, L. M., Freeman, G. J., Glimcher, L. H. (1993) Constitutive expression of B7 restores immunogenicity of tumor cells expressing truncated MHC class II molecules. *Proc. Natl. Acad. Sci. USA* 90: 5687–5690.
4. Blankenstein Th, Rowley DA, Schreiber, H. (1991) Cytokines and cancer: experimental systems. *Curr. Opin. Immunol.* 3: 694–698.
5. Blankenstein, T. (1996) Genetic approaches to cancer immunotherapy. *Rev. Physiol. Biochem. Pharmacol.* col (in press).
6. Boon, T., Cerottini, J. C., Van den Eynde, B., van der Bruggen, P., Van Pel, A. (1994) Tumor antigens recognized by T lymphocytes. *Annu. Rev. Immunol.* 12: 337–365.
7. Cavallo, F., Giovarelli, M., Gulino, A. et al. (1992) Role of neutrophils and CD4$^+$ T lymphocytes in the primary and memory response to nonimmunogenic murine mammary adenocarcinoma made immunogenic by IL-2 gene transfection. *J. Immunol.* 149: 3627–3635.
8. Cavallo, F., Di Pierro, F., Giovarelli, M., Gulino, A., Vacca, A., Stoppacciaro, A., Forni, M., Modesti, A., Forni, G. (1993) Protective and curative potential of vaccination with interleukin-2-gene-transfected cells from a spontaneous mouse mammary adenocarcinoma. *Cancer Res.* 53: 5067–5070.
9. Cavallo, F., Martin-Fontecha, A., Bellone, M., Heltai, S., Gatti, E., Tornaghi, P., Freschi, M., Forni, G., Dellabona, P., Casorati, G. (1995) Co-expression of B7-1 and ICAM-1 on tumors is required for rejection and the establishment of a memory response. *Eur. J. Immunol.* 25: 1154–1162.
10. Cayeux, S., Beck, C., Aicher, A., Dörken, B., Blankenstein, Th. (1995) Tumor cells cotransfected with interleukin 7 and B7.1 genes induce CD25 and CD28 on tumor infiltrating lymphocytes and are strong vaccines. *Eur. J. Immunol.* 25: 2325–2331.
11. Cayeux, S., Beck, C., Dörken, B., Blankenstein, T. (1996) Coexpression of interleukin 4 and B7.1 in murine tumor cells leads to improved tumor rejection and vaccine effect compared to single gene transfectant and a classical adjuvant. *Hum. Gene Ther.* 7: 525–529.
12. Chen, L., Ashe, S., Brady, W. A., Hellström, I., Hellström, K. E., Ledbetter, J. A.,

McGowan, P., Linsley, P. S. (1992) Costimulation of antitumor immunity by the B7 coun-terreceptor for the T lymphocyte molecules CD28 and CTLA-4. *Cell* 71: 1093–1102.

13. Chen, L., Linsley, P. S., Hellström, K. E. (1993) Costimulation of T cells for tumor im-munity. *Immunol. Today* 14: 482–486.

14. Chen, L., McGowan, P., Ashe, S., Johnston, Y., Li, Y., Hellström, I., Hellström, K. E. (1994) Tumor immunogenicity determines the effect of B7 costimulation on T cell-medi-ated tumor immunity. *J. Exp. Med.* 179: 523–532.

15. Colombo, M. P., Modesti, A., Parmiani, G. et al. (1992) Local cytokine availability elic-its tumor rejection and systemic immunity through granulocyte-T-lymphocyte cross-talk. *Cancer Res.* 52: 4853–4857.

16. Connor, J., Bannerji, R., Saito, S., Heston, W., Fair, W., Gilboa, E. (1993) Regression of bladder tumors in mice treated with interleukin 2 gene-modified tumor cells. *J. Exp. Med.* 177: 1127–1134.

17. Costello, R., Brailly, H., Mallet, F. et al. (1993) Interleukin-7 is a potent co-stimulus of the adhesion pathway involving CD2 and CD28 molecules. *Immunology* 80: 451–457.

18. Dorsch, M., Hock, H., Kunzendorf, U., Diamantstein, T., Blankenstein, Th. (1993) Macrophage colony-stimulating factor gene transfer into tumor cells induces macrophage infiltration but not tumor suppression. *Eur. J. Immunol.* 23: 186–190.

19. Dranoff, G., Jaffee, E., Lazenby, A. et al. (1993) Vaccination with irradiated tumor cells engineered to secrete murine granulocyte-macrophage colony-stimulating factor stimu-lates potent, specific, and long-lasting anti-tumor immunity. *Proc. Natl. Acad. Sci. USA* 90: 3539–3543.

20. Fearon, E. R., Pardoll, D. M., Itaya, T. et al. (1990) Interleukin 2 production by tumor cells bypasses T helper function in the generation of an antitumor response. *Cell* 60: 397–403.

21. Gansbacher, B., Zier, K., Daniels, B., Cronin, K., Bannerjy, R., Gilboa, E. (1990) Interleukin 2 gene transfer into tumor cells abrogates tumorigenicity and induces protec-tive immunity. *J. Exp. Med.* 172: 1217–1224.

22. Golumbek, P. T., Lazenby, A. J., Levitsky, H. I. et al. (1991) Treatment of established re-nal cancer by tumor cells engineered to secrete interleukin 4. *Science* 254: 713–716.

23. Hock, H., Dorsch, M., Diamantstein, T., Blankenstein, T. (1991) Interleukin 7 induces CD4$^+$ T cell-dependent tumor rejection. *J. Exp. Med.* 174: 1291–1298.

24. Hock, H., Dorsch, M., Kunzendorf, U., Qin, Z., Diamantstein, T., Blankenstein, Th. (1993a) Mechanisms of rejection induced by tumor cell targeted gene transfer of inter-leukin-2, interleukin-4, interleukin-7, tumor necrosis factor or interferon-gamma. *Proc. Natl. Acad. Sci. USA* 90: 2774–2778.

25. Hock, H., Dorsch, M., Kunzendorf, U. et al. (1993b) Vaccinations with tumor cells ge-netically engineered to produce different cytokines: effectivity not superior to a classical adjuvant. *Cancer Res.* 53: 714–716.

26. June, C. H., Bluestone, J. A., Nadler, L. M., Thompson, C. B. (1994) The B7 and CD28 receptor families. *Immunol. Today* 15: 321–331.

27. Karp, S. E., Farber, A., Salo, J. C. et al. (1993) Cytokine secretion by genetically modi-fied nonimmunogenic murine fibrosarcoma. Tumor inhibition by IL2 but not tumor necrosis factor. *J. Immunol.* 150: 896–908.

28. Krüger-Krasagakes, S., Li, W., Richter, G., Diamantstein, T., Blankenstein, Th. (1993) Eosinophils infiltrating interleukin 5 gene transfected tumors do not suppress tumor growth. *Eur. J. Immunol.* 23: 992–995.

29. Levitsky, H. I., Lazenby, A., Hayashi, R. J., Pardoll, D. M. (1994) *In vivo* priming of two distinct antitumor effector populations: The role of MHC class I expression. *J. Exp. Med.* 179: 1215–1224.

30. Linsley, P. S., Ledbetter, J. A. (1993) The role of the CD28 receptor during T cell response to antigen. *Annu. Rev. Immunol.* 11: 191–212.

31. Lotze, M. T., Rosenberg, S. A. (1991) Interleukin-2: Clinical applications. *In:* V. T. DeVita, S. Helman, S. A. Rosenberg (eds), *Biologic Therapy of Cancer, Principles and Pratice.* J.B. Lippincott Press, New York, pp. 159.

32. Oettgen, H. F., Old, L. J. (1991) The history of cancer immunotherapy. *In:* V. T. DeVita, S. Helman, S. A. Rosenberg (eds), *Biologic Therapy of Cancer, Principles and Practice.* J.B. Lippincott Press, New York, pp. 87–119.

33. Pardoll, D. (1993) Cancer vaccines. *Immunol. Today* 6: 310–316.

34. Pericle, F., Giovarelli, M., Colombo, M. P., Ferrari, G., Musiani, P., Modesti, A., Cavallo, F., Di Pierro, F., Novelli, F., Forni, G. (1994) An efficient Th2-type memory follows CD8[+] lymphocyte-driven and eosinophil-mediated rejection of a spontaneous mouse mammary adenocarcinoma engineered to release IL-4. *J. Immunol.* 153: 5659–5673.

35. Porgador, A., Tzehoval, E., Katy, V. E. et al. (1992) Interleukin 6 gene transfection into Lewis lung carcinoma tumor cells suppresses the malignant phenotype and confers immunotherapeutic competence against parental metastatic cells. *Cancer Res.* 52: 3679–3686.

36. Porgador, A., Bannerji, R., Watanabe, Y., Feldman, M., Gilboa, E., Eisenbach, L. (1993a) Anti-metastatic vaccination of tumor-bearing mice with two types of γIFN gene inserted tumor cells. *J. Immunol.* 150: 1458–1470.

37. Porgador, A., Bannerji, R., Tzehoval, E., Gilboa, E., Feldman, M., Eisenbach, L. (1993b) Anti-metastatic vaccination of tumor-bearing mice with IL-2 gene inserted tumor cells. *Int. J. Cancer* 53: 471–477.

38. Qin, Z., Krüger-Krasagakes, S., Kunzendorf, U., Hock, H., Diamantstein, T., Blankenstein Th (1993) Expression of tumor necrosis factor by different tumor cell lines results either in tumor suppression or augmented metastasis. *J. Exp. Med.* 178: 355–360.

39. Qin, Z., Noffz, G., Mohaupt, M. and Blankenstein, Th. (1997) Interleukin 10 prevents dendritic cell infiltration and vaccination with granulocyte-macrophage colony-stimulating factor gene modified tumor cells. *J. Immunol.* 159: 770–776.

40. Ramarathinam, L., Castle, M., Wu, Y., Liu, Y. (1994) T cell costimulation by B7/BB1 induces CD8 T cell-dependent tumor rejection: An important role of B7/BB1 in the induction, recruitment, and effector function of antitumor T cells. *J. Exp. Med.* 179: 1205–1214.

41. Schreiber, H. (1993) Tumor immunology. *In:* W. E. Paul (ed.), *Fundamental Immunology.* Raven Press, New York, pp. 1143–1178.

42. Steinman, R. M. (1991) The dendritic cell system and its role in immunogenicity. *Annu. Rev. Immunol.* 9: 271–296.

43. Townsend, S. E., Allison, J. P. (1993) Tumor rejection after direct costimulation of CD8+ T cells by B7-transfected melanoma cells. *Science* 259: 368–370.

44. Townsend, S. E., Su, F. W., Atherton, J. M., and Allison, J. P. (1994) Specificity and Longevity of Antitumor Immune Responses Induced by B7-transfected Tumors. *Cancer Res.* 54: 6477–6483.

45. Wu, L. C., Huang, A. Y. C., Jaffee, E. M, Levitzky, H. I. and Pardoll, D. M. (1995) A reassessment of the role of B7-1 expression in tumor rejection. *J. Exp. Med.* 182: 1415–1421.

5 Autologous and Allogeneic Tumor Cell Vaccines

E. Nößner, D. J. Schendel

5.1 Human Tumor Cell Vaccine

5.1.1 The Principle of Tumor Cell Vaccination

The basic principle underlying the development of tumor vaccines is that tumor cells express antigens that enable the immune system to distinguish them from their normal counterparts. Because of other deficiencies, tumor cells are ineffective at initiating specific immune responses. Genetic engineering of tumors can compensate for some deficiencies, thereby enhancing their potential to induce antitumor immunity. Immunization of cancer patients with genetically engineered tumor vaccines should activate lymphocytes that are able to eliminate residual, unmodified tumor cells which themselves would be incapable of inducing an immune response.

This contention is supported by studies demonstrating that the requirements for initiation of immune responses are both quantitatively and qualitatively distinct from those required for activated lymphocytes to mediate their effector functions [1,2]. As an example, few cell types can stimulate optimal immune responses following infection with virus; virus-infected dendritic cells are particularly efficient because they express high levels of virus-associated determinants at their surface which stimulate specific lymphocytes expressing complementary receptors. Additionally, they express other surface molecules and secrete cytokines that deliver supplementary activation signals to specific lymphocytes. An important component of the resultant anti-viral immunity is a population of activated, virus-specific cytotoxic T lymphocytes (CTL) that are able to recognize and eliminate any cell in the body that harbors virus, as long as it expresses the same antigens to which the CTL were primed. In contrast to the dendritic cells, other virus-infected target cells can be recognized even when they express much lower levels of virus-associated antigen and they need not express costimulatory molecules to be destroyed by activated CTL.

Different requirements for the induction of tumor-specific immune responses versus subsequent recognition and elimination of tumor cells by activated effector lymphocytes have also been observed in animal tumor models *in vivo* as well as by studying the lymphocyte-tumor cell interactions of cancer patients *in vitro*. While

T. Blankenstein (ed.) Gene Therapy
©1999, Birkhäuser Verlag Basel

antitumor antibody responses are generated during the course of tumor development [3], animal studies indicate that cell-mediated immune responses usually play a greater role in tumor rejection [4, 5]. Various cell-mediated effector mechanisms may participate in a tumor response but the induction of specific thymus-derived (T) lymphocytes is necessary to provide protection against outgrowth of tumor cells and to eliminate established metastases. Lymphocytes of both the CD4 and CD8 subsets can contribute to optimal protection [6–8]. T cells express specific receptors that can be triggered through vaccination causing them to clonally expand. This leads not only to the generation of effector cells that specifically attack tumor cells but also to the induction of memory cells that provide long-term protection against metastases.

5.1.2 Autologous and Allogeneic Tumor Cell Vaccines

Two types of genetically engineered tumor cell vaccines are being considered currently for cancer patients [9] (Fig. 5-1). The first involves the genetic modification of a patient's own tumor cells and is designated as "autologous vaccination". A tumor is excised and tumor cells are genetically modified *ex vivo* to enhance their immunogenicity. Modified tumor cells are reapplied intradermally or subcutaneously as a vaccine to the same individual. Modified cells can be cryopreserved for multiple rounds of vaccination. It is also possible to modify other cells of the patient, such as fibroblasts, which are admixed with unmodified tumor cells. Alternatively, modifying genes can be directly introduced into tumors in the patient if they are surgically accessible. All autologous vaccines are patient specific, thus the same procedures are repeated to generate a unique vaccine for each individual. The second approach is that of "allogeneic tumor cell vaccination" in which tumor cell lines, established from one or several patients, are used to vaccinate selected patients bearing the same type of tumor which expresses the same tumor antigen. These tumor cell lines are also genetically engineered *in vitro* to improve their immunogenicity.

T lymphocytes are of central importance for successful tumor vaccination, therefore, it is necessary to understand the molecular basis for their activation.

5.2 Molecular Basis for T Cell Recognition and the Development of Effector Function

5.2.1 Antigen Recognition: MHC Restriction and the Trimolecular Complex Involving MHC, Peptide and T Cell Receptor

Immune recognition by T lymphocytes occurs through surface interactions involving a specific T cell receptor (TCR) on the lymphocyte and an antigen on the target cell. Unlike antibodies, TCR do not recognize intact antigens. With few exceptions,

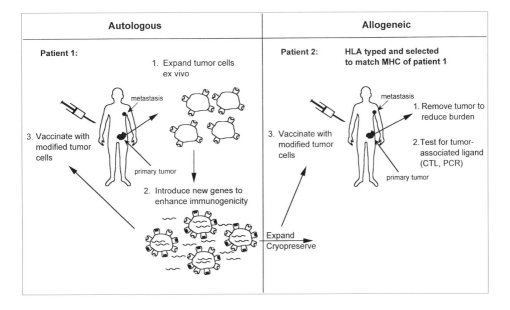

Figure 5-1. Tumor cell vaccination: In an autologous vaccination procedure, a patient's tumor is removed and tumor cells are expanded "ex vivo". New genes are introduced to enhance their immunogenicity. These modified cells are used as a vaccine for the patient from whom the tumor was originally derived. In an allogeneic setting, vaccine cells from one patient are used to vaccinate a second patient. This patient is selected to share MHC subtypes of patient 1 that present tumor-associated determinants. After removal of the primary tumor, patient 2 is immunized with vaccine cells derived from patient 1. In both settings it is hoped that specific immune responses are stimulated that will eliminate disseminated tumor cells.

they only recognize fragments of antigens, in the form of peptides, which are presented on the cell surface by molecules of the major histocompatibility complex (MHC). The ability of T cells to recognize antigen only when it is presented by MHC molecules is called "MHC restriction" [10]. There are two classes of MHC molecules, class I and class II, which are functionally distinguished by the type of antigen they bind and the subset of T cells with which they interact. The solution of the crystal structure of class I and class II molecules clarified the phenomenon of MHC restriction in molecular detail [11, 12]. The three-dimensional structures revealed that both types of molecule possess a binding groove into which a single peptide can be tightly fit. The peptide, thereby, becomes an integral part of the MHC structure providing stability as well as antigenic specificity. Most of the peptide is buried when the MHC complex is assembled, thus, the surface of the MHC/peptide complex is seen as a single contiguous surface by the TCR [13, 14].

MHC molecules are characterized by their highly polymorphic nature and many different variants have been defined which are encoded by different gene loci and alleles. Many of the amino acid variations that distinguish MHC alleles are localized in the binding groove where they modulate its shape and charge. Peptides capable of binding to a particular MHC molecule must fit to the configuration of that molecule's binding groove. The specificity, imparted by the mutual interaction of the peptide and the MHC molecule, is clearly demonstrated by the observation that two closely related class I molecules, which differ by only one amino acid in the binding groove, may not bind the same antigenic peptides [15]. Within the human population there are many different MHC alleles and an individual generally expresses six different class I molecules, each of which binds a different repertoire of peptides due to different amino acid composition of the peptide binding groove. As a consequence of MHC polymorphism and MHC restriction, a particular antigen, for example a tumor-associated peptide, may be presented to CTL in one individual expressing a class I molecule that allows binding; whereas in a second patient with other class I molecules, the peptide may not fit and cell surface presentation does not occur.

For the immune system to respond to antigenic stimulation, MHC/peptide complexes must be expressed at the surface of a target cell and the T cells must express a cognate TCR. T cells are educated to differentiate between self and foreign by the use of MHC molecules. Through a developmental process in the thymus, each individual T cell is selected to express a TCR that recognizes peptide only, if it is presented by the same MHC molecule to which it was exposed in the thymus [16].

5.2.2 The Dichotomy between MHC Molecules and T Lymphocytes

MHC class I and class II proteins have evolved to deal with different sets of antigens. Exogenous antigens such as bacteria, toxins, cellular debris and molecules shed from tumor cells are present in extracellular body fluids. Antigen presenting cells will take them up by phagocytosis, endocytosis or receptor mediated uptake mechanisms. Within the cell they are degraded into peptides and presented by class II molecules. Endogenous antigens which originate inside the cell, such as viruses, mutated proteins or oncogenic products, are degraded in the cytosol and presented by class I molecules. Class I molecules also present peptides derived by proteolysis of normal cellular proteins (self-peptides). Thus, the complete internal environment of a cell, consisting of self as well as foreign proteins, is reflected at the cell surface by peptides bound to class I molecules. Here, they are surveyed by T lymphocytes: Complexes containing normal self peptides generally are ignored while complexes containing foreign peptides activate T cell responses (Fig. 5-2).

The dichotomy between class I and class II molecules relates to their different roles in T cell activation. Class I molecules present peptide to CD8 positive CTL.

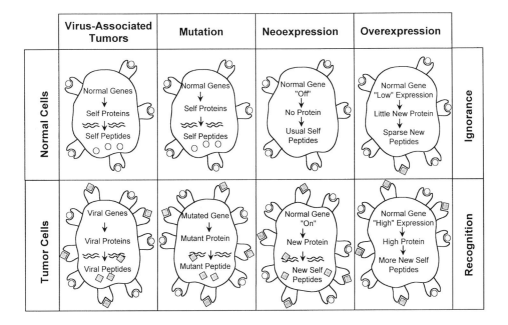

Figure 5-2. Sources of tumor-associated antigens. Tumor cells express not only tumor-associated determinants but also normal (self) determinants as MHC class I/peptide complexes at the cell surface. MHC class I/peptide complexes which contain normal self-peptides are ignored by T cells, while those MHC complexes presenting foreign or altered-self peptides will stimulate T cells expressing appropriate TCR.

These cells directly lyse target cells; therefore it is biologically appropriate that they recognize peptides derived from intracellular antigens that are presented by class I molecules. Moreover, since class I molecules are expressed by nearly all nucleated cells, CTL are able to recognize and destroy virtually any cell if it presents the appropriate MHC/peptide complex. On the other hand, class II molecules present peptides to CD4 positive T helper cells, the primary function of which is to secrete cytokines that promote activities of other lymphocytes, including B cells, macrophages and CTL. They play a dominant role, therefore, in orchestrating an immune response.

5.2.3   Activation of T Lymphocyte Function

Activation of CD8 positive CTL, which leads to their clonal expansion and differentiation into activated effector cells, requires two signals. The first signal is mediated by the engagement of the TCR with its cognate MHC/peptide complex on the target cell and determines the specificity of the response. The second signal is not antigen specific and generally involves the participation of CD4 positive T helper cells [1, 2]. The helper activity of CD4 positive T cells is stimulated by class II/peptide complexes and mainly involves the secretion of cytokines, such as IL-2, which are required for optimal activation of CD8 positive CTL. Alternatively, the second signal may be delivered through interaction of B7 molecules of the antigen presenting cell with CD28 molecules of the T cell [17, 18]. CTL which receive their second signal through the interaction of CD28 with B7 do not require cytokine help from CD4 positive T cells [19]. Engagement of the TCR with MHC class I/peptide complexes alone, that is delivery of signal 1 in the absence of signal 2, is one mechanism known to induce T cell anergy in cells making their first contact with antigen. The failure of many tumors cells to deliver second signals may allow them to escape immune recognition [8, 17].

5.3 The Molecular Basis of T Cell Responses to Tumors

5.3.1 Sources of Tumor-Associated Antigens

Antigens that activate tumor-specific, CD4 positive T helper cells are derived primarily from exogenous sources and are presented by class II molecules. Few class II restricted, tumor-specific peptides have been identified at the molecular level [20, 21]. Theoretically, such peptides could be processed from proteins that are shed into extracellular body fluids from the surface of tumor cells or released from inside when tumor cells die. Priming of CD8 positive, tumor-specific T cells requires presentation of peptides derived from the processing of proteins that are present in the cytosol of the tumor cell. At least four changes in tumor cells have been found to generate proteins that can serve as sources of tumor-associated peptides (Fig. 5-2) ([22] and references in Wölfel and Melief, this volume).

5.3.2 Virus-associated Peptides

Some viruses are associated with malignant transformation. As examples, human papilloma virus is correlated with cervical cancer, human hepatitis virus with liver cancer, and Epstein-Barr virus with certain B cell lymphomas and nasopharyngeal carcinoma [23]. When proteins encoded by viral genomes are synthesized in tumor cells they may be processed into peptides which bind to MHC class I molecules of some individuals, leading to their presentation at the cell surface. If proper CTL activation occurs the tumor cells can be destroyed.

5.3.3 Peptides Derived from Mutated Proteins

Tumors accumulate mutations that often give them growth advantages. Proteins encoded by mutated genes can serve as tumor antigens if peptides are processed from the region encompassing a mutation. CTL recognizing peptides of mutated p21-ras and p53 proteins [24, 25] as well as a mutant protein involved in cell cycle regulation [26] have been described for pancreatic carcinoma, breast carcinoma and melanoma, respectively. While expression of mutated oncogenes may be characteristic of many tumors, not every individual's immune system will be able to detect tumors cells via these changes, since peptides containing the mutation may not be generated by the cell or may not bind to the patient's MHC molecules.

5.3.4 Peptides Derived from Neoexpression of Normal Genes

The human genome encodes many genes, the function of which is limited to special cell types or to specific time points in development. For unknown reasons, such proteins can be newly expressed in a variety of tumor types. Examples of this category include several tumor antigens found in melanoma, tumors of the head and neck, non-small cell lung tumors and bladder cancer. Normal expression of these genes is limited to the testes (see Wölfel, this volume).

5.3.5 Peptides Derived from Proteins that are Overexpressed in Tumor Cells

Another source of tumor antigens are proteins which are present in normal cells at levels insufficient to induce responses. Tumors may express higher levels of these MHC/peptide complexes so that they are no longer ignored by lymphocytes. Examples of proteins that are overexpressed in tumor cells and recognized by specific T cells are the HER-2/neu proto-oncogene found in ovarian, breast, colon and renal cell cancers [27], the p53 protein present in many tumors and several proteins identified in melanoma and cells of the melanocytic lineage.

5.3.6 MHC/Peptide Complexes on Autologous and Allogeneic Vaccines

Two parameters regarding MHC/peptide complexes distinguish autologous and allogeneic vaccines. The first is the nature of the tumor peptides that can be targeted for immune recognition. In an autologous setting, T cells, educated to recognize peptides in the context of the six self-MHC class I molecules, are present in the lymphocyte repertoire and the patient is capable of responding to tumor-associated peptides presented by any of these different MHC molecules. Any alteration that leads to surface presentation of a tumor-associated peptide may activate specific T cells: Some genetic changes create peptides that are common to tumor cells of different patients, others may be unique to cells of a single tumor; nevertheless both types of peptide can contribute to the vaccine potential in the autologous setting. Hypothetically, this allows an intricate immune response to develop, which includes T cells with specificities for multiple MHC/peptide complexes. This multiplicity can help to limit immune selection of tumor variants that can escape immune detection *in vivo*.

Autologous tumor vaccination can be attempted with every patient from whom tumor cells are available, whereas allogeneic vaccination is restricted only to those patients who share at least one MHC molecule with the vaccine cells. Partial MHC matching is necessary to allow TCR, that were selected in the patient to detect peptides presented by self-MHC molecules, to recognize tumor-peptides presented by the matched MHC molecules of the allogeneic vaccine. Due to the high level of MHC polymorphism in the human population, full MHC matching of a patient with any vaccine cell is highly improbable. Therefore, only limited matching is feasible and development of tumor-specific lymphocytes with different specificities is limited. Peptides that can function in immunization are also limited: only common tumor-peptides that are expressed both by vaccine cells and tumor cells of a patient can be targeted by allogeneic vaccination.

The second parameter that distinguishes these two approaches relates to the immunogenicity of the tumor cells. In autologous tumor vaccines, the assumption is made that every tumor expresses tumor-associated peptides that can be recognized by T lymphocytes. Tumor-associated antigens that are recognized by T cells have been demonstrated for many melanomas and information about the expression of T cell-defined antigens for other tumors starts to accumulate [22]. However, it remains to be determined whether every individual tumor expresses antigens that can stimulate T cells. If such T cell antigens are lacking on tumors, a vaccination effect will not be achieved even if the tumor cell has undergone a sophisticated genetic engineering to compensate for other deficiencies.

In the allogeneic setting, well characterized tumor cell lines with defined MHC/tumor-peptide complexes can be selected for vaccine use. Nevertheless vaccine efficacy will still depend upon other variables discussed below.

5.3.7 Poor Immunogenicity of Spontaneous Tumors

Although most tumors are thought to be immunogenic, and tumor-specific CTL are found in peripheral blood or infiltrating the tumors of some patients [28–30], tumor cells generally are not eliminated. Several mechanisms may contribute to this failure. Firstly, the rate of tumor cell proliferation may outpace the ability of activated effector cells to destroy tumor cells. Alternatively, various forms of suppression, mediated by faulty expression of signal transduction molecules [31, 32] or killer cell inhibitory receptors [33] in lymphocytes or by soluble factors, such as TGFβ, prostaglandins [34] or IL-15 [33], may hinder an ongoing immune response.

Some patients may be incapable of even generating an initial antitumor response. In this situation, several factors may contribute. For example, only peptides that bind to an MHC class I molecule can be presented to the immune system. Therefore, if a patient's MHC molecules fail to bind a relevant tumor-associated peptide, immune recognition of the tumor cell cannot occur. Even if tumor-associated peptides would be able to bind to MHC molecules, the peptides may not gain access to MHC class I molecules, because proteins necessary for their transport to MHC molecules are poorly expressed. Both, lack and low level of expression of transport proteins, have been noted in various tumors [35, 36]. Tumors may also downregulate MHC class I expression or reduce expression of proteins which serve as sources of tumor-associated antigens [37, 38]. As a result, the density of MHC/peptide complexes at the surface of the tumor cell may be insufficient for immune recognition.

Even if MHC/peptide complexes are adequately expressed, lack of molecules involved in intercellular contact of T cells and tumor cells may limit successful immune recognition. One important intercellular interaction is mediated through an adhesion molecule, ICAM-1 and its partner molecule, LFA-1, expressed by a T cell. Tumor cells with low ICAM-1 expression are less susceptible to lysis by most CTL. Tumor cells may also shed their adhesion molecules and, thereby, inhibit intercellular interactions with T cells. This has been found for ICAM-1 molecules released by tumor cells. The levels of soluble ICAM-1 found in the sera of cancer patients correlated directly with tumor stage and inversely with patient survival (ref. in [39]).

Although failure of tumor cell recognition by the immune system is one reason for tumor cell survival, impaired induction of tumor-specific T cell responses is probably the main reason for immunological blindness. Tumor cells are generally poor at providing signal 2 for lymphocyte activation because they express few or no B7 molecules and they do not secrete substituting cytokines [8,17]. Many tumor cells cannot activate CD4 T helper cells because they do not express class II molecules. Consequently, immune cells encountering tumor cells may receive signal 1 in the absence of signal 2, inducing T cell anergy.

5.3.8 Limitations of Autologous and Allogeneic Tumor Cell Vaccines

A number of genetic modifications can be undertaken to alter these detrimental characteristics of tumor cells, thereby allowing development of more effective antitumor immunity following vaccination (see Blankenstein, this volume). Autologous and allogeneic tumor cells can be engineered with the same modifying genes but complex alterations involving introduction of multiple genes into the same cells are only feasible in established tumor lines. Despite such improvements, important problems may still limit the effectiveness of genetically engineered tumor vaccines.

Heterogeneity in antigen expression has often been described for individual tumors; thus one human melanoma may express multiple antigens, but not all cells of the tumor may express all of the antigens [29, 40]. Additionally, some tumor cells may downregulate or loose MHC expression and cannot be recognized by CTL. Vaccines that can activate multiple sets of T cells, specific for distinct MHC/peptide complexes, can reduce selection of tumor cells that escape destruction. Furthermore, if vaccines are applied early in an adjuvant setting, an effective immune response may be able to eliminate residual tumor cells before they become too heterogenous.

Until tumor antigens are identified at the molecular level, whole tumor cells are the only source of tumor antigens. Autologous vaccines offer the advantage that both common and unique tumor peptides can be presented and there is complete MHC matching between vaccine cells and the patient's residual tumor cells. It does, however, have severe limitations. Insufficient tumor tissue, difficulties in establishing the short term lines needed for genetic engineering *ex vivo* and ineffective gene transfer can limit the numbers of vaccine cells that are obtained. Furthermore, considerable effort is required to determine if each tumor cell vaccine is immunogenic following gene transfer.

Allogeneic vaccines are technically more feasible since only one or several well established lines must be genetically engineered. Their application, however, is restricted to patients matched for particular MHC molecules. Matching is severely limited by the high degree of MHC polymorphism. MHC matching also needs to be made at the molecular level which requires complex technology. For example, the prevalent HLA-A2 antigen, that is defined by standard serological methods used for transplantation matching, represents a group of 21 molecular subtypes, many of which vary in their peptide binding groove. In allogeneic vaccination, molecular matching for subtypes is needed to assure that peptides presented by the vaccine are also presented by the tumor cells of the patient. Selecting lines with prevalent MHC subtypes and using a small pool of tumor lines, each expressing a different set of prevalent MHC molecules, may overcome the limitations of MHC matching.

Since only partial matching is feasible, the allogeneic vaccine will express other MHC subtypes that can activate alloresponsive T cells similar to those induced in organ transplantation between unrelated individuals. This strong alloresponse may eliminate the vaccine cells before adequate tumor-specific responses have developed. On the other hand, alloreactive T cells may secrete cytokines that foster antitumor immunity [41].

Of the three forms of autologous tumor cell modification being evaluated currently, direct genetic engineering of the tumor in the patient may be the most simple; however it has several important disadvantages. The safety and specificity of gene delivery and the low efficiency of genetic modification achieved *in vivo* remain important considerations. In the future, these problems may be solved technically;

nevertheless important immunological obstacles will remain unresolved. Due to the necessity for surgical intervention to access the tumor, multiple rounds of vaccination are less feasible and this may limit the effectiveness of immunization. Furthermore, later boosting of memory responses which may be important to provide life-long protection against outgrowth of metastases is not possible. Finally, large tumors may produce suppressive factors; tumor removal may reduce such suppression and allow activated T cells to more efficiently destroy residual tumor cells. Based on these criteria, the direct genetic engineering of a tumor is most likely to find application in situations where complete removal of tumors is not possible.

While *ex vivo* modification of autologous tumor cells entails substantially more effort, important considerations regarding efficiency of gene transfer and safety can be resolved by applying selection procedures *in vitro* to isolate modified tumor cells and to analyze them for unexpected properties resulting from their genetic alteration. Individual variations in tumor size and cell viability will dictate whether or not adequate numbers of vaccine cells are available for multiple rounds of vaccination and boosting of memory responses. Modified fibroblast lines provide a source of cells that is less limiting. When admixed with tumor cells, they may function equally well to the use of cytokine secreting tumor cells but expression of costimulatory molecules, such as B7, alongside the MHC/tumor-peptide complexes on the same cell, may be essential to achieve optimal lymphocyte induction.

If appropriate tumor cell lines can be identified, allogeneic vaccines offer substantial advantages. Since they utilize established cell lines genetic engineering must be performed only once. Theoretically, unlimited numbers of tumor cells can be cultivated so that any vaccination dose and frequency can be tested and it is possible to make booster immunizations over long periods of time.

5.4 Concluding Remarks

Although tumor vaccination was first attempted by Coley in 1893, the development of tumor vaccines, employing genetically engineered cells, is in a stage of infancy. In the intervening years, major advances have been made in understanding the function of lymphocytes and how they can recognize tumors. Over the past 20 years, studies of animal models and of patients' responses to their cancers have yielded detailed insight into the nature of tumor-associated ligands that are recognized by specific T cells. The precise knowledge about tumor peptides associated with melanomas is now being used to develop optimal vaccine strategies [42,43]. For the remaining solid tumors, detailed information is still lacking; therefore major emphasis must be placed on understanding the molecular and cellular basis of antitumor immunity for other tumors. Regardless of the strategy selected for vaccination, variations in antigen expression by tumors of different types, among individual tumors of the same type, and even among cells of an individual tumor will play a critical role in determining the clinical success or failure of vaccination, even if optimal activation of specific lymphocytes has been achieved through immunization with genetically engineered tumor cells.

Acknowledgements

This work is supported by the Deutsche Forschungsgemeinschaft, DFG, Klinische Forschergruppe (Ho 1596/3-2).

References

1. Paul, W. E., Seder, R. A. (1994) Lymphocyte Responses and Cytokines. *Cell* 76: 241–251.
2. Grabbe, S., Beissert, S., Schwarz, T. et al. (1995) Dendritic cells as initiators of tumor immune responses: A possible strategy for tumor immunotherapy. *Immunol. Today* 16: 117–121.
3. Old, L. J. (1981) Cancer immunology: the search for specificity-GHA. Clowes Memorial lecture. *Cancer Res.* 41: 361–375.
4. Greenberg, P. D. (1991) Adoptive T cell therapy of tumors: Mechanisms operative in the recognition and elimination of tumor cells. *Adv. Immunol.* 49: 281–355.
5. Melief, C. J. (1992) Tumor eradication by adoptive transfer of cytotoxic T lymphocytes. *Adv. Cancer Res.* 58: 143–175.
6. Hamaoka, T., Fujiwara, H. (1987) Phenotypically and functionally distinct T-cell subsets in antitumor responses. *Immunol. Today* 8: 267–269.
7. Topalian, S. L. (1994) MHC class II restricted tumor antigens and the role of CD4+ T cells in cancer immunotherapy. *Curr. Opin. Immunol.* 6: 741–745.
8. Allison, J. P., Hurwitz, A. A., Leach, D. R. (1995) Manipulation of costimulatory signals to enhance antitumor T-cell responses. *Curr. Opin. Immunol.* 7: 682–686.
9. Roth, J. A., Cristiano, R. J. (1997) Gene therapy for cancer: what have we done and where are we going? In *J. Nat. Cancer Inst.* 89: 21–39.
10. Zinkernagel, R. M., Doherty, P. C. (1997) The discovery of MHC restriction. *Immunol. Today* 18: 14–17.
11. Bjorkman, P. J., Saper, M. A., Samraoui, B. et al. (1987) Structure of the human class I histocompatibility antigen, HLA-A2. *Nature* 329: 506–512.
12. Brown, J. H., Jardetzky, T. S., Gorga, J. C. et al. (1993) Three-dimensional structure of the human class II histocompatibility antigen HLA-DR1 [see comments]. *Nature* 364: 33–39.
13. Garcia, K. C., Degano, M., Stanfield, R. L. et al. (1996) An αβ T Cell Receptor Structure at 2.5 A and its Orientation in the TCR-MHC Complex. *Science* 274: 209–219.
14. Garcia, K. C., Degano, M., Pease, L. R. et al. (1998) Structural basis of plasticity in T cell receptor recognition of a self peptide MHC antigen. *Science* 279: 1166–1172.
15. McMichael, A. J., Gotch, F. M., Santos-Aguado, J. et al. (1988) Effect of mutations and variations of HLA-A2 on recognition of a virus peptide epitope by cytotoxic T lymphocytes. *Proc. Natl. Acad. Sci. USA* 85: 9194–9198.
16. Pawlowski, T. J., Staerz, U. D. (1994) Thymic education-T cells do it for themselves.

Immunol. Today 15: 205–209.

17. Guinan, E. C., Gribben, J. G., Boussiotis, V. A. et al. (1994) Pivotal role of the B7:CD28 pathway in transplantation tolerance and tumor immunity. *Blood* 84: 3261–3282.

18. Sperling, A. I., Bluestone, J. A. (1996) The complexities of T cell co-stimulation: CD28 and beyond. *Immunol. Rev.* 153: 155–182.

19. Harding, F. A., Allison, J. P. (1993) CD28-B7 interactions allow the induction of CD8+ cytotoxic T lymphocytes in the absence of exogenous help. *J. Exp. Med.* 177: 1791–1796.

20. Topalian, S. L. (1994) MHC class II restricted tumor antigens and the role of CD4+ T cells in cancer immunotherapy. *Curr. Opin. Immunol.* 6: 741–745.

21. Monach, P. A., Meredith, S. C., Siegel, C. T. et al. (1995) A unique tumor antigen produced by a single amino acid substitution. *Immunity* 2: 45–59.

22. Van den Eynde, B. J., Van der Bruggen P. (1997) T cell defined tumor antigens. *Curr. Opin. Immunol.* 9: 684–693.

23. Masucci, M. G. (1993) Viral immunopathology of human tumors. *Curr. Opin. Immunol.* 5: 693–700.

24. Melief, C. J. M., Kast, W. M. (1993) Potential immunogenicity of oncogene and tumor suppressor gene products. *Curr. Opin. Immunol.* 5: 709–713.

25. Cheever, M. A., Disis, M. L., Bernhard, H. et al. (1995) Immunity to oncogenic proteins. *Immunol. Rev.* 145: 33–59.

26. Wölfel, T., Hauer, M., Schneider, J. et al. (1995) A p16INK4a-insensitive CDK4 mutant targeted by cytolytic T lymphocytes in a human melanoma. *Science* 269: 1281–1284.

27. Brossart, P., Stuhler, G., Flad, T. et al. (1998) Her-2/neu-derived peptides are tumor-associated antigens expressed by human renal cell and colon carcinoma lines and are recognized by *in vitro* induced specific cytotoxic T lymphocytes. *Cancer Res.* 58: 732–736.

28. Anichini, A., Fossati, G., Parmiani, G. (1987) Clonal analysis of the cytolytic T-cell response to human tumors. *Immunol. Today* 8: 385–389.

29. Anichini, A., Mazzocchi, A., Fossati, G. et al. (1989) Cytotoxic T lymphocyte clones from peripheral blood and from tumor site detect intratumor heterogeneity of melanoma cells. *J. Immunol.* 142: 3692–3701.

30. Topalian, S. L., Solomon, D., Rosenberg, S. A. (1989) Tumor-specific cytolysis by lymphocytes infiltrating human melanomas. *J. Immunol.* 142: 3714–3725.

31. Zier, K., Gansbacher, B., Salvadori, S. (1996) Preventing abnormalities in signal transduction of T cells in cancer: The promise of cytokine gene therapy. *Immunol. Today* 17: 39–45.

32. Levey, D. L., Srivastava, P. K. (1995) T cells from late tumor-bearing mice express normal levels of p56lck, p59fyn, and CD3zeta despite suppressed cytolytic activity. *J. Exp. Med.* 182: 1029–1036.

33. Mingari, M. C., Moretta, A., Moretta, L. (1998) Regulation of KIR expression in human T cells: a safety mechanism that may impair protective T cell responses. *Immunol. Today* 19: 153–157.

34. Sulitzeanu, D. (1993) Immunosuppressive factors in human cancer. *Adv. Cancer Res.* 60: 247–267.

35. Seliger, B., Maeurer, M. J., Ferrone, S. (1997) TAP off – Tumors on. *Immunol. Today* 18: 292–299.

36. Seliger, B., Hohne, A., Knuth, A. et al. (1996) Analysis of the major histocompatibility complex class I antigen presentation machinery in normal and malignant renal cells:

Evidence for deficiencies associated with transformation and progression. *Cancer Res.* 56: 1756–1760.

37. Jaeger, E., Ringhoffer, M., Altmannsberger, M. et al. (1997) Immunoselection *in vivo*: independent loss of MHC class I and melanocyte differentiation antigen expression in metastatic melanoma. *Int. J. Cancer* 71: 142–147.

38. Jaeger, E., Ringhoffer, M., Karbach, J. et al. (1996) Inverse relationship of melanocyte differentiation antigen expression in melanoma tissues and CD8+ cytotoxic-T-cell responses: evidence for immunoselection of antigen-loss variants *in vivo*. *Int. J. Cancer* 66: 470–476.

39. Fleuren, G. J., Gorter, A., Kuppen, P. J. K. et al. (1995) Tumor heterogeneity and immunotherapy of cancer. *Immunol. Rev.* 145: 91–122.

40. Topalian, S. L., Kasid, A., Rosenberg, S. A. (1990) Immunoselection of a human melanoma resistant to specific lysis by autologous tumor-infiltrating lymphocytes. Possible mechanisms for immunotherapeutic failures. *J. Immunol.* 144: 4487–4495.

41. Mandelboim, O., Vadai, E., Feldman, M. et al. (1995) Expression of two H 2K genes, syngeneic and allogeneic, as a strategy for potentiating immune recognition of tumor cells. *Gene Ther.* 2: 757–765.

42. Jaeger, E., Bernhard, H., Romero, P. et al. (1996) Generation of cytotoxic T-cell responses with synthetic melanoma-associated peptides *in vivo*: implications for tumor vaccines with melanoma-associated antigens. *Int. J. Cancer* 66: 162–169.

43. Marchand, M., Weynants, P., Rankin, E. et al. (1995) Tumor regression responses in melanoma patients treated with a peptide encoded by gene MAGE-3. *Int. J. Cancer* 63: 883–885.

6 Identification of Tumor Antigens Defined by Cytolytic T Lymphocytes and Therapeutic Implications

Th. Wölfel

The hope that tumors can be destroyed by the autologous immune system, like foreign infectious agents, is almost as old as the discoveries of immune defense mechanisms [1]. Herein, the focus is on tumor antigens recognized by cytolytic T lymphocytes (CTL). The identification of CTL-defined tumor rejection antigens in animal models motivated the search for comparable antigens on human tumors. The current knowledge on murine and human tumor antigens recognized by the T cell system is reviewed, thereby excluding viral antigens [reviewed in: [2, 3]]. Further, the potential role of these antigens as targets of rejection responses and inducers of immune protection will be discussed.

6.1 Animal Tumor Models

6.1.1 Identification of Tumor-Specific Transplantation Antigens on Experimental Rodent Tumors

When murine tumors, induced by methylcholanthrene (MC) and ultraviolet light (UV), were surgically removed from the original hosts and transplanted into syngeneic animals they grew progressively and finally killed their new hosts. If transplanted tumors were removed, animals were protected against further challenge with the respective tumor. However, no protection against other syngeneic tumors was observed, even when the challenging tumors had been induced independently in a single animal. Observations that transplantation of normal skin did not induce protective immunity against tumors and reports on successful immunization of original tumor-bearing mice excluded that genetic impurities of inbred strains, causing expression of distinct minor histocompatibility antigens in individual mice, accounted for the immunogenicity of tumors [4–9]. These experiments unequivocally demonstrated that MCA- and also UV-induced murine tumors carry individual tumor-specific transplantation antigens (TSTAs). The variability of TSTAs appeared indefinite.

Efforts to induce immunity against spontaneous rodent tumors failed in general. This led to the conclusion, that the presence of TSTAs might be confined to experimentally induced tumors [10, 11]. However, heterogenization or xenogenization of seemingly non-immunogenic tumors induced protective immune responses to un-modified tumor [12–14]. A large proportion of tumors appear to express antigens that can be targets of rejection responses but nevertheless, these antigenic tumors are not able to elicit efficient immune responses without modifications. Essentially, these are either modification of tumor cells with strongly immunogenic antigens by mutagenisation [15], transfection with genes encoding alloantigens [13] and viral in-fection [16] or enhancement of the tumor cells´ immunostimulatory capabilities by transfection with genes encoding costimulatory molecules or cytokines [17].

6.1.2 Characterization of Murine Tumor Antigens recognized by T-Lymphocytes

Adoptive transfer and depletion experiments revealed that in protected animals tu-mor-specific immunity is mediated by T lymphocytes, with both CD8-positive cy-tolytic T lymphocytes (CTL) and CD4-positive T helper cells playing important roles [18, 19]. However, TSTAs did not evoke a humoral response. This was the main reason, why conventional biochemical approaches to identify the nature of TSTAs regularly failed as well as efforts to identify other purely T cell-defined anti-gens like minor histocompatibility or some viral antigens. Among questions arising were, whether or not TSTAs are completely independent from one another or if they are encoded by gene families, and whether their expression is related to the malig-nant phenotype. After decades of almost complete standstill, ten years ago progress arrived. Several T cell-defined antigens were identified, at least some of which seem to be functionally equivalent to TSTAs, especially by their ability to induce protec-tive immunity in naive animals.

Individually Distinct Antigens Associated with Heat-Shock Protein gp96

Pramod Srivastava and collegues observed that vaccination with heat-shock protein gp96, purified from MCA-induced tumors, led to protection against subsequent challenge, which was specific for the respective tumor and depended on CD8-posi-tive T lymphocytes [20–22]. Subsequent work demonstrated that gp96 was not antigenic itself but rather served as a carrier of pre-formed tumor-specific peptide antigens [23].

Identification of CTL Target Antigens in Murine Mastocytoma P815

After numerous failures to unravel the nature of T cell epitopes with biochemical means, the group of Thierry Boon at the Ludwig Institute for Cancer Research in

Brussels successfully applied a gene transfection approach to identify CTL-defined antigens present on murine mastocytoma P815 [24, 25]. The P815 tumor had been induced by MCA-treatment and is considered as weakly immunogenic. However, when syngeneic mice were immunized with so-called tum− variants of P815 cells, expressing neoantigens induced by *in vitro* mutagen treatment, Thierry Boon and collegues regularly saw cross-protection against parental tumor cells, named tum+. A number of tum− antigens was identified to represent mutated proteins [26–28]. Peptides containing mutated residues were processed from these proteins and presented on the cell surface by MHC class I molecules according to previous discoveries of Alan Townsend and colleges for viral epitopes [29]. The exchange of amino acids either improved peptide-binding to the presenting MHC molecule, thereby generating an aggretope, or formed new T cell epitopes within normally processed peptides.

Using the same approach as for tum− antigens, an antigen originally present on the P815 tumor was identified and named P1A. The observation that escape variants of P815, growing after a latency period in challenged immune mice, lacked expression of P1A suggested that it represented a potent rejection antigen. The P1A gene was not altered. P1A is expressed in tumor cells and testes, but little or none in other normal cell types tested [30–31]. It is still not clear, whether immunization with P1A protects mice against subsequent challenge with P815 tumor cells.

Point mutations, accounting for the specific immunity of tum− antigens, were supposed to represent TSTAs. Therefore, it came as a surprise that the first potential rejection antigen was not tumor-specific. P1A was found to be expressed also on other mastocytomas and is, therefore, to be called tumor-associated. Such an antigen, of course, could not explain tumor-specific immunity, originally observed after transplantation of independently induced experimental tumors of the same histological type (see 6.1.1). However, examples of antigens that presumably represent TSTAs, in the original sense, have been published during the last year.

Mutated Connexin 37 in Lewis Lung Carcinoma

Mandelboim and colleges reported on an antigen of murine Lewis lung carcinoma 3LL-D122. After acid extraction from MHC molecules of tumor cells, they purified and sequenced an octapeptide that was recognized by CTL specific for 3LL cells. The peptide was derived from a variant molecule of connexin 37 and comprised amino acids 52–59. At position 54, it contained glutamine (Q) instead of cysteine (C). Only the variant peptide bound stably to H-2 Kb but not the normal connexin homologue [32]. Connexin 37 is abundant in lungs and belongs to a family of transmembrane proteins that form intercellular gap junctions. It is unclear, whether or not the C54Q amino acid exchange also affected the function of connexin 37. All nucleotides of codon 54 were exchanged in variant connexin 37 allele of the 3LL tumor, which raised concern that the C54Q exchange reflected a natural polymorphism of connexin 37 and was not the result of a mutagenic event during tumorigenesis [33]. In the latter case, connexin 37-C54Q would represent a minor histocompatibility antigen. More recently, protection of mice from metastatic spread of 3LL-D122 as well as rejection of established micrometastases was observed after vaccination with the C54Q connexin 37 peptide [34].

Retroviral Antigen in Spontaneous Leukemia LEC

On the spontaneous murine leukemia LEC, originally thought to be non-immunogenic [10], a target antigen for syngeneic CTL was identified by transfecting a LEC cosmid library. The antigen LEC-A was found to be encoded by the *gag* gene of an endogenous defective retrovirus belonging to the intracisternal A particle (IAP) family. The LEC-A peptide was presented by H-2 Dk molecules. Expression of the antigenic LEC-A peptide antigen depended on the presence of a 3′ LTR (long terminal repeat) sequence. Obviously, transposition of the IAP sequence to a new genomic location caused unique expression in the LEC leukemia. This work also demonstrated that retrotransposons are to be included in the list of principal CTL-target antigens. Such retroelements are also part of the human genome [35].

Antigen Encoded by the c-akt Gene in a Radiation-Induced Leukemia

Another TSTA candidate antigen was reported for BALB/c radiation leukemia RLO1. Semigeneic CTL specifically recognize RLO1 cells. Peptides extracted from RLO1 cells were separated, purified by HPLC and tested for recognition by these CTL. After sequencing of fractions with sensitizing activity by Edman degradation, an octapeptide was finally identified. It is derived from the 5′-untranslated region of the c-*akt* oncogene. How the peptide-coding region is translated in RLO1 cells is apparently not known [36].

Mutated Ribosomal Protein L9

Against a panel of UV-induced tumors, syngeneic T cell clones were established *in vitro* shown to recognize unique antigens [9]. One such antigen on tumor 6132A was recently characterized. Intact proteins were extracted from 6132A tumor cells, fractionated by size and screened with a CD4+ T cell hybridoma using an immunoblot assay. Thereby, ribosomal protein L9 was purified. The cDNA sequence of L9 from 6132A tumor differed from L9 cDNA of normal cells of the original tumor-bearing mouse at one nucleotide, which caused an amino acid exchange at position 47 of protein L9. A 25-residue mutant peptide centered on position 47 stimulated the release of IL-2 by L9-reactive T cells over 1000-fold more effectively than did the wild-type peptide. Immunization with the mutant L9 peptide induced lymph node cells capable of mediating antigen-specific tumor rejection when transferred into SCID mice. Whether or not the mutation of L9 contributed to the malignant phenotype of 6132A is unclear [37].

Immune Responsiveness to Proteins Known to be Involved in Oncogenesis

Mutated ras and p53 would seem ideal candidates for T cell targeting of tumor cells because they are frequently mutated and – as for p53 – also overexpressed in animal as well as in human cancers. A number of groups succeeded in inducing T lymphocytes that recognize targets that were pulsed with peptides, whose sequences were derived from mutated ras and p53 proteins. However, it proved difficult to demonstrate lysis of cells expressing the respective mutated oncoproteins by CTL generated with peptide-pulsed cells [38]. In the following selected work performed in mouse models is described.

Skipper et al. immunized H2b mice with recombinant vaccinia viruses containing cDNA encoding either wild-type human N-ras or human N-ras with a glutamine to lysine mutation at position 61. From spleens of immunized animals, H2-Kb-restricted CTLs that lysed cells infected with recombinant vaccinia viruses and were specific for either mutated or wild-type peptides comprising residues 60–67 were isolated. Residue 61 turned out to be critical for T cell recognition. A second ras epitope was mapped to residues 152–159 and was recognized equally well by CTL raised against normal or mutant ras [39]. Peace and collegues also elicited a H2-Kb-restricted CTL response against a mutated segment of transforming p21ras protein. In contrast to the previous study, this was achieved by primary *in vitro* immunization with synthetic peptides that contained one of the most common mutations of residue 61 consisting of a substitution of leucine for glutamine. The ras peptide-induced CTL specifically lysed syngeneic fibroblasts, transfected with a recombinant plasmid encoding p21c-Ha-ras, carrying the same single amino acid substitution [40]. Although these studies clearly demonstrated the principal immunogenicity of the ras epitope encompassing residue 61, none of the ras-specific CTL, derived in these studies, lysed untransfected cells endogenously expressing mutated ras. Whether vaccination with ras peptides can protect against tumor challenge or even cure tumor-bearing animals was not reported so far.

Noguchi et al. analyzed the syngeneic T-cell response to mutated and normal p53 of murine sarcoma Meth A [41]. The p53 of Meth A is known to carry three distinct missense point mutations. Mice were immunized with 24 peptides spanning these mutational sites and with the corresponding wild-type peptides. Only one of the peptides induced cytotoxic effector cells consistently, and this was detectable only after *in vivo* immunization with peptide, emulsified in incomplete Freund´s adjuvant, and subsequent to *in vitro* sensitization with peptide-pulsed syngeneic cells. This peptide was designated 234CM. Mouse mastocytoma cells, transfected with plasmids coding for the 234CM epitope, were lysed by 234CM-specific CTL after pretreatment with interferon gamma. Growth of Meth A was suppressed in mice immunized with 234CM in incomplete Freund´s adjuvant. The latter observation was the only *in vivo* correlate for the T-cell response against the 234-region of p53. The authors failed to detect 234-reactive CD8+ or CD4+ T lymphocytes in mice immunized with Meth A cells that were highly resistent against to Meth A, which indicated that mutated p53 *per se* does not elicit transplantation resistance, but might, nevertheless, induce protective immunity after appropriate immunization.

6.1.3   Principal Lessions from Animal Experiments

Regarding the work described in 1.1, clearly, the immunogenicity of murine tumors and the variability of T-cell defined antigens became apparent only through transplantation experiments and preventive immunization on naive syngeneic animals. Progressive growth of the model tumors did not result in an effective immune response. It proved even difficult to immunize primary hosts against a tumor that had been previously surgically removed and that was known to be immunogenic in naive syngeneic animals [42]. These considerations allow to predict difficulties in demonstrating the existence and the molecular nature of tumor rejection antigens on human tumors.

Some of the antigens mentioned in 6.1.2 are derived from unaltered and some are derived from mutant proteins. It is conceivable that many unique TSTAs are mutant proteins because physical and chemical carcinogens, inducing tumors bearing unique antigens, are mutagens, and because treatment of tumors with mutagens generated immunogenic mutant proteins [43]. The appearance of the latter can be independent of tumor transformation [44]. Alternatively, mutagen treatment might result in *de novo* expression or overexpression of unaltered proteins. But in the latter situation, cross-protectivity should have been induced by vaccination with at least some experimental tumors, which according to the literature was not the case [44, 45], even when tumors shared a common antigen [46]. Mutant antigenic epitopes should be more immunogenic than normal, though overexpressed antigens, since immune responses against the latter should be inhibited by mechanisms that induce tolerance towards self. However, this does not exclude that transferring immunity or enhancing an immune response against self determinants, either overexpressed in tumor cells or expressed only on certain tissues, can have an advantageous effect on tumor growth, as seen in some animal models [47, 48].

6.2 Antigens on Human Tumors Recognized by Autologous T Lymphocytes

The existence of murine tumor rejection antigens could only be proven by *in vivo* transplantation experiments. Obviously this is not applicable to human tumors. Therefore, those who intend to identify potential rejection antigens on human tumors depend exclusively on *in vitro* experiments. The advent of techniques to generate and to maintain clonal T lymphocyte populations from blood and tissues was the basis of these studies [49, 50].

6.2.1 Derivation of CTL Lysing Autologous Tumor Cells

The general availability of the T cell – growth factor, named interleukin-2 (IL-2), greatly facilitated reproducible results in the derivation and maintenance of mono-specific T lymphocyte population, reactive with autologous tumor cells from blood, tumor tissues and draining lymph nodes of tumor patients. Over the last decade, numerous groups succeeded in isolating, propagating and analyzing CTL that recognize autologous tumor cells. This was carried out mainly in melanoma but also in other tumors [reviewed in [51]]. Restricting HLA molecules for CTL recognition were regularly identified by blocking experiments. In some melanoma models, a hierarchy of restricting HLA alleles was observed [52, 53].

Using allogeneic HLA-compatible tumor lines for typing, tumor-reactive CTL exhibited a wide cross-reactivity, irrespective of histological descendence of the targets [54], were limited in recognition to a certain tissue type [55], preferentially lysed tumor cells but not normal cells of the same tissue origin [56], or apparently only lysed autologous tumor cells [53, 56]. When several independent tumor-reactive CTL clones were available in individual models, it was repeatedly seen that these CTL recognize different antigens simultaneously expressed on autologous tumor cells [57–60].

6.2.2 Identification of Antigens on Human Tumors that are Recognized by Autologous T-Lymphocytes

When stable, tumor-reactive CTL lines or clones became available, efforts were initiated to characterize their respective target antigens and thus, the discovery of the principles of T cell recognition was inevitable. CTL do not recognise intact proteins but do recognise short peptides processed from any cytoplasmic or membrane-bound, nuclear or extranuclear protein and bound to HLA class I and β2-microglobulin in a trimolecular complex (reviewed in [61]) (Fig. 6.1 schematically summarizes the essential steps required for HLA I-restricted antigen presentation).

319

Thierry Boon's group used a genetic approach, that has led to the identification of CTL-defined murine tumor antigens (reviewed in [31]). This approach was based on the transfection of cosmid libraries from tumor cell lines into antigen-negative recipient cells and subsequent screening of transfected cell populations for the presence of antigen-expressing transfectants. Screening assays were performed with antigen-specific CTL clones. Cosmid rescue strategies finally allowed cloning and, thereby, identification of genes encoding CTL-defined antigens [31]. During the recent years, this time and labour consuming approach was replaced by applying expression cloning of cDNA [62–64]. Transfectants were screened for expression of antigens with cytokine assays, based on the ability of tumor-reactive T lymphocytes to release TNF-alpha [65–69], IFN-gamma [7–72] or GM-CSF [73, 74] after antigen contact.

Approximately ten years ago it was discovered that T lymphocytes recognize short peptides presented in grooves of MHC molecules on the cell surface [29, 75]. H. -G. Rammensee and his group sequenced those peptide pools detached from single MHC alleles with conventional Edman degradation. They found that peptides binding to certain MHC molecules share typical sequence motifs [76]. The knowledge of such motifs greatly facilitated the search for CTL epitopes either within proteins in which genes had been cloned by transfection (see above), or within proteins that seemed to be of particular interest due to their association with the malignant phenotype of cancer cells. To the latter type of proteins belong mutated ras, bcr-abl and p53 (see below). Peptides with typical HLA binding motifs were regularly iden-

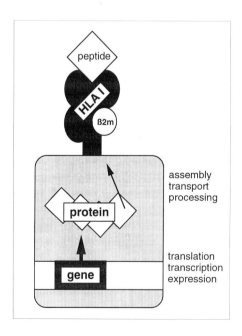

Figure 6.1: Presentation of peptide antigens by HLA class I molecules as the result of a multistep process (schematically).

tified within those proteins and were also shown to elicit T cell responses *in vitro*. However, in many cases it was difficult or impossible to demonstrate that that responders of *in vitro* sensitizaton against peptide-loaded targets recognize cells expressing the respective proteins, which would prove that these peptides are actually processed intracellularly. Improved biochemical techniques, combining HPLC (reversed-phase high-performance liquid chromatography) and tandem mass spectrometry, allowed in some instances to isolate and sequence T cell epitopes directly from peptide pools detached from MHC molecules of tumor cells [77].

6.2.2.1 Antigens Encoded by Genes MAGE[1], BAGE AND GAGE

Antigens encoded by MAGE, BAGE AND GAGE genes were all detected as targets of autologous blood-derived CTL by Thierry Boon´s group in a single human melanoma model derived from patient MZ2 [58, 67, 78–82]. Their sequences are unrelated and the function of their proteins is unclear at present. Instead MAGE, BAGE and GAGE have several other features in common, because of which BAGE and GAGE were named with the suffix AGE in analogy to MAGE (reviewed in [83]): They each belong to unrelated families of previously unknown genes. In adult tissues they are expressed only in tumor tissues and in testes but not in other normal tissues. Aside from melanoma, a variable proportion of bladder, head and neck, lung (NSC), mammary and prostatic carcinomas and sarcomas expresse "AGE" genes. No expression was found in leukemias and lymphomas and rarely in colorectal and renal carcinomas. Tumors expressing MAGE were reported to almost invariably also express members of the BAGE and GAGE gene families [82, 67]. MAGE, BAGE and GAGE are about two to three times more frequently expressed in melanoma metastases than in primary melanomas [84]. In addition, these genes are expressed more than fourfold in invasive bladder carcinomas than in superficial lesions [82, 67]. Taken together, these observations suggest that expression of the "AGE" genes is similarly regulated and is associated with a more aggressive tumor phenotype.

Known MAGE-, BAGE- and GAGE-derived peptides and their presenting HLA molecules are summarized in Table 6.1. Taking into account the frequency of antigen expression and the allele frequencies of presenting HLA molecules, it was calculated that tumors of aproximately 51% of melanoma patients and 49% of non small cell lung cancer patients carry at least one of the known antigenic epitopes encoded by MAGE, BAGE and GAGE genes [67]. Notably, due to the lack of suitable antibodies expression of these genes in tumors is still only verified by PCR amplification of reverse transcribed RNA, which does not provide information about the heterogenity and distribution of "AGE" proteins in tissues.

---

[1] MAGE = melanoma antigen-encoding gene

Table 6-1. Human tumor antigens recognized by autologous T lymphocytes

Gene/Protein	peptide	amino acid position[1]	peptide-presenting HLA molecule	reference
A) Shared antigens				
MAGE-1	EADPTGHSY		HLA-A1	(78)
	SAYGEPRKL		HLA-Cw16	(79)
MAGE-3	EVDPIGHLY		HLA-A1	(81,174)
	FLWGPRALV		HLA-A2.1	(80)
BAGE	AARAVFLAL		HLA-Cw16	(82)
GAGE-1/-2	YRPRPRRY		HLA-Cw6	(67)
tyrosinase	MLLAVLYCL	1–9	HLA-A2.1	(88)
	YMNGTMSQV	369–377	HLA-A2.1	(88)
	YMDGTMSQV[2]	369–377	HLA-A2.1	(94)
	SEIWRDIDF	192–200	HLA-B44.3	(96)
	AFLPWHRLF	206–214	HLA-A24	(95)
	?		HLA-DR	(175)
gp100/Pmel17	YLEPGPVTA[2]	280–288	HLA-A2.1	(77)
	LLDGTATLRL	457–466	HLA-A2.1	(173)
	ITDQVPFSV	209–217	HLA-A2.1	(102)
	KTWGQYWQV	154–162	HLA-A2.1	(102)
	VLYRYGSFSV	476–485	HLA-A2.1	(102)
Melan-A/MART-1	AAGIGILTV	27–35	HLA-A2.1	(103)
	ILTVILGVL2	32–40	HLA-A2.1	(106)
	AEEAAGIGIL(T)	24–34	HLA-B45.1	(unpublished)
gp75/TRP1	MSLQRQFLR		HLA-A31	(73)
her2/neu	IISAVVGIL		HLA-A2.1	(176)
muc-1	PDTRP[3]		non-restricted	(111)
B) mutated antigens				
Mum-1	EEKLIVVLF		HLA-B44.2	(68)
CDK4-R24C	ACDPHSGHFV		HLA-A2.1	(69)

[1]amino acid positions are given when distinct epitopes have been identified
[2]biochemically identified natural peptides
[3]polypeptide core contained in 20 amino acid long tandem repeats

6.2.2.2 Melanocyte Differentiation Antigens

As mentioned above, CTL were isolated from blood or melanoma tissues that recognized autologous melanoma cells and also a high proportion of allogeneic melanomas. Regularly, those CTL were restricted by HLA-A2.1 [85–87, 53]. At least some of them also lysed normal melanocytes [55], which suggested that they are targeted against antigens specific to the melanocyte lineage. During the past three years, a number of these antigens have been characterized:

Tyrosinase

Tyrosinase was the first shared melanoma antigen to be characterized by applying COS transfection and cDNA cloning [65]. Tyrosinase contributes to the synthesis of the melanin precursor molecule dihydroxyphenylalanine (DOPA) and is localized within melanosomes. Tyrosinase is not related to tyrosine hydroxylase which also produces DOPA in adrenal cells and adrenergic neurons. We, initially, identified two distinct tyrosinase peptides targeted by CTL of different patients [88] (see Table 6.1). Both are binding to HLA-A2.1. Peptide MLLAVLYCL, recognized by CTL of patient LB24, is derived from the signal sequence of tyrosinase. Signal sequence peptides prevail among peptides bound to the HLA-A2.1 molecules of mutant cell line T2, which is deficient for the expression of TAP molecules transporting peptides into the endoplasmic reticulum [89–91]. Thus, the presence of a signal-derived tumor antigen should be independent of TAP function and should therefore deprive the melanoma cells from escaping T cell responses through loss of these transporter molecules [92]. Peptide YMNGTMSQV, recognized by blood-derived CTL of patient SK29(AV), corresponds to amino acid region 369–377 of the tyrosinase protein.

Tyrosinase peptides 1–9 and 366–377 both contain known consensus motifs for binding to HLA-A2.1. However, in peptide recognition assays, they had to be applied at concentrations several orders of magnitude higher than usually observed for HLA-A2.1-binding peptides. For peptide MLLAVLYCL, this could be explained by its rather low binding affinity to HLA-A2.1. Its corresponding naturally processed peptide might differ in length, because most of the signal sequence peptides identified among natural ligands eluted from HLA-A2.1 molecules were longer than nine amino acids and characteristically had a rather low-affinity to HLA-A2 [93]. In contrast, tyrosinase peptide YMNGTMSQV seemed to bind to HLA-A2 as strongly as viral peptides of known high binding affinity [88]. An explanation for its inferior target-sensitizing ability has, meanwhile, been found. Cox and collegues separated peptides, eluted from HLA-A2.1 molecules of a melanoma cell line, by HPLC and identified, in one of the HPLC fractions, a peptide that proved to be almost identical to tyrosinase peptide 368–376. Its sequence was determined as YMDGTMSQV

[77]. Thus, the aspartic acid (N) residue of the synthetic peptide was replaced by an asparagine (D) in the eluted peptide, which resulted from a minor chemical modification in the course of intracellular peptide processing or transport as discovered by that same group [94]. CTL of patient SK29(AV) indeed recognized the natural ligand YMDGTMSQV at about 1000-fold lower concentrations than its synthetic homologue, that had been predicted from the coding DNA sequence.

T lymphocytes directed against tyrosinase were isolated from blood or tumor-infiltrating lymphocytes of further patients. Their target antigens were peptides associated with HLA-A24 [71, 95], HLA-B44 [96] and HLA-DR [97]. The fine-specificity of anti-tyrosinase T lymphocytes appears quite heterogeneous when analyzing T cell responses against tyrosinase in different individuals, as opposed to T cell responses against other melanocyte-associated antigens (see below). T lymphocyte reactivity against tyrosinase has been observed not only in melanoma patients but also in healthy individuals [98, 99].

gp100

The transmembrane glycoprotein gp100 and its closely related homologue Pmel17 are specific for the melanocytic lineage. Their function is unknown, although it was postulated that they contribute to melanin biosynthesis [100]. Monoclonal antibodies NKI-beteb, HMB50 and HMB45 against gp100/Pmel17 are widely used for diagnostic purposes. Two years ago, it was reported that gp100 is also targeted by a melanoma-derived HLA-A2.1-restricted CTL line, derived from tumor infiltrating lymphocytes [101]. Peptide LLDGTATLRL recognized by this TIL line comprised gp100 residues 457–466 [102]. By tandem mass spectrometry Cox et al. identified a gp100/Pmel17 epitope as a natural melanoma peptide, recognized by high-affinity CTLs from five different melanoma patients. Its sequence was determined as YLEPGPTVTA and corresponds to residues 280–288 of gp100 [77]. According to the literature, almost half of HLA-A2-restricted melanoma-reactive TIL lines, derived from different patients, are directed against gp100. As HLA-A2.1-associated target epitopes altogether five peptides were identified so far [102]. They are processed from distinct protein regions (Tab. 6.1). Two of them seem to be dominant. Aside from peptide YLEPGPTVTA (residue 280–288) [77], peptide KTWGQYWQV (residues 154–162) also appears to be targeted by CTL in more than half of the patients that were tested [102].

Melan-A/MART-1

The antigen most commonly recognized by anti-melanoma CTL is Melan-A or MART-1, as it has also been named. Neither its gene nor its protein sequence was known before. Similarly to tyrosinase and gp100/Pmel17, Melan-A/MART-1 is expressed in almost all melanoma samples and and in about half of long-term cultured melanoma cell lines. Most remarkably, HLA-A2.1-restricted Melan-A/MART-1 reactive CTL of all patients analyzed, so far, are directed against a single epitope, which is in contrast to the interindividual diversity of lymphocyte reactivity observed with tyrosinase and gp100/Pmel17. Nine out of ten melanoma-reactive TIL lines from HLA-A2-positive patients lysed target cells incubated with overlapping Melan-A/MART-1 peptides AAGIGILTV (residues 27–35) and EAAGIGILTV (residues 26–35) at concentrations below 1 µg/ml [103]. While in this original report, peptide AAGIGILTV was recognized at 10- to 100-fold lower concentrations than EAAGIG-ILTV, we have observed the reverse for anti-Melan-A/MART-1 CTL in melanoma patient SK29(AV). CTL from the blood of this patient also responded to a Melan-A/MART-1 peptide processed from the same region and presented by HLA-B45.1 (unpublished). This immunodominant epitope region is located within a hydrophobic putative transmembrane domain of the Melan-A/MART-1 protein [66, 72].

In independently analyzed melanoma models, naturally processed Melan-A/MART-1 peptides were localized in several neighbouring HPLC fractions [104, 105]. A peptide in one of these fractions was identified by tandem mass spectrometry as ILTVILGVL. It is partially overlapping with the above mentioned epitope. Melanoma-reactive CTL clones and lines from different patients, as well as freshly isolated TILs, recognized this peptide [106].

gp75/TRP-1

Sera from some melanoma patients contain antibodies against the gp75 protein [107], which is involved in melanin synthesis [108] and most abundantly expressed in melanocytic and melanoma cells [109]. The amino acid sequence of gp75, also named tyrosinase related protein 1 (TRP-1), has about 40–45% homology to tyrosinase, gp100 and TRP-2.

By cDNA expression cloning, it was shown that gp75 cDNA is also targeted by cytolytic T lymphocytes. The particular TIL-derived CTL line used for screeening of the cDNA library was restricted by HLA-A31 [73]. The peptide recognized by these CTL is listed in Table 6-1.

Common Features

Tyrosinase, gp100, Melan-A/MART-1 and gp75 are normal antigens, commonly expressed in melanomas and melanocytes but not in any other tissue, whereas, melanocytes are located not only in the skin but also in the choroid of the eye and in the leptomeninx. In immunoselection experiments with CTL, we have repeatedly observed concomitant reduction or loss of tyrosinase and Melan-A/MART-1 expression on a melanoma cell line [53], which indicates a common regulation of gene expression. According to PCR analyses, approximately 50–70% of melanoma cell lines score positive for the expression of these antigens, while almost all melanoma tissues are positive. However, as for the "AGE"-type antigens the advent of antibodies for immunohistochemistry will provide us with information about the phenotypic heterogeneity of tumor tissues. Considering the frequency of antigens and their presenting HLA alleles, melanomas, in almost half of the patients, carry T cell epitopes derived from melanocyte differentiation antigens.

T-lymphocytes against melanocyte antigens have also been found in normals [98, 99, 110]. The question, whether anti-melanocyte T lymphocyte reactivity in melanoma patients differs functionally or quantitatively from healthy individuals, has not been answered yet. Because these antigens seem to represent normal autoantigens, the identification of antigens shared by melanomas but not by melanocytes is eagerly awaited [56].

6.2.2.3 Mucin MUC-1 (reviewed in [111])

Epithelial cell mucin, MUC-1, can be targeted to tumor cells by T lymphocytes. Mucins are high molecular weight glycoproteins produced by normal ductal epthelial cells. Their physiological function is to lubricate ductal surfaces and to protect epithelium from damage in the ductal lumen. MUC-1 is the only transmembrane molecule among mucins. Its expression is polarized and restricted to the ductal lumen on normal epithelium. However, on tissues of epithelial tumors, its expression is enhanced and no longer polarized and in addition, it is incompletely glycosylated on tumor cells. These changes associate with the malignant phenotype and contribute to the immunogenicity of MUC-1 on tumor cells. MUC-1 is expressed in pancreatic, breast and ovarian carcinomas, prostate tumors, lung adenocarcinomas, salivary gland tumors and multiple myelomas.

MUC-1 recognition by T cells is not HLA-restricted. T cells appear uniformly to recognize a polypeptide core sequence (PTRP) contained in a tandem repeat peptide specific to MUC-1, but different from mucins i.e. those expressed in colon epithelium. It was proposed that repeating epitopes interact directly with T cell receptors, which means that MUC-1 is recognized in its native form and does not require presentation by HLA molecules. Like "AGE"-type antigens and melanocyte-associated antigens, MUC-1 is a normal cellular protein. However, for two reasons, MUC-1 is mainly considered as a tumor-specific antigen. On normal tissues MUC-1 is hidden from access by the immune system by its polarized expression and the T cell epitopes are masked by glycosylation. Therefore, due to its independence from HLA expression with respect to recognition by T cells, MUC-1 seems to be an attractive target for T cell based immunotherapy.

In chimpanzees, vaccination with MUC-1-transfected antigen-presenting cells induced measurable frequencies of MUC-1 reactive CTL in peripheral blood and high frequencies of mucin-specific CTL in draining lymph nodes. In a phase I trial, breast, colon and pancreatic cancer patients were vaccinated with a 105 amino acid long peptide containing five tandem repeats. No adverse effects were observed after these vaccinations and ongoing trials continue to test their efficacy.

6.2.2.4 Mutated Melanoma Proteins Targeted by Autologous CTL

As soon as it was realized that T lymphocytes recognize peptide fragments processed from proteins of virtually any intracellular location, it was speculated that the T cell system can survey the organism not only for the presence of foreign, i.e. viral, proteins, but also for self proteins carrying point mutations [112]. Point mutations generating T cell epitopes were found after mutagen treatment in mouse mastocytoma P815 (see 1.2). For years, only unaltered self proteins were discovered as target antigens for autologous T lymphocytes on human tumors. However, recently, two mutated melanoma antigens were described.

Antigen MUM-1 [68]

Five distinct antigens were recognized on melanoma cells of patient LB33. One of them, originally named LB33-B, was recognized by CTL in association with HLA-B44.1 and was identified by cDNA expression cloning. The sequence of the cDNA encoding LB33-B did not appear similar to any known gene and was found to be ubiquitously expressed. Its peptide-coding region is located across an exon-intron junction. A comparison with homologous cDNA from normal cells of the same patient demonstrated the presence of a point mutation in the intron part of the peptide-coding region. This point mutation led to an amino acid exchange (serine → isoleucine) at position 5 of the antigenic peptide (see Tab. 6.1). Both peptides bound equally well to HLA-B44.1, but only the mutated peptide was recognized by CTL. The mutation was found in two independently generated LB33-melanoma cell lines and in a tumor sample from patient LB33, indicating that the mutation had occured *in vivo*. Expression and translation of the peptide-coding region spanning an exon-intron border resulted from differential mRNA splicing. Alternate splicing was not specific to the LB33 tumor cells and, therefore, independent of the mutation. The gene encoding LB33-B was named MUM-1 (= melanoma ubiquitous mutated). A tumor cell variant, that had lost expression of MUM-1, did not show a phenotypic change [60], suggesting that its mutation did not contribute to the tumor transformation process.

Antigen CDK-R24C [69]

Among melanoma-reactive CTL, generated from the blood of patient SK29(AV) over a period of almost ten years, some clones recognized melanocyte differentiation antigens and, therefore, cross-reacted with a high proportion of allogeneic HLA-A2-positive melanoma lines. Other CTL clones seemed to lyse only autologous melanoma cells. Based on immunoselection, we distinguished between two distinct private antigens in this particular model, originally named SK29-B and -C [53, 57]. Using the COS transfection technique we cloned cDNA encoding antigen SK29-C. It was identified as a mutated cyclin-dependent kinase 4. The mutant allele was present in autologous cultured melanoma cells and metastatic tissue but was not found in the patient´s lymphocytes. It carries a UV-specific point mutation causing an arginine (R) to cysteine (C) exchange at residue 24. Cysteine 24 is part of the antigenic peptide processed from CDK4. It does not constitute the CTL epitope, but rather improves binding to HLA-A2.1. Aside from its role in immune recognition, the R24C mutation also prevents binding of the inhibitors p16INK4a and p15INK4b but does not affect the ability of this CDK4 mutant to bind cyclin D and form a functional kinase. Deregulated CDK4/cyclin D complexes may acquire oncogenic potential. This view is supported by the frequent inactivation of the p16INK4 gene in certain tumor types including melanoma [113, 114]. Therefore, it seems plausible to assume that expression of the CDK4-R24C allele contributed to the malignant phenotype in the SK29 melanoma. In addition, the mutated site of CDK4-R24C generates a tumor-specific target antigen for autologous CTL. Antigens with such dual functions were described as targets of potent tumor rejection responses *in vivo* in virally induced murine tumor models [115]. Aside from the SK29 melanoma, the CDK4-R24C allele was found in 1/28 melanomas analyzed so far.

In patient SK29(AV), the CDK4-R24C allele was generated by a somatic mutation. Recently, the same CDK4 mutation was found in two families with hereditary melanoma. The mutated allele was detected in 11/11 melanoma patients, but only in 2/17 unaffected family members. The authors of this study conclude that the germline R24C mutation in CDK4 generated a dominant oncogene [116].

6.2.2.5 Exploiting the T-Lymphocyte Response to Oncogenetic Proteins in Humans

Human tumor antigens described above, were identified as targets of tumor-reactive CTL isolated from blood or tumor tissues. As mentioned on page 320, discovering the rules for T cell recognition of protein epitopes, enabled immunologists to predict candidate T cell epitopes within given proteins. Mutated and/or overexpressed oncogenic and tumor-suppressor proteins are hoped-for candidate tumor rejection antigens due to several common features:

a They are shared by a high proportion of individuals.

b Their presence is associated with malignant transformation, maintainance of the malignant phenotype and tumor progression, and because of this, loss of expression is not likely to occur with these proteins.

c There is evidence for interaction of the humoral immune system, including CD4 responsiveness, with at least some of these proteins in cancer patients, which increases the likelihood that these antigens can be targeted successfully by T cell vaccines and T cell therapy.

Oncogenic proteins as potential targets for the immune system are listed in the following and have been reviewed recently [117, 118]:

a The p21 ras protein most commonly carries mutations in codons 12 and 61. These mutations determine its oncogenic effect and occur in a high proportion of several tumor types, like pancreatic carcinoma (90%), follicular carcinomas of the thyroid (60%), colon carcinomas (50%) and myeloid leukemias (40%).

b The bcr-abl fusion protein is expressed in more than 95% of chronic myelogenous leukemias (CML) and in about 25% of adult acute lymphocytic leukemia (ALL). It is generated by translocation of the human c-abl proto-oncogene from chromosome 9 to a specific breakpoint cluster (bcr) region on chromosome 22. This translocation is consistent and usually forms one of two new fusion genes encoding a 210 kD fusion protein with an abnormal tyrosine kinase activity essential for leukemogenesis.

c Mutated and overexpressed p53 is commonly observed in a great variety of human cancers. Inactivation of p53 most frequently results from mutation of one of the p53 alleles and deletion of the other. As opposed to ras, p53 mutations are not restricted to certain codons and, thus, are not predictable.

d HER-2/neu, a member of the epidermal growth factor receptor family, which is often overexpressed on malignant cells. Overexpression is due to gene amplification in many human malignancies, like carcinomas of the breast, ovary, uterus, stomach, and the lung. Moreover, at least in breast cancer patients overexpression is associated with a more aggressive tumor phenotype. It is a non-mutated self-protein. Tumor-reactive CTL specific for HER2/neu have been repeatedly isolated from cancer patients [119–121].

It proved possible to evoke in animals by vaccination T lymphocyte responses against non-viral oncogenic proteins [122]. It is, however, unclear, whether or not such peptide-specific immunity generates protectivity against tumors expressing the respective protein. From blood of tumor patients CTL that specifically recognized the sensitizing peptide were isolated after stimulation with cells loaded with HLA-binding peptides from oncogenic proteins. However, those CTL often did not lyse cells expressing their original proteins. From the candidate proteins listed here, HER-2/neu is the only one that has been identified as the target of tumor-reactive CTL isolated from patients. While CTL against HER-2/neu are directed against a normal protein, tumour reactive CTL against mutated epitopes on oncogneic proteins listed above have not been reported so far [123–124].

6.3 Evidence for the Relevance of T Cell-Defined Human Tumor Antigens *in vivo*

Tumor progression *per se* does not exclude potential immunogenicity of tumors, as we learnt from mouse models (see 6.1.1). Indeed, all of the antigens in Table 6.1 have been identified on tumor cells derived from progressive metastases. In addition, TIL populations, propagated *in vitro* from tumor infiltrates of progressive tumors, were reproducibly shown contain CTL subsets that can lyse autologous tumor cells [125]. Among the explanations for this seeming discrepancy are the following:

a Although adequately expressing peptide/HLA-complexes, tumor cells lack the expression of costimulatory molecules, like B7-1, needed for interaction with CD28 on T lymphocytes as a first step towards induction of a T cell response. Inadequate costimulation even renders T cells anergic [126–128].

b Tumor cells or infiltrating lymphocytes secrete immunosuppressive cytokines like transforming growth factor β (TGF-β) and IL-10 [129, 130].

c Tumors carrying weakly immunogenic antigens do not induce a rejection response, but, nevertheless, attract lymphocytes to the tumor site. This hypothesis of "tumor immunostimulation" was presented by R.T. Prehn [42].

The target antigens of autologous T lymphocyte responses on human tumors, as listed in Table 6.1, were identified exclusively by applying artefactual *in vitro* technology. Therefore, *a priori* it cannot be predicted, if these antigens are able to elicit *in vivo* tumor regression or protective immunity against tumor cells expressing them. This is in contrast to TSTAs in murine models, that were identified in the course of transplantation experiments as outlined above (see 1.3). Most of the antigens in Table 6.1 are autoantigens. Immune reactivity against them is likely to be controlled by mechanisms of thymic or peripheral tolerance induction [131, 132]. However, a series of independent observations point to the interaction between the immune system with tumor cells *in vivo* and suggest that at least some patients might profit from it.

Depigmentation or vitiligo of the skin is believed to result from an autoimmune response to melanocytes [133]. Serum antibodies against tyrosinase were found in these conditions [134]. Clinical studies on melanoma patients report the appearance of vitiligo to correlate with improved prognosis [135–137] and responsiveness to chemoimmunotherapy [138]. Similarly, the presence of autoantibodies against the ganglioside GM2 correlates with prolonged disease-free interval and improved survival [139]. However, it is unclear, whether immune reactivity towards normal melanocyte structures is by itself causing a rejection of melanoma lesions, or whether it is merely the bystander effect of a rejection response against unknown tumor antigens.

From tissue of a spontaneously regressing melanoma CTL that lysed autologous melanoma cells were isolated and were restricted by HLA-B14 [140], which offers direct evidence for immunsurveillance *in vivo*. Indirect evidence comes from observations indicating antigen loss by immunoselection *in vivo*, as reported in a melanoma model [60] and in cervical cancer [141].

All these observations support the notion of interaction between the T cell system and tumor cells *in vivo*. However, it will be difficult, if not impossible, to prove, in an individual patient, that a given antigen serves as a rejection antigen. For antigens identified so far and in the future, this can only be achieved in prospective immuntherapeutic trials as outlined in the following section.

6.4   Implications of T Cell-Defined Antigens on Immunotherapy

Tumor rejection and protectivity against murine tumors is predominantly conferred by the T cell system. Developing rational strategies for T-cell-based immunotherapy in humans depended very much on the identification of T cell target antigens and is now within reach. Recognition by T lymphocytes is very sensitive. An estimated 200 peptide/HLA-complexes are sufficient for recognition by CTL [142]. Presentation of a given peptide on the cell surface, however, is the end point of protein processing and transport involving many independently regulated steps [143] and is, therefore, especially susceptible to disturbances. Indeed, tumor progression is accompanied by loss or reduction of total HLA class I expression or selective losses of HLA I alleles by different mechanisms [reviewed in [144, 145]]. Investigations that determine the precise frequency of these events and their underlying cause are awaited and will be greatly facilitated by improved typing reagents and techniques. Aside from loss of target antigen expression, reduction or loss of HLA expression will be a major obstacle to T-cell-based immunotherapy and should be carefully assessed as a predictive parameter in clinical studies. However, it should be considered that tumor antigens, shed from the cell surface or released after cell death, can be processed and presented by infiltrating cells, which can then be targeted by tumor reactive T cells and thereby cause destruction of tumor tissues. Also still hypothetical, this concept seems attractive, because it would overcome defective processing or transport mechanisms in tumor cells.

Therapeutic Vaccination

The term vaccination was inaugurated for preventing infectious disease. Individuals without previous contact to an infectious agent are immunized with inactivated viruses, bacteria or their protein components in order to prevent infection. The situation will be different in tumor patients, who have already had a history of interaction with their own tumor over an unknown period of time [146]. For tumor patients one may apply the term therapeutic vaccination. The principal challenge is to induce a destructive anti-tumor T cell response, where the immune system, obviously, had failed before. Therapeutic vacccination certainly requires a deeper understanding of the immunological network as well as tolerance induction and has to meet higher demands on vaccine constructs and vaccination schedules than for disease prevention. Therefore, it ought to be kept in mind that all present efforts in that field should be regarded as contribution to basic science and should be designed to provide a maximum amount of information on parameters that can be associated with success or failure, according to present knowledge. Advantages of developing therapeutic vaccine strategies are obvious, especially because of their tolerability. This applies not only to malignant, but also to persistent infectious disease [147].

Animal studies demonstrated that even when tumors are already established, vaccination with gene-modified tumor cells, tumor peptides or peptide-pulsed presenting cells can cure a high percentage of syngeneic animals [34, 148–150]. These experiments cannot be immediately extrapolated to humans. Among other reservations, these animals were not primary tumor hosts. But also in human malignant disease, observations, in different tumor types, have been reported that point to the efficiency of therapeutic vaccination with defined antigens:

a   Kwak et al. induced in a small number of patients with B-cell lymphoma after conventional therapy T-cell and antibody responses against the surface-immunglobulin idiotype of autologous tumors. In two patients they observed a partial response of remaining lymphomas [151].

b   B. Mukherji and collegues immunized HLA-A1-positive melanoma patients with advanced disease, whose tumors expressed the MAGE-1 gene, intradermally with autologous antigen presenting cells pulsed with the synthetic MAGE-1 peptide EADPTGHSY. The investigators reported induction of T cells lysing autologous tumor cells and specific for the MAGE-1 peptide both at the immunization site and at distant metastatic sites [152].

c   In a pilot study, 16 patients with advanced disease, who were HLA-A1-positive and whose tumors expressed MAGE-3, were recruited [153]. The protocol intended to inject patients three times, at monthly intervals, subcutaneously with 100 or 300 µg of MAGE-3 peptide EVDPIGHLY, but without adjuvant. Of twelve melanoma patients, who were injected with the peptide, three developed mild inflammatory reactions at cutaneous or superficial lymph node metastatic sites requiring no further treatment. This was the only side effect, which did not correlate with visible tumor regression. Six of the melanoma patients were withdrawn, after one or two injections, because of rapid disease progression. Of six patients, who were immunized three times, regressions of multiple metastases occured. Regressions involved cutaneous, subcutaneous, and in one of the patients, lung metastases. Although tumor responses, seen even in the absence of an adjuvant, suggest that vaccination with MAGE-3 peptide may be effective in a significant proportion of patients, the authors emphasize that these results are to be regarded as preliminary and require confirmation in a larger group of patients.

d Five patients with adenocarcinoma of the pancreas participated in another pilot vaccination study. Pancreatic adenocarcinomas often harbor mutations in codon 12 of K-ras. The ras mutations in tumors of the participants had been identified by PCR amplification and sequencing analysis. Peptides encompassing amino acids 5–21 of the K-ras protein were synthesized and contained different amino acids at position 12, corresponding to the mutation in the individual tumor. For vaccination, patients were infused repeatedly with large quantities of peripheral blood mononuclear cells, pulsed with synthetic peptides carrying the individual mutation. In two of the five patients a transient ras-specific T cell response was observed following vaccination. The rationale was that part of the peptide-pulsed cells will reach lymphoid organs, where induction of a T cell response and thereby breaking of a preexisting tolerance to mutant ras is supposed to take place. No major therapeutic response was seen, which could be explained by the patients´ endstage disease resulting, also, in detoriation of the immune system [154, 155]. In continuation of this pilot study, the authors are now recruiting patients with minimal residual disease [156].

These examples of pioneering vaccination studies are heterogeneous and were performed in very small cohorts. Nevertheless, they motivate to further explore therapeutic vaccination. There is a wide variety of vaccine designs, such as:

- peptides admixed to adjuvants [34], conjugated to lipids or associated with liposomes [157],
- macrophages or dendritic cells pulsed with peptides [149, 150, 152, 156],
- intact proteins with adjuvants [158] and
- recombinant vaccinia viruses [39] or recombinant BCG [159].

At this stage it is not possible to predict, which of these strategies will prove optimal and – not to forget – whether or not the antigens identified so far will be appropriate immunogens. However, some principles have evolved during the last years and are briefly discussed in the following.

Peptides versus Whole Proteins

Choosing peptides for vaccination is not only tempted by the relative ease of producing peptides in sufficient quantity and quality. If T cell reactivity to dominant epitopes of self proteins is deleted by selection in the thymus, but reactivity against subdominant epitopes is ignored [160], immunization with peptides, containing only subdominant epitopes, would bypass tolerance to whole protein, as indicated by animal experiments (Melief, herein) [117].

Peptide-based vaccination is limited to patients expressing appropriate HLA molecules. In addition, individuals might respond to different epitopes, as appears to be the case with T cell responses to tyrosinase and gp100 (see 6.2.2.2). Widening the applicability of peptide vaccination would require the identification of many more T cell epitopes, although there is experimental evidence that vaccination with a single peptide not only elicits an immune response to the immunizing peptide, but also to other determinants contained in the same protein. The latter has been termed "determinant spreading" of the immune response [117, 161]. Mixing different peptides potentially circumvents individual preferences and would, in addition, counteract, to some degree, the threat of antigen-loss during tumor progression. However, vaccinating mixtures of peptides can at present not be generally recommended. Single peptides are able to both induce and tolerize T cells, depending on the mode of application (Melief, herein) [162]. Moreover, overstimulation of T cell responses might lead to exhaustion [163, 164] or apoptosis [165]. Since these effects are hard to predict, there are very good reasons to confine treatment to single peptides in initial studies, in the course of which, optimal delivery and dosage has to be determined. When choosing peptides for vaccination trials, the possibility of posttranslational modification also has to be taken into account [94].

Whole proteins are certainly more difficult to supply in sufficient quantities than peptides but would offer some advantages over peptides. Intact proteins contain all possible epitopes and, thus, are applicable regardless of the HLA type of patients. The presence of T helper epitopes might greatly facilitate the generation of a cytolytic T cell response [166]. Vaccination with whole proteins *per se* does not preclude the generation of a class I-restricted CTL response. There is increasing evidence that a subset of antigen-presenting cells acquires and presents exogenous antigens on HLA class I molecules [167]. Animal models demonstrated that vaccination with purified proteins can elicit strong enough CTL responses to provide protection [158].

Normal versus Mutated Antigens

As outlined in 6.1.3, vaccination of animals with tumor cells led to protection against the immunizing tumor but did not lead to cross-protection against other tumors of the same histotype or against tumors known to express antigens shared with the immunizing tumor. According to these experiments, shared antigens obviously do not elicit protective immunity. However, there is, to date, no systematic animal study really addressing this issue, which would require the immunization of animals with shared antigens in various ways (see vaccine designs listed above) and testing for immune protectivity in different tumor models and tumor types. In human cancer, shared antigenic epitopes are the only type to be applicable in a significant proportion of patients. Moreover, both in animals and in humans, immune responses to normal autoantigens, also expressed on tumors, have been associated with improved survival (see 6.3).

Mutated antigens are restricted to tumor tissues. It is assumed that high affinity T lymphocyte responses can be induced more readily against such antigens, because they have been generated comparably late during lifetime, and immunity against them is not subjected to thymic tolerance induction. However, depending on the duration of tumor history, tolerance induction might have evolved over years and might eventually prove as difficult to be circumvented as for normal autoantigens. Only in two instances have mutated CTL target antigens been identified in non-viral human tumors, and with considerable efforts [68] [69]. It cannot be predicted now, which proportion of tumors carry such antigens and whether technical progress will ever allow their more regular and more rapid identification. The existence of tumor-specific antigens argues for using autologous tumor tissue as a source for vaccines. Vaccinations with either autologous tumor cells, i.e. modified by viruses [168] or gene transfection (Forni and Blankenstein, herein), autologous peptide-loaden heat shock proteins [169] or unfractionated peptide pools eluted from tumor tissues [150] all provide a chance to immunize against unknown tumor-specific antigens.

Adoptive Transfer of T Lymphocytes

In animal models, T lymphocytes were able to eradicate tumors after adoptive transfer *in vivo*. Transferred lymphocytes were either specifically reacting with tumor tissues [115] or were directed against a tumor-associated antigen also expressed at lower levels on normal cells [47]. In human disease, the efficiency of adoptive transfer of antigen-specific T lymphocytes was most convincingly demonstrated in EBV-associated lymphoma [170] and in the treatment of relapsing chronic myeloid leukemia [171]. A myeloma idiotype-specific T cell response was successfully transferred to a patient by bone marrow transplantation. This patient´s healthy sibling donor had, previously, been immunized with the particular myeloma immunoglobulin [172]. Donor immunization with tumor-specific antigens is, therefore, regarded as a new concept for generating or enhancing the graft-versus-tumor effect of marrow grafts.

It was reported that administration of TIL populations, recognizing melanocyte differentiation antigens *in vivo*, resulted in tumor regression [71, 173]. However, as the authors themselves pointed out, it has to be considered that IL-2 was infused in parallel with T lymphocytes and that those TIL populations were not clonal. The identification of CTL target peptides offers the opportunity to expand peptide-specific CTL *in vitro* for adoptive therapy. Answers to whether or not such CTL are effective *in vivo* and how durable responses would be, are eagerly awaited. In general, wide applicability of adoptive transfer in T cell-based immunotherapy will certainly be hindered by costs and labor, associated with production of appropriate T cell populations.

6.5 Concluding Remarks

During the recent years, methods have been developed that will allow the identification of many more target antigens of anti tumor lymphocyte responses. Among these will hopefully be antigens with most of "ideal" characteristics, such as antigens,

- whose expression is restricted to tumor tissues,
- that are expressed on a high proportion of tumors,
- whose expression is stable,
- whose epitopes are presented by frequent HLA alleles and
- that are immunogenic in most individuals.

Aside from the search for new antigens, the development of vaccine adjuvants and delivery systems are areas of major interest. The reasonable aim of therapeutic immunizations will be to complement conventional therapies like surgery or chemotherapy (Cao, herein) in those patients, whose tumors have not yet lost their antigenicity.

References

1. Ehrlich, P (1909) Ueber den jetzigen Stand der Karzinomforschung. *Ned. Tijdschr. Geneeskd.* 1: 273–290.
2. Masucci, M. G (1993) Viral immunopathology of human tumors. *Curr. Opin. Immunol.* 5: 693–700.
3. Melief, C. J., W. M. Kast (1995) T-cell immunotherapy of tumors by adoptive transfer of cytotoxic T lymphocytes and by vaccination with minimal essential epitopes. *Immunol. Rev.* 145: 167–77.
4. Gross, L (1943) Intradermal immunization of C3H mice against a sarcoma that originated in an animal of the same line. *Cancer Res.* 3: 326–333.
5. Foley, E. J (1953) Antigenic properties of methylcholanthrene-induced tumors in mice of the strain of origin. *Cancer Res.* 13: 835–837.
6. Prehn, R. T., J. M. Main (1957) Immunity to methylcholanthrene-induced sarcomas. *J. Nat. Cancer Inst.* 18: 769–778.
7. Klein, G., Sjogren, H. O., Klein, E., Hellström, K. E (1960) Demonstration of resistance against methylcholanthrene-induced sarcomas in the primary autochtonous host. *Cancer Res.* 20: 1561–1572.
8. Kripke, M. L (1974) Antigenicity of murine skin tumors induced by ultraviolet light. *J. Nat. Cancer Inst.* 53: 1333.
9. Ward, P. L., H. Koeppen, T. Hurteau, H. Schreiber (1989) Tumor antigens defined by cloned immunological probes are highly polymorphic and are not detected on autologous normal cells. *J. Exp. Med.* 170: 217–32.
10. Hewitt, H. B., E. R. Blake, A. S. Walder (1976) A critique of the evidence for active host defence against cancer, based on personal studies of 27 murine tumours of spontaneous origin. *Brit. J. Cancer* 33: 241–259.
11. Middle, J. G., M. J. Embleton (1981) Naturally arising tumors in the inbred WAB/Not

rat strain. II. Immunogenicity of transplanted tumors. *J. Nat. Cancer Inst.* 67: 637–643.

12. Van Pel, A., F. Vessiére, T. Boon (1983) Protection against two spontaneous mouse leukemias conferred by immunogenic variants obtained by mutagenesis. *J. Exp. Med.* 157: 1992–2001.

13. Hui, K. M., T. Sim, T. T. Foo, A. A. Oei (1989) Tumor rejection mediated by transfection with allogeneic class I histocompatibility gene. *J. Immunol.* 143: 3835–3843.

14. Grohmann, U., R. Bianchi, M. C. Fioretti, F. Fallarino, L. Binaglia, C. Uyttenhove, A. Van Pel, T. Boon, P. Puccetti (1995) CD8+ cell activation to a major mastocytoma rejection antigen, P815AB: requirement for tumor helper peptides in priming for skin test reactivity to a P815AB-related peptide. *Eur. J. Immunol.* 25: 2797–2802.

15. Boon, T (1983) Antigenic tumor cell variants obtained with mutagens. *Adv. Cancer Res.* 39: 121–151.

16. Schirrmacher, V (1992) Immunity and metastasis: *In situ* activation of protective T cells by virus modified cancer vaccines. *Cancer Surv.* 13: 129–154.

17. Colombo, M. P., G. Forni (1994) Cytokine gene transfer in tumor inhibition and tumor therapy. *Immunol. Today* 15: 48–51.

18. Melief, C. J. M (1992) Tumor eradication by adoptive transfer of cytotoxic T lymphocytes. *Adv. Cancer Res.* 58: 143–175.

19. Greenberg, P. D (1991) Adoptive T cell therapy of tumors: mechanisms operative in the recognition and elimination of tumor cells. *Adv. Immunol.* 49: 281–355.

20. Srivastava, P. K., A. B. DeLeo, L. J. Old (1986) Tumor rejection antigens of chemically induced tumors of inbred mice. *Proc. Natl. Acad. Sci. USA* 83: 3407–3411.

21. Srivastava, P. K., R. G. Maki (1991) Stress-induced proteins in immune response to cancer. *Curr. Top. Microbiol. Immunol.* 167: 109–123.

22. Uduno, H., D. L. Levey, P. K. Srivastava (1994) Cellular requirements for tumor-specific immunity elicited by heat shock proteins: tumor rejection antigen gp96 primes CD8+ T cells *in vivo. Proc. Natl. Acad. Sci.* 91: 3077–3081.

23. Suto, R., P. K. Srivastava (1995) A mechanism for the specific immunogenicity of heat shock protein-chaperoned peptides. *Science* 269: 1585–1588.

24. Wölfel, T., Van Pel, A., De Plaen, E., Lurquin, C., Maryanski, J. and Boon, T (1987) Immunogenic (tum⁻) variants obtained by mutagenesis of mouse mastocytoma P815. VIII. Detection of stable transfectants expressing a tum⁻ antigen with a cytolytic T cell stimulation assay. *Immunogenetics* 26: 178–187.

25. De Plaen, E., Lurquin, C., Van Pel, A., Mariame, B., Szikora, J. -P., Wölfel, T., Sibille, C., Chomez, P. and Boon, T (1988) Immunogenic (tum⁻) variants of mouse tumor P815: Cloning of the gene of tum-antigen P91A and identification of the tum⁻ mutation. *Proc. Natl. Acad. Sci. USA* 85: 2274–2278.

26. Lurquin, C., A. Van Pel, B. Mariame, E. De Plaen, J. -P. Szikora, C. Janssens, M. J. Reddehase, J. Lejeune, T. Boon (1989) Structure of the gene of tum-transplantation antigen P91A: the mutated exon encodes a peptide recognized with Ld by cytolytic T cells. *Cell* 58: 293–303.

27. Sibille, C., P. Chomez, C. Wildmann, A. Van Pel, E. De Plaen, J. L. Maryanski, V. de Bergeyck, T. Boon (1990) Structure of the gene coding for tum⁻ transplantation antigen P198: a point mutation generates a new antigenic peptide. *J. Exp. Med.* 172: 35–45.

28. Szikora, J. -P., A. Van Pel, V. Brichard, M. André, N. Van Baren, P. Henry, E. De Plaen, T. Boon (1990) Structure of the gene of tum-transplantation antigen P35B: presence of a point mutation in the antigenic allele. *EMBO J.* 9: 1041–1050.

29. Townsend, A. R. M., Rothbard, J., Gotch, F. M., Bahadur, G., Wraith, D. and A. J. McMichael (1986) The epitopes of influenza nucleoprotein recognized by cytotoxic T lymphocytes can be defined with short synthetic peptides. *Cell* 44: 959–968.

30. Van den Eynde, B., B. Lethe, A. Van Pel, E. De Plaen, T. Boon (1991) The gene coding for a major tumor rejection antigen of tumor P815 is identical to the normal gene of syngeneic DBA/2 mice. *J. Exp. Med.* 173: 1373–1384.

31. Boon, T., P. Coulie, M. Marchand, P. Weynants, T. Wölfel, V. Brichard (1994) Genes coding for tumor rejection antigens: perspectives for specific immunotherapy. *Biologic therapy of cancer updates* 4: 1–14.

32. Mandelboim, O., G. Berke, M. Fridkin, M. Feldman, M. Eisenstein, L. Eisenbach (1994) CTL induction by a tumour-associated antigen octapeptide derived from a murine lung carcinoma. *Nature* 369: 67–71.

33. Pardoll, D. M (1994) A new look for the 1990s. *Nature* 369: 357–358.

34. Mandelboim, O., E. Vadai, M. Fridkin, A. Katz-Hillel, M. Feldman, G. Berke, L. Eisenbach (1995) Regression of established murine carcinoma metastases following vaccination with tumour-assiated antigen peptides. *Nat. Med.* 1: 1179–1183.

35. de Bergeyck, V., E. De Plaen, P. Chomez, T. Boon, A. Van Pel (1994) An intracisternal A-particle sequence codes for an antigen recognized by syngeneic cytolytic T lymphocytes on a mouse spontaneous leukemia. *Eur. J. Immunol.* 24: 2203–2212.

36. Uenaka, A., T. Ono, T. Akisawa, H. Wada, T. Yasuda, E. Nakayama (1994) Identification of a unique antigen peptide pRL1 on BALB/c RLo->1 leukemia recognized by cytotoxic T lymphocytes and its relation to the *Akt* oncogene. *J. Exp. Med.* 180: 1599–1607.

37. Monach, P. A., S. C. Meredith, C. T. Siegel, H. Schreiber (1995) A unique tumor antigen produced by a single amino acid substitution. *Immunity* 2: 45–59.

38. Melief, C. J. M., W. M. Kast (1993) Potential immunogenicity of oncogene and tumor suppressor gene products. *Curr. Opin. Immunol.* 5: 709–713.

39. Skipper, J., H. J. Stauss (1993) Identification of two cytotoxic T lymphocyte-recognized epitopes in the Ras protein. *J. Exp. Med.* 177: 1493–8.

40. Peace, D. J., J. W. Smith, W. Chen, S. -G. You, W. L. Cosand, J. Blake, M. A. Cheever (1994) Lysis of ras oncogene-transformed cells by specific cytolytic T lymphocytes elicited by primary *in vitro* immunization with mutated ras peptide. *J. Exp. Med.* 179: 473–479.

41. Noguchi, Y., Y.-T. Chen, L. -J. Old (1994) A mouse mutant p53 product recognized by CD4+ and CD8+ T cells. *Proc. Natl. Acad. Sci. USA* 91: 3171–3175.

42. Prehn, R. T (1993) Tumor immunogenicity: how far can it be pushed? *Proc. Natl. Acad. Sci. USA* 90: 4332–4333.

43. Boon, T., J.-P. Szikora, E. De Plaen, T. Wölfel, A. Van Pel (1989) Cloning and characterization of genes coding for tum⁻ transplantation antigens. *J. Autoimmun.* 2 (Supplement): 109–114.

44. Hostetler, L. W., H. N. Ananthaswamy, M. L. Kripke (1986) Generation of tumor-specific transplantation antigens by UV radiation can occur independently of neoplastic transformation. *J. Immunol.* 137: 2721–2725.

45. Basombrio, M. A (1970) Search for common antigenicities among twenty-five sarcomas induced by methylcholanthrene. *Cancer Res.* 30: 2458–2462.

46. Ramarathinam, L., M. Castle, Y. Wu, Y. Liu (1994) T cell costimulation by B7/BB1 induces CD8 T cell-dependent tumor rejection: an important role of B7/BB1 in the induction, recruitment, and effector function of antitumor T cells. *J. Exp. Med.* 179:

1205–1214.
47. Hu, J., W. Kindsvogel, S. Busby, M. C. Bailey, Y. Y. Shi, P. D. Greenberg (1993) An evaluation of the potential to use tumor-associated antigens as targets for antitumor T cell therapy using transgenic mice expressing a retroviral tumor antigen in normal lymphoid tissues. *J. Exp. Med.* 177: 1681–90.
48. Hara, I., Y. Takechi, A. N. Houghton (1995) Implicating a role for immune recognition of self in tumor rejection: passive immunization against the Brown locus protein. *J. Exp. Med.* 182: 1609–1614.
49. Morgan, D. A., Ruscetti, F. W., Gallo, R. C (1976) Selective *in vitro* growth of T lymphocytes from normal human bone marrows. *Science* 193: 1007.
50. Gillis, S., Smith, K (1977) Long term culture of tumor specific cytotoxic T cells. *Nature* 268: 154.
51. Knuth, A., T. Wölfel, K. H. Meyer zum Büschenfelde (1992) T cell responses to human malignant tumors. *Cancer Surv.* 13: 39–52.
52. Crowley, N. J., T. L. Darrow, M. A. Quinn-Allen, H. F. Seigler (1991) MHC-restricted recognition of autologous melanoma by tumor-specific cytotoxic T cells. Evidence for restriction by a dominant HLA-A allele. *J. Immunol.* 146: 1692–1699.
53. Wölfel, T., M. Hauer, E. Klehmann, V. Brichard, B. Ackermann, A. Knuth, T. Boon, K. - H. Meyer zum Büschenfelde (1993) Analysis of antigens recognized on human melanoma cells by A2-restricted cytolytic T lymphocytes (CTL). *Int. J. Cancer* 55: 237–244.
54. Bernhard, H., J. Karbach, T. Wölfel, P. Busch, S. Störkel, M. Stöckle, C. Wölfel, B. Seliger, C. Huber, K. -H. Meyer zum Büschenfelde, A. Knuth (1994) Cellular immune response to human renal-cell carcinomas: definition of a common antigen recognized by HLA-A2-restricted cytotoxic T-lymphocyte (CTL) clones. *Int. J. Cancer* 59: 837–842.
55. Anichini, A., C. Maccalli, R. Mortarini, S. Salvi, A. Mazzocchi, P. Squarcina, M. Herlyn, G. Parmiani (1993) Melanoma cells and normal melanocytes share antigens recognized by HLA-A2-restricted cytotoxic T cell clones from melanoma patients. *J. Exp. Med.* 177: 989–998.
56. Anichini, A., R. Mortarini, C. Maccalli, P. Squarcina, K. Fleischhauer, L. Mascheroni, G. Parmiani (1996) Cytotoxic T cells directed to tumor antigens not expressed on normal melanocytes dominate HLA-A2.1-restricted immune repertoire to melanoma. *J. Immunol.* 156: 208–217.
57. Knuth, A., T. Wölfel, E. Klehmann, T. Boon, K. H. Meyer zum Büschenfelde (1989) Cytolytic T-cell clones against an autologous human melanoma: Specificity study and definition of three antigens by immunoselection. *Proc. Natl. Acad. Sci. USA* 86: 2804–2808.
58. Van den Eynde, B., P. Hainaut, M. Herin, A. Knuth, C. Lemoine, P. Weynants, P. Van der Bruggen, R. Fauchet, T. Boon (1989) Presence on a human melanoma of multiple antigens recognized by autologous CTL. *Int. J. Cancer* 44: 634–640.
59. Ioannides, C. G., R. S. Freedman, C. D. Platsoucas, S. Rashed, Y. P. Kim (1991) Cytotoxic T cell clones isolated from ovarian tumor-infiltrating lymphocytes recognize multiple antigenic epitopes on autologous tumor cells. *J. Immunol.* 146: 1700–1707.
60. Lehmann, F., M. Marchand, P. Hainaut, P. Pouillart, X. Sastre, H. Ikeda, T. Boon, P. G. Coulie (1995) Differences in the antigens recognized by cytolytic T cells on two successive metastases of a melanoma patient consistent with immunoselection. *Eur. J. Immunol.* 25: 340–347.
61. Unanue, E. R (1995) The 1995 Albert Lasker Medical Research Award. The concept of

antigen processing and presentation. *Jama* 274: 1071–3.

62. Seed, B., A. Aruffo (1987) Molecular cloning of the CD2 antigen, the T-cell erythrocyte receptor, by a rapid immunoselection procedure. *Proc. Natl. Acad. Sci. USA* 84: 3365–3369.

63. Karttunen, J., N. Shastri (1991) Measurement of ligand-induced activation in single viable T cells using the *lacZ* reporter gene. *Proc. Natl. Acad. Sci. USA* 88: 3972–3976.

64. Karttunen, J., S. Sanderson, N. Shastri (1992) Detection of rare antigen-presenting cells by the *lacZ* T-cell activation assay suggests an expression cloning strategy for T-cell antigens. *Proc. Natl. Acad. Sci. USA* 89: 6020–6024.

65. Brichard, V., A. Van Pel, T. Wölfel, C. Wölfel, E. De Plaen, B. Lethé, P. Coulie, T. Boon (1993) The tyrosinase gene codes for an antigen recognized by autologous cytolytic T lympocytes on HLA-A2 melanomas. *J. Exp. Med.* 178: 489–495.

66. Coulie, P., V. Brichard, A. Van Pel, T. Wölfel, J. Schneider, C. Traversari, S. Mattei, E. De Plaen, C. Lurquin, J. -P. Szikora, J. -C. Renauld, T. Boon (1994) A new gene coding for a differentiation antigen recognized by autologous cytolytic T lymphocytes on HLA-A2 melanomas. *J. Exp. Med.* 180: 35–42.

67. Van den Eynde, B., O. Peeters, O. De Backer, B. Gaugler, S. Lucas, T. Boon (1995) A new family of genes coding for an antigen recognized by autologous cytolytic T lymphocytes on a human melanoma. *J. Exp. Med.* 182: 689–698.

68. Coulie, P. G., F. Lehmann, B. Lethé, J. Herman, C. Lurquin, M. Andrawiss, T. Boon (1995) A mutated intron sequence codes for an antigenic peptide recognized by cytolytic T lymphocytes on a human melanoma. *Proc. Natl. Acad. Sci. USA* 92: 7976–7980.

69. Wölfel, T., M. Hauer, J. Schneider, M. Serrano, C. Wölfel, E. Klehmann-Hieb, E. De Plaen, T. Hankeln, K. -H. Meyer zum Büschenfelde, D. Beach (1995) A p16INK4a-insensitive CDK4 mutant targeted by cytolytic T lymphocytes in a human melanoma. *Science* 269: 1281–1284.

70. Kawakami, Y., S. Eliyahu, C. H. Delgado, P. F. Robbins, L. Rivoltini, S. L. Topalian, T. Miki, S. A. Rosenberg (1994) Cloning of the gene coding for a shared human melanoma antigen recognized by autologous T cells infiltrating into tumor. *Proc. Natl. Acad. Sci. USA* 91: 3515–3519.

71. Robbins, P. F., E. -G. M., Y. Kawakami, S. A. Rosenberg (1994) Recognition of tyrosinase by tumor-infiltrating lymphocytes from a patient responding to immunotherapy. *Cancer Res.* 54: 3124–6.

72. Kawakami, Y., S. Eliyahu, C. H. Delgado, P. F. Robbins, L. Rivoltini, S. L. Topalian, T. Miki, S. A. Rosenberg (1994) Cloning of the gene coding for a shared human melanoma antigen recognized by autologous T cells infiltrating into tumor. *Proc. Natl. Acad. Sci. USA.* 91: 3515–9.

73. Wang, R. F., P. F. Robbins, Y. Kawakami, X. Q. Kang, S. A. Rosenberg (1995) Identification of a gene encoding a melanoma tumor antigen recognized by HLA-A31-restricted tumor-infiltrating lymphocytes. *J. Exp. Med.* 181: 799–804.

74. Robbins, P. F., M. El Gamil, Y. F. Li, S. L. Topalian, L. Rivoltini, K. Sakaguchi, E. Appella, Y. Kawakami, S. A. Rosenberg (1995) Cloning of a new gene encoding an antigen recognized by melanoma-specific HLA-A24-restricted tumor-infiltrating lymphocytes. *J. Immunol.* 154: 5944–50.

75. Bjorkman, P. J., M. A. Saper, B. Samraoui, W. S. Bennett, J. L. Strominger, D. C. Wiley (1987) The foreign antigen binding site and T cell recognition regions of class I histocompatibility antigens. *Nature* 329: 512–518.

76. Rammensee, H. -G., T. Friede, S. Stevanovic (1995) MHC ligands and peptide motifs: first listing. *Immunogenetics* 41: 178–228.

77. Cox, A. L., J. Skipper, Y. Chen, R. A. Henderson, T. L. Darrow, J. Shabanowitz, V. H. Engelhard, D. F. Hunt, C. L. Slingluff Jr (1994) Identification of a peptide recognized by five melanoma-specific human cytotoxic T cell lines. *Science* 264: 716–719.

78. van der Bruggen, P., C. Traversari, P. Chomez, C. Lurquin, E. De Plaen, B. Van den Eynde, A. Knuth, T. Boon (1991) A gene encoding an antigen recognized by cytolytic T lymphocytes on a human melanoma. *Science* 254: 1643–1647.

79. van der Bruggen, P., J. -P. Szikora, P. Boel, C. Wildmann, M. Somville, M. Sensi, T. Boon (1994) Autologous cytolytic T lymphocytes recognize a MAGE-1 nonapeptide on melanomas expressing HLA-Cw*1601. *Eur. J. Immunol.* 24: 2134–2140.

80. van der Bruggen, P., J. Bastin, T. Gajewski, P. G. Coulie, P. Boel, D. S. C., C. Traversari, A. Townsend, T. Boon (1994) A peptide encoded by human gene MAGE-3 and present-ed by HLA-A2 induces cytolytic T lymphocytes that recognize tumor cells expressing MAGE-3. *Eur. J. Immunol.* 24: 3038–43.

81. Gaugler, B., B. Van den Eynde, P. van der Bruggen, P. Romero, J. J. Gaforio, E. De Plaen, B. Lethé, F. Brasseur, T. Boon (1994) Human gene MAGE-3 codes for a antigen recog-nized on a melanoma by autologous cytolytic T lymphocytes. *J. Exp. Med.* 179: 921–930.

82. Boel, P., C. Wildmann, M. L. Sensi, R. Brasseur, J. -C. Renauld, P. Coulie, T. Boon, P. van der Bruggen (1995) BAGE: a new gene encoding an antigen recognized on human melanomas by cytolytic T lymphocytes. *Immunity* 2: 167–175.

83. Van Pel, A., P. van der Bruggen, P. G. Coulie, V. G. Brichard, B. Lethé, B. Van den Eynde, C. Uyttenhove, J. -C. Renauld, T. Boon (1995) Genes coding for tumor antigens recog-nized by cytolytic T lymphocytes. *Immunol. Rev.* 145: 229–249.

84. Brasseur, F., a. others (1995) Expression of MAGE genes in primary and metastatic cu-taneous melanoma. *Int. J. Cancer* 63: 375–380.

85. Anichini, A., Fossati, G. and Parmiani, G (1987) Clonal analysis of the cytolytic T cell re-sponse to human tumors. *Immunol. Today* 8: 385–389.

86. Crowley, N. J., C. L. Slingluff, T. L. Darrow, H. F. Seigler (1990) Generation of human autologous melanoma-specific cytotoxic T-cells using HLA-A2 matched allogeneic melanomas. *Cancer Res.* 50: 492–498.

87. Viret, C., F. Davodeau, Y. Guilloux, J. -D. Bignon, G. Semana, R. Breathnach, F. Jotereau (1993) Recognition of shared melanoma antigen by HLA-A2-restricted cytolytic T cell clones derived from human tumor-infiltrating lymphocytes. *Eur. J. Immunol.* 23: 141–146.

88. Wölfel, T., A. Van Pel, V. Brichard, J. Schneider, B. Seliger, K. -H. Meyer zum Büschenfelde, T. Boon (1994) Two tyrosinase nonapeptides recognized on HLA-A2 melanomas by autologous cytolytic T lymphocytes. *Eur. J. Immunol.* 24: 759–764.

89. Chen, Q., P. Hersey (1992) MHC-restricted responses of CD8+ and CD4+ T-cell clones from regional lymph nodes of melanoma patients. *Int. J. Cancer* 51: 218–24.

90. Stam, N. J., T. N. Vroom, P. J. Peters, E. B. Pastoors, H. L. Ploegh. submitted.HC-A2.

91. Spies, T., V. Cerundolo, M. Colonna, P. Cresswell, A. Townsend, R. De Mars (1992) Presentation of viral antigen by MHC class I molecules is dependent on a putative pep-tide transporter heterodimer. *Nature* 355: 644–646.

92. Restifo, N. P., F. Esquivel, Y. Kawakami, J. W. Yewdell, J. J. Mule, S. A. Rosenberg, J. R. Bennink (1993) Identification of human cancers deficient in antigen-processing. *J. Exp. Med.* 177: 265–272.

93. Wei, M. L., P. Cresswell (1992) HLA-A2 molecules in an antigen-processing mutant cell contain signal sequence-derived peptides. *Nature* 356: 443–446.
94. Skipper, J. C. A., R. C. Hendrickson, P. H. Gulden, V. Brichard, A. Van Pel, Y. Chen, J. Shabanowitz, T. Wölfel, J. Slingluff, G. L., T. Boon, D. F. Hunt, V. H. Engelhard (1996) An HLA-A2-restricted tyrosinase antigen on melanoma cells resulted from posttranslational modification and suggests a novel pathway for processing of membrane proteins. *J. Exp. Med.* 183: 527–534.
95. Kang, X., Y. Kawakami, G. M. el, R. Wang, K. Sakaguchi, J. R. Yannelli, E. Appella, S. A. Rosenberg, P. F. Robbins (1995) Identification of a tyrosinase epitope recognized by HLA-A24-restricted, tumor-infiltrating lymphocytes. *J. Immunol.* 155: 1343–8.
96. Brichard, V. G., J. Herman, A. Van Pel, C. Wildman, B. Gaugler, T. Wölfel, T. Boon, B. Lethé (1996) A tyrosinase nonapeptide presented by HLA-B44 is recognized on a human melanoma by autologous cytolytic T lymphocytes. *Eur. J. Immunol.* 26: 224–230.
97. Topalian, S. L (1994) MHC class II restricted tumor antigens and the role of CD4+ T cells in cancer immunotherapy. *Curr. Opin. Immunol.* 6: 741–745.
98. Visseren, M. J., E. A. van, d. V. E. van, M. E. Ressing, W. M. Kast, P. I. Schrier, C. J. Melief (1995) CTL specific for the tyrosinase autoantigen can be induced from healthy donor blood to lyse melanoma cells. *J. Immunol.* 154: 3991–8.
99. Herr, W., J. Schneider, A. W. Lohse, K. -H. Meyer zum Büschenfelde, T. Wölfel (1996) Detection and quantification of blood-derived CD8+ T lymphocytes secreting tumor necrosis factor alpha in response to HLA-A2.1-binding melanoma and viral peptide antigens. *J. Immunol. Meth.* 191: 131–142.
100. Kwon, B. S., C. Chintamaneni, C. A. Kozak, N. G. Copeland, D. J. Gilbert, N. Jenkins, D. Barton, U. Francke, Y. Kobayashi, K. K. Kim (1991) A melanocyte-specific gene, Pmel 17, maps near the silver coat color locus on mouse chromosome 10 and is in a syntenic region on human chromosome 12. *Proc. Natl. Acad. Sci. USA.* 88: 9228–32.
101. Bakker, A. B. H., M. W. J. Schreurs, A. J. de Boer, Y. Kawakami, S. A. Rosenberg, G. J. Adema, C. G. Figdor (1994) Melanocyte lineage-specific antigen gp100 is recognized by melanoma-derived tumor-infiltrating lymphocytes. *J. Exp. Med.* 179: 1005–1009.
102. Kawakami, Y., S. Eliyahu, C. Jennings, K. Sakaguchi, X. Kang, S. Southwood, P. F. Robbins, A. Sette, E. Appella, S. A. Rosenberg (1995) Recognition of multiple epitopes in the human melanoma antigen gp100 by tumor-infiltrating T lymphocytes associated with *in vivo* tumor regression. *J. Immunol.* 154: 3961–3968.
103. Kawakami, Y., S. Eliyahu, K. Sakaguchi, P. F. Robbins, L. Rivoltini, J. R. Yannelli, E. Appella, S. A. Rosenberg (1994) Identification of the immunodominant peptides of the MART-1 human melanoma antigen recognized by the majority of HLA-A2-restricted tumor infiltrating lymphocytes. *J. Exp. Med.* 180: 347–52.
104. Slingluff, C. L., A. L. Cox, R. A. Henderson, D. F. Hunt, V. H. Engelhard (1993) Recognition of human melanoma cells by HLA-A2.1-restricted cytotoxic T lymphocytes is mediated by at least six shared peptide epitopes. *J. Immunol.* 150: 2955–2963.
105. Wölfel, T., J. Schneider, K. -H. Meyer zum Büschenfelde, H. -G. Rammensee, O. Rötzschke, K. Falk (1994) Isolation of naturally processed peptides recognized by cytolytic T lymphocytes (CTL) on human melanoma cells in association with HLA-A2.1. *Int. J. Cancer* 57: 413–418.
106. Castelli, C., W. J. Storkus, M. J. Maeurer, D. M. Martin, E. C. Huang, B. N. Pramanik, T. L. Nagabhushan, G. Parmiani, M. T. Lotze (1995) Mass spectrometric identification of a naturally processed melanoma peptide recognized by CD8+ cytotoxic T lympho-

cytes. *J. Exp. Med.* 181: 363–8.

107. Mattes, M. J., T. M. Thomson, L. J. Old, K. O. Lloyd (1983) A pigmentation associated, differentiation antigen of human melanoma defined by a precipitating antibody in human serum. *Int. J. Cancer* 32: 717.

108. Jimenez-Cervantes, C., F. Solano, T. Kobayashi, K. Urabe, V. J. Hearing, J. A. Lozano, J. C. Garcia-Borron (1994) A new enzymatic function in the melanogenic pathway. *J. Biol. Chem.* 269: 17993–18001.

109. Thomsen, T. M., F. X. Real, S. Murakami, C. Cordon-Cardi, L. J. Old, A. N. Houghton (1988) differentiation antigens of melanocytes and melanoma: analysis of melanosome and cell surface markers of human pigmented cells with monoclonal antibodies. *J. Invest. Dermatol.* 90: 459–466.

110. Bakker, A. B., G. Marland, A. J. de Boer, R. J. Hujbens, E. H. Danen, G. J. Adema, C. G. Figdor (1995) Generation of anti-melanoma cytotoxic T lymphocytes from healthy donors after presentation of melanoma-associated antigen-derived epitopes by dendritic cells *in vitro. Cancer Res.* 15: 5330–5334.

111. Finn, O. J., K. R. Jerome, R. A. Henderson, G. Pecher, N. Domenech, J. Magarian-Blander, S. M. Barratt-Boves (1995) MUC-1 epithelial tumor mucin-based immunity and cancer vaccines. *Immunol. Rev.* 145: 61–89.

112. Rajan, T. V (1987) Is there a role for MHC class I antigens in the elimination of somatic mutants? *Immunol. Today* 8: 171–172.

113. Kamb, A., N. A. Gruis, J. Weaver-Feldhaus, Q. Liu, K. Harshman, S. V. Tavtigian, E. Stockert, R. S. Day III, B. E. Johnson, M. H. Skolnick (1994) A cell cycle regulator potentially involved in genesis of many tumor types. *Science* 264: 436–440.

114. Mori, T., K. Miura, T. Aoki, S. Mori, Y. Nakamura (1994) Frequent somatic mutation of the MTS1/CDK4I (multiple tumor suppressor/cyclin-dependent kinase 4 inhibitor) gene in esophageal squamous cell carcinoma. *Cancer Res.* 54: 3396–3397.

115. Kast, W. M., R. Offringa, P. J. Peters, A. C. Voordouw, R. H. Meloen, A. J. van der Eb, C. J. M. Melief (1989) Eradication of adenovirus E1-induced tumors by E1A-specific cytotoxic lymphocytes. *Cell* 59: 603–614.

116. Zuo, L., J. Weger, B. Yang, A. M. Goldstein, M. A. Tucker, N. Hayward, N. C. Dracopoli (1996) Germline mutations in the p16INK4A binding domain of CDK4 in familial melanoma. *Nat. Genet.* 12: 97–99.

117. Cheever, M. A., M. L. Disis, H. Bernhard, J. R. Gralow, S. L. Hand, E. S. Huseby, H. L. Qin, M. Takahashi, W. Chen (1995) Immunity to oncogenic proteins. *Immunol. Rev.* 145: 33–59.

118. Nijman, H. W., d. B. S. Van, M. P. Vierboom, J. G. Houbiers, W. M. Kast, C. J. Melief (1994) p53, a potential target for tumor-directed T cells. *Immunol. Lett.* 40: 171–8.

119. Ioannides, C. G., B. Fisk, D. Fan, W. E. Biddison, J. T. Wharton, C. A. O'Brian (1993) Cytotoxic T cells isolated from ovarian malignant ascites recognize a peptide derived from the HER-2/neu proto-oncogene. *Cell. Immunol.* 151: 225–234.

120. Peoples, G. E., I. Yoshino, C. C. Douville, J. V. Andrews, P. S. Goedegebuure, T. J. Eberlein (1994) TCR V beta 3+ and V beta 6+ CTL recognize tumor-associated antigens related to HER2/neu expression in HLA-A2+ ovarian cancers. *J. Immunol.* 152: 4993–9.

121. Fisk, B., B. Chesak, M. S. Pollack, J. T. Wharton, C. G. Ioannides (1994) Oligopeptide induction of a cytotoxic T lymphocyte response to HER-2/neu proto-oncogene *in vitro. Cell. Immunol.* 157: 415–27.

122. Chen, W., D. J. Peace, D. K. Rovira, S. -G. You, M. A. Cheever (1992) T-cell immunity to the joining region of p210BCR-ABL protein. *Proc. Natl. Acad. Sci. USA* 89: 1486–1472.

123. Houbiers, J. G. A., H. W. Nijman, S. H. Van der Burg, J. W. Drijfhout, P. Kenemans, C. J. H. Van de Velde, A. Brand, F. Momburg, W. M. Kast, C. J. M. Melief (1993) *In vitro* induction of human cytotoxic T lymphocyte responses against peptides of mutant and wild type p53. *Eur. J. Immunol.* 23: 2072–2077.

124. Fossum, B., T. I. Gedde-Dahl, J. Breivik, J. A. Eriksen, A. Spurkland, E. Thorsby, G. Gaudernack (1994) p21-ras-peptide-specific T-cell responses in a patient with colorectal cancer. CD4+ and CD8+ T cells recognize a peptide corresponding to a common mutation (13Gly→Asp). *Int. J. Cancer* 56: 40–45.

125. Spagnoli, G. C., C. Schaefer, T. E. Willimann, T. Kocher, A. Amoroso, A. Juretic, M. Zuber, U. Luscher, F. Harder, M. Heberer (1995) Peptide-specific CTL in tumor infiltrating lymphocytes from metastatic melanomas expressing MART-1/Melan-A, gp100 and Tyrosinase genes: a study in an unselected group of HLA-A2.1-positive patients. *Int. J. Cancer* 64: 309–15.

126. Townsend, S. E., J. P. Allison (1993) Tumor rejection after direct costimulation of CD8+ T cells by B7-transfected melanoma cells. *Science* 259: 368–370.

127. Becker, J. C., T. Brabletz, C. Czerny, C. Termeer, E. B. Brocker (1993) Tumor escape mechanisms from immunosurveillance: induction of unresponsiveness in a specific MHC-restricted CD4+ human T cell clone by the autologous MHC class II+ melanoma. *Int. Immunol.* 5: 1501–8.

128. Lanzavecchia, A (1993) Identifying strategies for immune intervention. *Science* 260: 937–944.

129. Luscher, U., L. Filgueira, A. Juretic, M. Zuber, N. J. Luscher, M. Heberer, G. C. Spagnoli (1994) The pattern of cytokine gene expression in freshly excised human metastatic melanoma suggests a state of reversible anergy of tumor-infiltrating lymphocytes. *Int. J. Cancer* 57: 612–9.

130. Becker, J. C., C. Czerny, E. B. Brocker (1994) Maintenance of clonal anergy by endogenously produced IL-10. *Int. Immunol.* 6: 1605–12.

131. von Boehmer, H (1992) Thymic selection: a matter of life and death. *Immunol. Today* 13: 454–8.

132. Arnold, B., G. Schönrich, G. J. Hämmerling (1993) Multiple levels of peripheral tolerance. *Immunol. Today* 14: 12–14.

133. Naughton, G. K., M. Eisinger, J. C. Bystryn (1983) Antibodies to normal human melanocytes in vitiligo. *J. Exp. Med.* 158: 246–251.

134. Song, Y. H., E. Connor, Y. Li, B. Zorovich, P. Balducci, N. Maclaren (1994) The role of tyrosinase in autoimmune vitiligo. *Lancet* 344: 1049–52.

135. Nordlund, J. J., J. M. Kirkwood, B. M. Forget, G. Milton, D. M. Albert, A. B. Lerner (1983) Vitiligo in patients with metastatic melanoma: a good prognostic sign. *J. Amer. Acad. Derm.* 9: 689–696.

136. Bystryn, J. -C., R. Darrell, R. J. Friedman, A. Kopf (1987) Prognostic significance of hypopigmentation in malignant melanoma. *Arch. Dermatol.* 123: 1053–1055.

137. Duhra, P., A. Ilchyshin (1991) Prolonged survival in metastatic malignant melanoma associated with vitiligo. *Clin. Exp. Dermatol.* 16: 303–305.

138. Richards, J. M., N. Mehta, K. Ramming, P. Skosey (1992) Sequential chemoimmunotherapy in the treatment of metastatic melanoma. *J. Clin. Oncol.* 10: 1338–1343.

139. Livingston, P. O., G. Y. C. Wong, S. Adluri, Y. Tao, M. Padavan, R. Parente, C. Hanlon, M. Jones Calves, F. Helling, G. Ritter, H. F. Oettgen, L. J. Old (1994) Improved survival in stage III melanoma patients with GM2 antibodies: a randomized trial of adjuvant vaccination with GM2 ganglioside. *J. Clin. Oncol.* 12: 1036–1044.

140. Mackensen, A., G. Carcelain, S. Viel, M. -C. Raynal, H. Michalaki, F. Triebel, J. Bosq, T. Hercend (1994) Direct evidence to support the immunosurveillance concept in a human regressive melanoma. *J. Clin. Invest.* 93: 1397–1402.

141. Ellis, J. R. M., P. J. Keating, J. Baird, E. F. Hounsell, D. V. Renouf, M. Rowe, D. Hopkins, M. S. Duggan-Keen, J. S. Bartholomew, L. S. Young, P. L. Stern (1995) The association of an HPV 16 oncogene variant with HLA-B7 has implications for vaccine design in cervical cancer. *Nat. Med.* 1: 464–470.

142. Christinck, L. R., M. A. Luscher, B. H. Barber, D. B. Williams (1991) Peptide binding to class I MHC on living cells and quantitation of complexes required for CTL lysis. *Nat.* 352: 67–70.

143. Neefjes, J. J., F. Momburg (1993) Cell biology of antigen presentation. *Curr. Opin. Immunol.* 5: 27–34.

144. Garrido, F., T. Cabrera, A. Concha, S. Glew, F. Ruiz-Cabello, P. L. Stern (1993) Natural history of HLA during tumor development. *Immunol. Today* 14.

145. Ferrone, S., F. M. Marincola (1995) Loss of HLA class I antigens by melanoma cells: molecular mechanisms, functional significance and clinical relevance. *Immunol. Today* 16: 487–494.

146. Hruban, R. H., d. R. P. van, Y. S. Erozan, D. Sidransky (1994) Brief report: molecular biology and the early detection of carcinoma of the bladder – the case of Hubert H. Humphrey. *N. Engl. J. Med.* 330: 1276–8.

147. Cohen, J (1994) Vaccines get a new twist. *Science (Research News)* 264: 503–505.

148. Golumbek, P. T., A. J. Lazenby, H. I. Levitsky, L. M. Jaffee, H. Karasuyama, M. Baker, D. M. Pardoll (1991) Treatment of established renal cancer by tumor cells engineered to secret interleukin-4. *Science* 245: 713–716.

149. Majordomo, J. I., T. Zorina, W. J. Storkus, L. Zitvogel, C. Celluzzi, L. D. Falo, C. J. Melief, S. T. Ildstad, W. M. Kast, A. B. Deleo, M. T. Lotze (1995) Bone marrow-derived dendritic cells pulsed with synthetic tumour peptides elicit protective and therapeutic antitumour immunity. *Nat. Med.* 1: 1297–1302.

150. Zitvogel, L., J. I. Mayordomo, T. Tjandrawan, A. B. DeLeo, M. R. Clarke, M. T. Lotze, W. J. Storkus (1996) Therapy of murine tumors with tumor peptide-pulsed dendritic cells: dependence on T cells, B7 costimulation, and T helper cell-associated cytokines. *J. Exp. Med.* 183: 87–97.

151. Kwak, L. W., M. J. Campbell, D. K. Czerwinski, S. Hart, R. A. Miller, R. Levy (1992) Induction of immune responses in patients with B-cell lymphoma against the surface-immunglobulin idiotype expressed by their tumors. *N. Engl. J. Med.* 327: 1209–1215.

152. Mukherji, B., N. G. Chakraborty, S. Yamasaki, T. Okino, H. Yamase, J. R. Sporn, S. K. Kurtzman, M. T. Ergin, J. Ozols, J. Meehan, F. Mauri (1995) Induction of antigen-specific cytolytic T cells *in situ* in human melanoma by immunization with synthetic peptide-pulsed autologous antigen presenting cells. *Proc. Natl. Acad. Sci. USA* 92: 8078–8082.

153. Marchand, M., P. Weynants, E. Rankin, F. Arienti, F. Belli, G. Parmiani, N. Cascinelli, A. Bourlond, R. Vanwijck, Y. Humblet, J. -L. Canon, C. Laurent, J. -M. Naeyaert, R. Plagne, R. Deraemaeker, A. Knuth, E. Jäger, F. Brasseur, J. Herman, P. G. Coulie, T.

Boon (1995) Tumor regression responses in melanoma patients treated with a peptide encoded by gene MAGE-3. *Int. J. Cancer* 63: 883–885.

154. Finke, J. H., A. H. Zea, J. Stanley, D. L. Longo, H. Mizoguchi, R. R. Tubbs, R. H. Wiltrout, J. J. O'Shea, S. Kudoh, E. Klein, a. l. et (1993) Loss of T-cell receptor zeta chain and p56lck in T-cells infiltrating human renal cell carcinoma. *Cancer Res.* 53: 5613–6.

155. Nakagomi, H., M. Petersson, I. Magnusson, C. Juhlin, M. Matsuda, H. Mellstedt, J. L. Taupin, E. Vivier, P. Anderson, R. Kiessling (1993) Decreased expression of the signal-transducing zeta chains in tumor-infiltrating T-cells and NK cells of patients with colorectal carcinoma. *Cancer Res.* 53: 5610–2.

156. Gjertsen, M. K., A. Bakka, J. Breivik, I. Saeterdal, B. G. Solheim, O. Soreide, E. Thorsby, G. Gaudernack (1995) Vaccination with mutant ras peptides and induction of T-cell responsiveness in pancreatic carcinoma patients carrying the corresponding RAS mutaton. *Lancet* 346: 1399–1340.

157. Alving, C. R., V. Koulchin, G. M. Glenn, M. Rao (1995) Liposomes as carriers of peptide antigens: induction of antibodies and cytotoxic T lymphocytes to conjugated and unconjugated peptides. *Immunol. Rev.* 145: 5–31.

158. Falo Jr., L. D., M. Kovacsovics-Bankowski, K. Thompson, K. L. Rock (1995) Targeting antigen into the phagocytic pathway *in vivo* induces protective tumour immunity. *Nat. Med.* 1: 649–653.

159. Hanson, M. S., C. V. Lapcevich, S. L. Haun (1995) Progress on development of the live BCG recombinant vaccine vehicle for combined vaccine delivery. *Ann. N.Y. Acad. Sci.* 754: 214–21.

160. Sercarz, E. E., P. V. Lehmann, A. Ametani, G. Benichou, A. Miller, K. Moudgil (1993) Dominance and crypticity of T cell antigenic determinants. *Annu. Rev. Immunol.* 11: 729-.

161. Jansson, L., P. Diener, A. Engström, T. Olsson, R. Holmdahl (1995) Spreading of the immune response to different myelin basic protein peptides in chronic experimental autoimmune encephalomyelitis in B10.RIII mice. *Eur. J. Immunol.* 25: 2195–2200.

162. Aichele, P., K. Brduscha-Riem, R. M. Zinkernagel, H. Hengartner, H. Pircher (1995) T cell priming versus T ell tolerance induced by synthetic peptides. *J. Exp. Med.* 182: 261–266.

163. Moskophidis, D., F. Lechner, H. Pircher, R. M. Zinkernagel (1993) Virus persistence in acutely infected immunocompetent mice by exhaustion of antiviral cytotoxic effector T cells. *Nature* 362: 758-.

164. Kearney, E. R., K. A. Pape, D. Y. Loh, M. K. Jenkins (1994) Visualization of peptide-specific T cell immunity and peripheral tolerance induction *in vivo*. *Immunity* 1: 327.

165. Suhrbier, A., S. R. Burrows, A. Fernan, M. F. Lavin, G. D. Baxter, D. J. Moss (1993) Peptide epitope induced apoptosis of human cytotoxic T lymphocytes. *J. Immunol.* 150: 2169.

166. Widman, C., P. Romero, J. L. Maryanski, G. Corradin, D. Valmori (1992) T helper epitopes enhance the cytotoxic response of mice immunized with MHC class I-restricted malaria peptides. *J. Immunol. Method.* 155: 95–99.

167. Rock, K. L (1996) A new foreign policy: MHC class I molecules monitor the outside world. *Immunol. Today* 17: 131–137.

168. Schlag, P., M. Manasterski, T. Gerneth, P. Hohenberger, M. Dueck, C. Herfarth, W. Liebrich, V. Schirrmacher (1992) Active specific immunotherapy with Newcastle-dis-

ease-virus-modified autologous tumor cells following resection of liver metastases in colorectal cancer. *Cancer Immunol. Immunother.* 35: 325–330.

169. Srivastava, P. K., H. Udono, N. E. Blachere, Z. Li (1994) Heat shock proteins transfer peptides during antigen processing and CTL priming. *Immunogenetics* 39: 93–8.

170. Papadopoulos, E. B., M. Ladanyi, D. Emanuel, S. Mackinnon, F. Boulad, M. H. Carabasi, H. Castro-Malaspina, B. H. Childs, A. P. Gillio, T. N. Small, J. W. Young, N. A. Kernan, R. J. O'Reilly (1994) Infusions of donor leukocytes to treat Epstein-Barr virus-associated lymphoproliferative disorders after allogeneic bone marrow transplantation. *N. Engl. J. Med.* 330: 1185–1191.

171. Porter, D. L., M. S. Roth, C. McGarigle, J. L. M. Ferrara, J. H. Antin (1994) Induction of graft-versus-host disease as immunotherapy for relapsed chronic myeloid leukemia. *N. Engl. J. Med.* 330: 100–106.

172. Kwak, L. W., D. D. Taub, P. L. Duffey, W. I. Bensinger, E. M. Bryant, C. W. Reynolds, D. L. Longo (1995) Transfer of myeloma idiotype-specific immunity from an actively immunised marrow donor. *Lancet* 345: 1016–1020.

173. Kawakami, Y., S. Eliyahu, C. H. Delgado, P. F. Robbins, K. Sakaguchi, E. Appella, J. R. Yannelli, G. J. Adema, T. Miki, S. A. Rosenberg (1994) Identification of a human melanoma antigen recognized by tumor-infiltrating lymphocytes associated with *in vivo* tumor rejection. *Proc. Natl. Acad. Sci. USA.* 91: 6458–62.

174. Celis, E., V. Tsai, C. Crimi, R. DeMars, P. A. Wentworth, R. W. Chesnut, H. M. Grey, A. Sette, H. M. Serra (1994) Induction of anti-tumor cytotoxic T lymphocytes in normal humans using primary cultures and synthetic peptides. *Proc. Natl. Acad. Sci. USA* 91: 2105–2109.

175. Topalian, S. L., L. Rivoltini, M. Mancini, N. R. Markus, P. F. Robbins, Y. Kawakami, S. A. Rosenberg (1994) Human CD4+ T cells specifically recognize a shared melanoma-associated antigen encoded by the tyrosinase gene. *Proc. Natl. Acad. Sci. USA.* 91: 9461–5.

176. Peoples, G. E., P. S. Goedegebuure, J. V. Andrews, D. D. Schoof, T. J. Eberlein (1993) HLA-A2 presents shared tumor-associated antigens derived from endogenous proteins in ovarian cancer. *J. Immunol.* 151: 5481–91.

7 Peptide Vaccination

R. E. M. Toes, F. Ossendorp, E. I. H. van der Voort, E. Mengedé, R. Offringa,
C. J. M. Melief

T Lymphocytes are capable of eradication of both virus- and non-virus induced tumor cells [1, 2]. The highly diverse repertoire of T cells and the increased knowledge of the recognized ligands make it possible to design protocols to specifically activate tumor-specific T cells *in vivo*. By T cell receptor interaction, T lymphocytes can recognize MHC molecules, presenting antigen-derived peptide fragments. The identification of tumor-associated antigens and antigenic peptides has firmly established the existence of tumor specific T cell targets [3].

Here, we will discuss the use of antigenic peptide based vaccines for induction of tumor-specific immunity as well as the potential drawbacks of usage of synthetic peptides leading to unwanted tolerization for tumor antigens.

7.1   Protective Immunity Induced by Peptide Vaccination

One of the protocols designed to specifically activate tumor-specific T cells *in vivo* is based on vaccination with defined peptides derived from tumor-associated antigens. The efficacy of *in vivo* induction of protective T cell mediated immunity, generated via peptide immunization, was demonstrated firstly in two pathogenic virus models [4, 5]. The first data, concerning induction of immunity against tumor cells by peptide immunization, came from a model system in which peptides corresponding to sequences of polyoma virus tumor antigens were used as immunogens [6]. In these studies, immunity against polyoma tumors was established but was not as efficient as that induced by immunization with polyoma virus and did not result in complete rejection of a small inoculum of polyoma tumor cells. The demonstration that a peptide based vaccination protocol was successful against the outgrowth of a lethal dose of tumor cells came from studies in which a cytotoxic T lymphocyte (CTL)-epitope containing peptide corresponding to sequences of the early region 7 (E7) gene product of human papilloma virus type 16 (HPV 16) was used. Peptide vaccination protected mice against a subsequent challenge of HPV 16 transformed syngeneic tumor cells [7]. CTL induced by the immunization protocol were not only able to lyse peptide-loaded target cells, but also HPV 16-transformed tumor cells. Interestingly, polyclonal CTL cultures, obtained after immunization with the HPV

16-transformed tumor cells, were not able to recognize the HPV 16 E7-encoded CTL epitope [8]. Also, several independently derived CTL clones, generated against the tumor cells, did not recognize the E7-encoded CTL epitope, suggesting that the latter CTL epitope is not immunodominant. These data imply that successful anti-tumor vaccination can be achieved with non-immunodominant CTL epitopes an show and additional value of peptide immunization over other vaccination types, such as tumor cells and possibly whole proteins harbouring several (sub)dominant T cells epitopes. Thus, an advantage of vaccination with peptides over other vacci-nation strategies is that peptides can selectively stimulate the T cell specificity of choice. Peptide immunization has also been applied successfully in several other tu-mor models [9–11]. Moreover, vaccination with a peptide, derived from a mutated connexin 37 gap-junction protein present on a murine lung carcinoma, prevented metastatic spread from a primary tumor that was allowed to develop for 30 days before surgical exicision. Both peptide, given in incomplete Freund's adjuvant, (IFA) and peptide loaded on RMA-S cells, reduced metastatic growth in mice carrying pre-established metastases. Tumor-specific immunity was primarily mediated by CD8+ T cells [12].

Independent studies show that presentation of antigenic peptides on specialized antigen-presenting cells is another strategy to induce an efficient therapeutic im-mune response, since these cells express essential co-stimulatory molecules next to the antigenic ligand [13]. In several different tumor models, treatment of animals, bearing established macroscopic tumors with dendritic cells, prepulsed with the re-spective tumor-specific peptides, resulted in tumor regression and eradication in more than 80% of cases [13, 14].

Not only vaccination with peptides representing tumor-derived CTL epitopes can induce protective (and therapeutic) anti-tumor immunity but also, T helper epi-topes can be used to induce long-lasting anti-tumor immunity. Vaccination with a tumor-specific T helper epitope, comprising synthetic peptide in IFA, induces pro-tective immunity against Murine Leukemia Virus (MuLV) induced lymphoma. The epitope is present in the MuLV env-gp70 gene product that is expressed by the tu-mor cells. The induced protection is long-lasting, since about 70% of the vaccinat-ed mice survive tumor challenge for more than one year, in contrast to mice immu-nized with a non-related (OVA) T helper peptide that die within one month. *In vi-vo* depletion of CD4+ cells with anti-CD4 monoclonal antibodies completely abro-gate the protective effect, strongly indicating the importance of tumor-(MuLV)-spe-cific T helper cells in the immune control of these lymphomas. Peptide specific, MHC class II restricted CD4+ T cells can be isolated from vaccinated mice but, paradoxically, were not able to directly recognize the tumor cells since these lack MHC class II expression. The mechanism of protection is still under investigation, but cross-priming by APC seems crucial and CD8+ effector T cells are likely to be involved.

Phase I/II clinical trials, using minimal CTL epitopes derived from several tumor-associated antigens, are currently in progress and the first indications that such peptide-based intervention schemes induce anti-tumor immunity have been described [15, 16].

7.2 Adverse Effects of Peptide Vaccination

Peptide antigens can also down-regulate the T cell immune response. In several models, with T cell receptor-transgenic mice, continuous and systemic exposure to a high dose of peptide was shown to lead to a selective depletion of most peripheral T cells bearing the relevant transgenic T cell receptor [17–20]. This depletion appeared to be the result of thymic clonal elimination, as well as peripheral loss of reactive T cells.

We have shown that peptide immunization can also induce functional deletion of tumor-specific CTL, leading to an enhanced tumor outgrowth of tumors expressing the relevant tumor-associated antigen. Murine syngeneic tumors, expressing the adenovirus type 5 early region 1B (Ad5E1B) tumor-associated antigen, present an H-2Db-restricted CTL epitope to the immune system [21]. Unexpectedly, vaccination of B6 mice with the synthetic Ad5E1B peptide-epitope in IFA, the same formulation that was used succesfully to induce protective immunity in the HPV16-system, did not result in protection against a subsequent challenge with Ad5E1B-expressing tumor cells. On the contrary, Ad5E1B-peptide vaccination promoted the growth of these tumors [22]. Moreover, protective immunity, induced by vaccination with irradiated tumor cells, was abolished by subsequent vaccination of the mice with the Ad5E1B peptide. The observed effects were strictly antigen-specific. Immunization with other H-2Db-binding peptides did not result in enhanced tumor growth, while vaccination with the Ad5E1B-peptide did not lead to an enhanced outgrowth of other, unrelated tumors. This was most clearly demonstrated with the use of Ad5E1B-expressing tumor cells harbouring a mutated Ad5E1B gene, of which the E1B epitope-encoding region was modified to encode the HPV16 E7-epitope. Vaccination with the Ad5E1B peptide did not enhance growth of these tumor cells, while protective immunity against a challenge with these cells was obtained by vaccination with the HPV 16 E7-peptide. Therefore, the contrasting effects of peptide vaccination, as observed in the Ad5E1 and HPV16 tumor systems, relate to properties of the peptide epitopes and not to differences between the tumor cells. The immunosuppressive effect appeared to be long-lasting, since injection of the Ad5E1B-peptide in tumor cell vaccinated animals, three weeks before tumor challenge, still abrogated the protective effects induced by tumor cell vaccination. Immunosuppression is observed after injection of this peptide in IFA or in complete Freund's adjuvant (CFA), and across a broad concentration range of peptide down to 1 µg [22]. Importantly, this phenomenon is paralleled by a decrease of Ad5E1B-specific CTL activity, indicating that deletion of specific CTL-activity explains the enhanced outgrowth of Ad5E1 tumor cells [22].

Similar effects have been found for a second Ad5E1-encoded tumor antigen. Vaccination of B6 mice with an Ad5E1A-peptide in IFA or CFA resulted in an enhanced outgrowth of Ad5E1A-transformed tumor cells, whereas, tumor cell vaccination protected the animals against tumor growth. The inability of Ad5E1A peptide-vaccinated mice to reject Ad5E1A-transformed tumors was peptide-and tumor-specific, was induced rapidly by a single injection of a low dose of peptide and was paralleled by a decrease in the Ad5E1A-specific CTL activity [23]. The immuno-suppressive effect of Ad5E1A-peptide vaccination was also observed in T cell receptor-transgenic mice that express an Ad5E1A-specific T cell receptor in that spleen cells of vaccinated mice are no longer able to lyse Ad5E1A-expressing tumors *in vitro*. Thus the enhanced outgrowth of Ad5E1A-expressing tumors after vaccination with the Ad5E1A-peptide is associated with the disappearance of Ad5E1A-specific CTL activity.

7.3 Peptides Inducing CTL Tolerance are Distributed Systemically

The reasons why vaccination with the two latter peptides in IFA leads to a functional deletion of the relevant CTL that is associated with enhanced tumor outgrowth (whereas immunization with the HPV16-E7-peptide in IFA induces CTL activity and protection against a challenge with an otherwise lethal dose of HPV16-transformed tumor cells) are not known. However, our current experimental data indicate that these differences in outcome, of peptide vaccination, might relate to different physical behaviours of the Ad5E1- and HPV16 E7-peptides *in vivo*. Intravenous injection of Ad5E1A- or Ad5E1B-specific CTL clones into nude mice, together with subcutaneous administration of interleukin-2 and of the relevant peptide in IFA (or CFA), revealed that the combination of CTL clone and cognate peptide was lethal to the mice [22, 23] The mice died, within 16 to 48 h, of severe lung congestion, most likely due to local, peptide-induced activation of Ad5E1-specific CTL that have been trapped in the capillary bed of the lung. Mice that received Ad5E1A-specific CTL clone in combination with Ad5E1B-peptide, or *vice versa*, did not show any signs of pathogenesis, indicating the antigen-specificity of these effects. Strikingly, the combination of HPV16 E7-peptide, a peptide capable of inducing protective immunity [7] and HPV16 E7-specific CTL clone did not lead to any pathogenic effects in this setting. Although these findings might be explained by differences in cytokine secretion- or migration pattern between the CTL clones, we favor the hypothesis that the HPV16 E7-peptide is retained more locally. The rapid and widespread distribution of the Ad5E1-peptides from the local depot of IFA/CFA causes pathogenic effects in mice that have simultaneously received activated Ad5E1-specific CTL. Systemic exposure of the Ad5E1-peptides might also explain the observed immunosuppression through clonal exhaustion or abberant triggering of Ad5E1-specific T cells, in contrast to HPV16 E7-specific T cells. Similar observations were made in a lymphocytic choriomeningitis virus (LCMV) model. Repetitive and systemic intraperitoneal injections to mice, of high doses of peptide derived from LCMV, caused inactivation of LCMV-specific CTL-activity, whereas, subcutaneous administration of the same LCMV peptide in IFA induced protective immunity [24]. Thus, systemic distribution of antigenic peptides might cause specific deletion of CTL activity rather than induction of protective CTL responses.

7.4 Conclusion

Although specific T cell tolerance induction by peptide immunization might be the desired goal in the treatment of T cell-mediated autoimmune diseases, it can severely hamper the application of peptide-based vaccines in anti-tumor immune intervention. Until now, it was not known whether the adverse effects on tumor outgrowth, after peptide vaccination in IFA, is the exception or the rule. In any case, these findings indicate both the potency of peptide vaccination and the delicate balance between immunization and inactivation of the T cell response upon specific triggering of the T cell receptor.

New vaccination formulations must be developed in order to induce controlled and reliable anti-tumor immunity by peptide vaccination. In this respect, immunization with T helper peptides and immunization with peptide-loaded dendritic cells seem to have good prospects. Alternatively, vaccination strategies based upon recombinant viral vectors, such as recombinant adenoviruses, expressing whole tumor-associated antigens or several defined T cell epitopes in a string-bead fashion, might be used to evoke effective anti-cancer immunity [21]. Thus, with the development of successful vaccination strategies, peptide- and T-cell epitope based immune intervention in cancer will eventually prove its applicability.

References

1. Greenberg, P. D. (1991) Adoptive T cell therapy of tumors: Mechanisms operative in the recognition and elimination of tumor cells. *Adv. Immunol.* 49: 281–355.
2. Melief, C. J. M. (1992) Tumor eradication by adoptive transfer of cytotoxic T lymphocytes. *Adv. Cancer Res.* 58: 143–175.
3. Boon, T., Cerottini, J. -C., van den Eynde, B., van der Bruggen, P. van Pel, A. (1994) Tumor antigens recognized by T lymphocytes. *Annu. Rev. Immunol.* 12: 337–365.
4. Schulz, M., Zinkernagel, R. M. Hengartner, H. (1991) Peptide-induced antiviral protection by cytotoxic T cells. Proc. Nat. *Acad. Sci. USA* 88: 991–993.
5. Kast, W. M., Roux, L., Curren, J., Blom, H. J. J., Voordouw, A. C., Meloen, R. H., Kolakofski, D., Melief, C. J. M. (1991) Protection against lethal Sendai virus infection by

in vivo priming of virus-specific cytotoxic T lymphocytes with an unbound peptide. *Proc. Natl. Acad. Sci. USA* 88: 2283–2287.

6. Reinholdsson-Ljunggren, G., Ramqvist, T., Åhrlund-Richter, L. Dalianis, T. (1992) Immunization against polyoma tumors with synthetic peptides derived from the sequences of middle- and large- T antigens. *Int. J. Cancer* 50: 142–146.

7. Feltkamp, M. C. W., Smits, H. L., Vierboom, M. P. M., Minnaar, R. P., De Jongh, B. M., Drijfhout, J. W., Ter Schegget, J., Melief, C. J. M., Kast, W. M. (1993) Vaccination with a cytotoxic T lymphocyte epitope-containing peptide protects against a tumor induced by human Papillomavirus type 16-transformed cells. *Eur. J. Immunol.* 23: 2242–2249.

8. Feltkamp, M. C. W., Vreugdenhil, G. R., Vierboom, M. P. M., Ras, E., Van der Burg, S. H., Ter Schegget, J. Melief, C. J. M. Kast, W. M. (1995) CTL raised against a subdominant epitope offered as a synthetic peptide eradicate human papillomavirus type 16-induced tumors. *Eur. J. Immunol.* 25: 2638–2641.

9. Schild, H., Norda, M., Deres, K., Falk, K., Rötschke, O., Wiesmüller, K,-H., Jung, G. Rammensee, H. -G. (1991) Fine specificity of cytotoxic T lymphocytes primed *in vivo* either with virus or synthetic lipopeptide vaccine or primed *in vitro* with peptide. *J. Exp. Med.* 174: 1665–1668.

10. Minev, B. R., McFarland, B. J., Spiess, P. J. Rosenberg, S. A. Restifo, N. P. (1994) Insertion signal sequence fused to minimal peptides elicits specific CD8+ T-cell responses and prolongs survival of thymoma-bearing mice. *Cancer Res.* 54: 4155–4161.

11. Vitiello, A., Ishioka, G., Grey, H. M., Rose, R., Farness, P., LaFond, R., Yuan, L., Chisari, F. V., Furze, J., Bartholomeuz, R.Chesnut, R. W. (1995) Development of a lipopeptide-based therapeutic vaccine to treat chronic HBV infection. I. Induction of a primary cytotoxic T lymphocyte response in humans. *J. Clin. Invest.* 95: 341–349.

12. Mandelboim, O., Vadai, E., Fridkin, M., Katz-Hillel, A., Feldmand, M., Berke, G. Eisenbach, L. (1995) Regression of established murine carcinoma metastases following vaccination with tumour-associated antigen peptides. *Nat. Med.* 1: 1179–1183.

13. Celluzzi, C. M., mayordomo, J. I., Storkus, W. J., Lotze, M. T., Falo, L. D. (1996) Peptide-pulsed dendritic cells induce antigen-specific, CTL-mediated protective immunity. *J. Exp. Med.* 183: 283–287.

14. Mayordomo, J. I., Zorina, T., Storkus, W. J., Zitvogel, L. Celluzzi, C., Falo, L. D., Melief, C. J. M., Ildstad, S. T., Kast, W. M., Deleo, A. B. Lotze, M. T. (1995) Bone marrow-derived dendritic cells pusled with with synthetic tumour peptides elicit protective and therapeutic antitumour immunity. *Nat. Med.* 1: 1297–1302.

15. Marchand, M., Weynants, P., Rankin, E., Arienti, F., Belli, F. Parmiani, G., Cascinelli, N., Bourlond, A., Vanwijck, R., Humblet, Y., Canon, J. -L., Laurent, C., Naeyaert, J. -M., Plagne, R., Deraemaeker, R., Knuth, A., Jäger, E., Brasseur, F., Herman, J., Coulie, P. G. Boon, T. (1995) Tumor regression responses in melanoma patients treated with a peptide encoded by gene MAGE-3. *Int. J. Cancer* 63: 883–885.

16. Mukherji, B., Chakraborty, N. G., Yamasaki, S., Okino, T., Yamase, H., Sporn, J. R. Kurtzman, S. K., Ergin, M. T., Ozols, J., Meehan, J. Mauri, F. (1995) Induction of antigen-specific cytolytic T cells *in situ* in human melanoma by immunization with synthetic peptide-pulsed autologous antigen presenting cells. *Proc. Natl. Acad. Sci. USA* 92: 8078–8082.

17. Mamalaki, C., Tanaka, Y., Corbella, P., Chandler, P., Simpson, E. Kioissis, D. (1993) T cell deletion follows chronic antien specific T cell activation *in vivo*. *Int. Immunol.* 5: 1285–1292.

18. Huang, L., Soldevilla, G., Leeker, M., Flavell, R. Crispe, I. N. (1994) The liver eliminates T cells undergoing antigen-triggered apoptosis. *Immunity* 1: 741–749.
19. Kearney, E. R., Pape, K. A., Loh, D. Y. Jenkins, M. K. (1994) Visualization of peptide-induced T cell immunity and peripheral tolerance induction *in vivo*. *Immunity* 1: 327–339.
20. Singer, G. C. Abbas, A. K. (1994) The fas antigen is involved in peripheral but not in thymic deletion of T lymphocytes in T cell receptor transgenic mice. *Immunity* 1: 366–371.
21 Toes, R. E. M., Offringa, R., Blom, H. J. J., Brandt, R. M. P., Van der Eb, A. J., Melief, C. J. M., Kast, W. M. (1995) An adenovirus type 5 early region 1B-encoded CTL epitope-mediating tumor eradication by CTL clones is down-modulated by an activated dras oncogene. *J. Immunol.* 154: 3396–3405.
22. Toes, R. E. M., Blom, R. J. J., Offringa, R., Kast, W. M., Melief, C. J. M. (1996) Functional deletion of tumor-specific cytotoxic T lymphocytes induced by peptide immunization can lead to the inability to reject tumors; *submitted*.
23. Toes, R. E. M., Offringa, R., Blom, R. J. J., Melief, C. J. M., Kast, W. M. (1996) Peptide vaccination can lead to enhanced tumor growth through specific T cell tolerance induction. *Proc. Natl. Acad. Sci. USA*
24. Aichele, P., Brduscha-Riem, K., Zinkernagel, R. M., Hengartner, H. Pircher, H. (1995) T cel priming versus T cell tolerance induced by synthetic peptides. *J. Exp. Med.* 182: 261–266.

8 Combination Effects of Gene Modified Tumor Cell Vaccines and Chemotherapy

X. Cao

8.1 Introduction

Cytokine gene therapy has been proposed as a new potential prospect in the management of cancer. Several approaches to cytokine gene therapy of cancer have been developed. One promising approach is the transfection of cytokine genes into tumor cells to prepare new types of cancer vaccines, which exert an antitumor effect mainly through active immunotherapy. It has been shown that vaccination with the cytokine gene-modified tumor cells can induce systemic immunity against a subsequent challenge with parental tumor cells at a distant site (Blankenstein et al., 1991a). But most of such animal studies, in which the effectiveness of cytokine gene-modified tumor cells as cancer vaccines was evaluated, suffer from drawbacks that limit their relevance to clinical application. One of these drawbacks is to immunize healthy animals with live cytokine gene-modified tumor cells to resist a subsequent challenge with parental tumors, rather than use the inactivated, cytokine gene-modified tumor cells to treat tumor-bearing mice. Because the primary goal of such studies is to generate more effective tumor cell vaccine and to eliminate the already existing tumor, the data obtained from these experiments have yielded restricted information about the effectiveness of such a approach in the treatment of established, especially metastatic, tumors.

Several studies have been undertaken to investigate the therapeutic effects of interleukin-2 (IL-2), interleukin-4 (IL-4), interleukin-6 (IL-6), interferon-alpha (IFN-alpha) or granulocyte-macrophage colony-stimulating factor (GM-CSF) gene-modified tumor cells on the already established tumors in the mouse tumor models (Cao et al., 1994; Dranoff et al., 1993; Golumbek et al., 1991; Porgador et al., 1994; Vieweg et al., 1994). Although the reduction of tumor progression was observed, the antitumor response induced by the vaccination with cytokine gene-modified tumor cells was usually not strong enough to eradicate the established tumors. Thus, attempts have to be made to improve the therapeutic efficacy of the cytokine gene transfer into tumor cells. For example, some investigators have reported that cotransfection of two cytokine genes into the same tumor cells, mixture of two tumor cell populations (each transfected with one cytokine gene), or cotransfection of cytokine gene with costimulator B7 gene into tumor cells can induce immune respons-

T. Blankenstein (ed.) Gene Therapy
©1999, Birkhäuser Verlag Basel

es more effectively and achieve better antitumor effects (Cayeux et al., 1995; Ohe et al., 1994; Rosenthal et al., 1994).

It is well known that the immune functions, including both humoral and cellular antitumor immunity, are depressed in the tumor-bearing host, especially at the advanced stage. There are many mechanisms by which tumor cells can nonspecifically interfere with the expression of immunity in the host. For examples, some tumor cells can release soluble factors that directly suppress immunological reactivity and the presence of suppressor T (Ts) cells during the tumor growth may down-regulate the host antitumor response. It has been shown by a number of laboratory experiments that CD4+ Ts, generated from the mice with advanced tumor, can suppress adoptive T-cell-mediated tumor regression in recipient mice. Preferential elimination of CD4+ Ts from the advanced tumor-bearing mice can result in an augmented level of immunity and in immunologically-mediated complete tumor regression (Awwad and North, 1988a; 1988b; Bear, 1986; Dye and North, 1981; North and Awwad, 1990; North and Dye, 1985). Interestingly, studies by Rakhmilevich et al. showed that intravenous injection of irradiated P815 tumor cells can induce a state of tumor-specific unresponsiveness, associated with the presence of CD4+ Ts (Rakhmilevich et al., 1993). Thus, tumor-induced immunosuppression, such as the presence of Ts, gives an obstacle to the activation of antitumor immunity in the tumor-bearing host. So, the reverse of the immunosuppression by the preferential elimination of Ts may contribute to the more efficient activation of an antitumor response by the tumor vaccine. On the basis of the findings from the previous studies which have demonstrated that low-dose cyclophosphamide (Cy) can selectively deplete some suppressor T-cell functions in murine and human subjects, we employed an experimental tumor model with B16 F10 melanoma pulmonary metastasis to determine whether or not the antitumor response could be induced more efficiently by the administration of low-dose Cy, prior to the vaccination with the inactivated, cytokine gene-modified tumor cells. In view of the fact that simultaneous administration of cytokine, such as IL-1, IL-2, with adjuvant activity may help the vaccine to activate the host immune response more effectively, we also investigated whether the therapeutic effects on the already established tumor could be improved by the treatment with cytokine gene-modified tumor cells in combination with low-dose Cy and low-dose IL-2, except that IL-1 was used as adjuvant in the combined therapy when using IL-2 gene-modified tumor cells.

8.2   Immunopotentiating Activity of Low-Dose Cy

Cy is one of the most widely used chemotherapeutic agents in cancer treatment. Now, it is well known that Cy has an antitumor effect when used either at high doses as a potent cytotoxic drug or at low-doses as an effective immunomodulator. The immunomodulating activity of Cy was demonstrated firstly by Maguire and Ettore in 1967 (Maguire and Ettore, 1967). They used a guinea pig model to show that Cy could augment the induction and expression of allergic contact dermatitis. Their experiment suggested that Cy might augment cell-mediated immunity. But this phenomenon did not attract much attention until the 1970s. Then, some researchers reported that Cy, especially at low-doses, could augment immune functions such as enhancement of delayed-type hypersensitivity (DTH) reaction or augmentation of antibody production (Askenase et al., 1975; Mitsuoka et al., 1976). However, this immunomodulatory activity of Cy may be schedule-dependent and dose-dependent. For example, administration of Cy, one to three days before antigen stimulation is effective in the induction of the immune response but Cy administration after immunization is usually immunosuppressive. Although Cy-mediated immunopotentiation has been achieved in experimental animals with a remarkably wide range of doses from 60 to 2100 mg/m2, lower doses of Cy (200–400 mg/m2, or 15 to 40 mg/kg) may be more effective. Some investigators demonstrated that low-dose Cy was effective in treating tumor-bearing mice through activation of the host immune responses (Hengst et al., 1980; Hengst et al., 1981; Nomi et al., 1985; Ray et al., 1981). It was very interesting to show that a single low-dose of Cy (15 mg/kg) could cure 92% of mice bearing large (≥ 20 mm) mature (15-day growth period) MOPC-315 plasmacytomas but only 10% of mice bearing small (nonpalpable) immature (four-day growth period) tumors (Hengst et al., 1980). Next, it was shown that the curative effect of this dosage of Cy was not due solely to its direct antitumor effects and required the contribution of T-cell-dependent antitumor immunity (Hengst et al., 1981; Culo et al., 1993). In 1980s, a variety of animal studies had confirmed that the *in vivo* antitumor effect of low-dose Cy is immunologically mediated and that the immunopotentiating effect of low-dose Cy is most probably due to its selective elimination of suppressor T-cells (Awward and North, 1989; North, 1982; Tutee et al., 1994; Wise et al., 1988).

In 1982, Berd et al. extended the immunopotentiating effects of low-dose Cy observed in mice to humans (Berd et al., 1982). They reported that Cy, at a dose of 1000 mg/m2, which is the dose most commonly used to achieve oncostatic effects, can augment human immune responses. Administration of Cy at that dose, 3 days before sensitization with the primary antigen, keyhole limpet hemocyanin (KLH), resulted in significant potentiation of the DTH response to KLH. However the antibody to KLH was not augmented by the pretreatment with Cy. On the basis of animal studies, in which doses of Cy as low as 60 mg/m2 augmented immunity and the *in vitro* human studies in which extremely low concentrations of the active metabolite, 4-hydroperoxy-Cy, augmented lymphocyte function, Berd et al. selected a low dose of Cy (300 mg/m2) to treat 18 patients with advanced metastatic cancer. They found that pretreatment with low-dose Cy not only resulted in significant augmentation of DTH, but also resulted in augmentation of the antibody response (Berd et al., 1984). Then, many clinical studies, in which low-dose Cy was used to treat cancer patients based on its immunopotentiation, were conducted. The results of representative clinical studies are presented in Table 8-1. It is confirmed by some clinical trials that low-dose Cy can reduce suppressor T cell activity in advanced cancer patients and exhibits the potent *in vivo* immunopotentiating activities. In these clinical trials, low-dose Cy was used as an immunomodulator to treat cancer patients either in combination with autologous tumor cell vaccine or with recombinant IL-2. Although the therapeutic effects of the combined use of low-dose Cy plus IL-2, on the different types of cancer patients, were controversial, the positive antitumor response was usually observed in melanoma patients receiving the combined therapy. In particular, superior clinical responses, including complete and partial responses in the active specific immunotherapy of metastatic melanoma patients, could be observed after treatment with low-dose Cy prior to vaccination with autologous tumor cell vaccine (Berd et al., 1984; Hoon et al., 1990).

8.3 Immunotherapy of Established Tumor by IL-2 Gene-Modified Tumor Vaccine in Combination with Low-Dose Cy and IL-1

Interleukin-1 (IL-1) is a multifunctional cytokine which has immunopotentiating effects. It has been shown that IL-1 can cause numerous types of pathophysiological and metabolic damage in murine tumor systems. Several studies have reported that IL-1 has an *in vitro* cytotoxic effect on some types of tumor cell lines and an *in vivo* antitumor effect in murine tumor models. In order to activate antitumor immunity, more markedly, by IL-2 gene-modified tumor vaccine, we selected IL-1 as adjuvant based on the following observations: Firstly, IL-1 is a highly effective systemic adjuvant for active specific immunotherapy of cancer. In the past, cancer vaccines were usually used together with adjuvant in order to activate host antitumor immune responses more efficiently because some potent adjuvants, such as muramyl dipeptide, tuftsin and lipopolysaccharide have been found to induce IL-1 production and may exert their adjuvant effect via IL-1. Thus, McCune and Marquis investigated the potential of IL-1 to serve as an adjuvant in a murine tumor model with a weakly immunogenic lung cancer. They found that IL-1 could improve vaccine effectiveness significantly and the enhancing effect was both IL-1 dose dependent and duration dependent (McCune and Marquis, 1990). Secondly, there are synergistic biological activities between IL-1 and IL-2. IL-1 amplifies T-cell activation, which is very important in the active specific immunotherapy, by inducing IL-2 production and IL-2 receptor expression, particularly in conjunction with antigens or mitogens. IL-1 and IL-2 can also increase NK activity synergistically. IL-1 was used once in combination with IL-2 to treat the mice with established Friend leukemia and a synergistic antitumor effect was achieved (Ciolli et al., 1991). It has been shown that combined IL-1/IL-2 treatment results in a synergistic effect on both the myelostimulatory activity and the rescue after Cy-induced myelosuppression (Proietti et al., 1993).

Table 8-1. Clinical trials of low-dose Cy in the treatment of cancer patients

Patients	Regimens	Conclusions	References
18 patients with advanced metastatic cancer, including 8 with melanoma, 8 with colorectal cancer, 1 with lung adenocarcinoma	Administration of two doses (1,000 mg and 300 mg/m² of Cy three days prior to sensitization with the primary antigen: keyhole limpet hemocyanin, (KLH) or 1-chior-2,4-dinitrobenzene(DNCB)	At both doses the acquisition of DTH to KLH was augmented, but antibody was generated only at lower dose of Cy	Berd et al. (1982) Berd et al. (1984)
19 patients with metastatic melanoma	Administration of 300 mg/m² Cy three days before injection of autologous melanoma cell vaccine	The DTH responses of Cy-pretreated patients to autologous tumor cells were significantly greater than those of control patients (vaccine only)	Berd et al. (1986)
26 patients with metastatic renal cell carcinoma	Administration of different doses (100, 500, or 1000 mg/m²) of Cy 24 h prior to the first of six weekly immunizations with irradiated autologous tumor cells mixed with Corynebacterium parvum	4 of 15 patients developed DTH to autologous tumor cells and 3 of these 4 had a clinical response. Only 1 of 11 patients with negative DTH achieved remission. These data indicated that acquisition of DTH to autologous tumor cells is associated with remission	Sahasrabudhe et al. (1986)
45 cancer patients including 37 with melanoma , 7 with colorectal carcinoma, 1 with breast adenocarcinoma	Administration of 300 mg/m² Cy three days before the 15 patients were sensitized with either dinitrochloro-benzene or KLH, and 30 patients were injected i.d. of irradiated autologous tumor cells mixed with Bacillus Calmette-Guerin	Administration of low-dose Cy to these patients reduced nonspecific T-suppressor function without selective depletion of the CD8+ T cells	Berd and Mastrangelo (1987)
35 patients with metastatic melanoma	Administration of 300 mg/m² Cy three days before injection of autologous melanoma cell vaccine	Patients receiving Cy plus vaccine developed DTH more significantly. There were 3 complete remission, 1 partial remission and 2 minor responses in evaluaTable 33 patients. Toxicity of the protocol was minimal	Berd and Mastrangelo (1988)

361

Table 8-1. (continued)

Patients	Regimens	Conclusions	References
42 patients with metastatic melanoma	Administration of 300 mg/m^2 Cy three days before intradermal injection of irradiated autologous melanoma cells mixed with *Baccillus Calmette-Guerin*	The cancer patients were treated with low-dose Cy plus an autologous tumor vaccine. Low-dose Cy could cause depletion of CD4$^+$ ZH4$^+$ suppressor-inducer T cells	Berd and Mastrangelo (1988)
24 patients with disseminated melanoma	Administration of 350 mg/m^2 Cy three days prior to daily i.v. injection of IL-2 (3.6×10^6 IU/m^2 Cetus U/m^2, i.e. 21.6× 10^6 IU/m^2) for five days on 2 successive weeks. The schedule was repeated at least twice at 1-week intervals	25% (6/24) patients who received more than one 2-week cycle of Cy plus IL-2 had a remission, one complete and five partial, with minor responses in eight others (33.3%). Toxicity was tolerable	Mitchell et al. (1988)
41 patients with stage II and III melanoma	Administration of varying dosages of Cy (300, 150, or 75 mg/m^2) three days prior to each treatment with an allogeneic melanoma cell vaccine 3 times in a 4-week interval and then every fourth week	In each trial group there were patients who had major reduction in suppressor cell activity ($>50\%$). Overall, the greatest reduction in suppressor cell activity occurred in patients receiving 300 mg/m^2 Cy compared to the other Cy dosages or vaccine alone	Hoon et al. (1990)
67 patients including 39 with melanoma, 15 with renal cancer, 13 with breast cancer	Administration of 350 mg/m^2 Cy three days prior to daily i.v. injection of IL-2 (21.6×10^6 IU/m^2) for five days on 2 successive weeks	26% (10/39) of evaluable patients with melanoma had major clinical responses. No effect was noted in 15 patients with renal cancer. Regressions of breast cancer were found in a shortened trial with 13 patients	Mitchell (1992)

Table 8-1. (continued)

Patients	Regimens	Conclusions	References
66 patients with advanced cancer resistant to standard therapy	Administration of 350 mg/m² Cy on day 1 followed by i.v. infusion of IL-2 on day 4-8 and 11–15. The doses of IL-2 ranged from 6.0 to 36.0 symbol 18×10^6 IU/m². Each treatment cycle consisted of 21 days	The combination of low-dose Cy and IL-2 is tolerable in most patients, but pretreatment with low-dose Cy prior to the administration of IL-2 did not enhance antitumor efficacy v.s. that previously reported with IL-2 alone	Verdi et al. (1992)
25 patients with advanced malignancies of varying types	Administration of varying dosages of Cy (300, 600, 1200 mg/m²) three days prior to i.v. injection of IL-2 (30×10^6 IU/m²) thrice weekly for 6 weeks	The group of patients receiving low-dose Cy (300 mg/m²) and IL-2 produced the highest, the most sustained levels of LAK and NK activity (P<0.05) when compared with the patients receiving IL-2 alone or those receiving the higher dosages of Cy	Abrams et al. (1993)
16 patients with advanced renal cell cancer	Patients received four cycles consisting of Cy (500mg/m²) three days prior to daily i.m. injections of alpha-IFN (3×10^6 U) and continuous infusion of (18×10^6 IU) IL-2 for five days. The cycle interval was three weeks	Pretreatment with low-dose Cy did not contribute to an increased major response rate of IL-2 plus IFN-alpha	Wersall et al. (1993)
13 patients with metastatic renal cancer	Administration of 350 mg/m² Cy three days prior to daily i.v. injections of IL-2 (21.6×10^6 IU/m²) for five days on two consecutive weeks. Treatment cycles were repeated every 21 days	The study was discontinued because of significant toxicity and lack of observed response	Quan et al. (1994)

Among the different types of cytokine gene transfer into tumor cells, IL-2 gene-modified tumor cells may be the most frequently studied vaccine by many researchers using different tumor models. Generally speaking, IL-2 gene-modified tumor cells exhibit both the decreased tumorigenicity and increased immunogenicity and can induce an effective systemic immune response to resist a subsequent challenge with wild-type tumor cells or to cure the established tumors. Different antitumor effects of vaccination with IL-2 gene-modified tumor cells were reported. In a very important experiment in which 22 molecules were examined, the vaccination with IL-2 gene-modified B16 F10 melanoma cells was found to be less effective (Dranoff et al., 1993). For further potentiation of the host antitumor immunity, induced by the IL-2 gene-modified vaccine, prepared from the IL-2 gene-transfected B16 F10 melanoma cells, we proposed a combined therapy in which IL-1 and low-dose Cy were also used in addition to the vaccine (Cao et al., 1994). As shown in Table 8-2, administration of IL-1 or low-dose Cy (20 mg/kg) alone has no *in vivo* antitumor effect on our experimental tumor model with pulmonary metastases. In contrast to the experimental results of Dranoff et al. (Dranoff et al., 1993), we found that the vaccination with inactivated, IL-2 gene-modified B16 F10 melanoma cells alone could reduce the pulmonary metastases more significantly than that of control vaccine. The anti-metastatic effect was better when the IL-2 gene-modified vaccine was used with IL-1 or low-dose Cy. However, the most significant therapeutic effect was achieved when the vaccine, IL-1 and low-dose Cy were all administered at the same time. Most (6/8) of tumor-bearing mice survived for more than 90 days after treatment with the combined therapy. Our data showed that the therapeutic efficacy of IL-2 gene-modified vaccine can be increased by the combination with IL-1 and low-dose Cy via the more effective activation of specific and non-specific antitumor immunity in the tumor-bearing host.

8.4 Immunotherapy of Established Tumor by IL-3, IL-4, or TNF Gene-Modified Tumor Vaccine in Combination with Low-Dose Cy and Low-Dose IL-2

In the combined therapy, low-dose IL-2 (2000 U, three times a day for 3 days, the cycle interval was 7 days) was selected on the basis of the following facts: Firstly, low-dose IL-2 may represent a useful adjuvant that can help the tumor vaccine to induce immune responses more efficiently (Cao et al., 1995b). It has been demonstrated that low-dose IL-2 may enhance T cell-mediated immune response against weakly immunogenic tumors and can make immunodeficient non-responders become responsive to vaccination (Meuer et al., 1989). For example, Harada et al. reported that the increased antitumor activity brought about by eliciting tumor-specific CTL *in vivo*, can be achieved when an i.p. injection of inactivated tumor cells and a subsequent consecutive i.p. injection of low-dose IL-2 (2500 U IL-2 twice for

Table 8-2 The number of pulmonary metastases in the tumor-bearing mice after treatment with inactivated, IL-2 gene-modified tumor vaccine in combination with low dose Cy and IL-2

Groups	Number of metastases (mean ± SD)
A (Hanks)	135.2 ± 22.8
B (IL-1)	129.4 ± 19.5
C (Cy)	131.8 ± 17.8
D (IL-1+Cy)	108.2 ± 21.5[a]
E (B16-Neo)	101.3 ± 23.3[a]
F (B16-IL-2)	40.7 ± 21.8[b]
G (B16-IL-2+IL-1)	32.4 ± 6.9[c]
H (B16-IL-2+Cy)	35.2 ± 8.9[c]
I (B16-IL-2+IL-1+Cy	23.7 ± 8.3[c]

C57BL/6 mice were injected i.v. with 2×10^5 wild-type B16 F10 melanoma cells. Three days later, tumor-bearing mice were divided into the groups as stated above. The following treatments were repeated twice at 7-day interval. In group A (Hanks), mice were given an i.p. injection of Hanks solution (once a day for 5 days). In group B (IL-1), mice were given i.p. injection of recombinant human IL-1 alpha (360 ng, once a day for 5 days). In group C (Cy) mice were given an i.p. injection of Cy (20 mg/kg). In group D (IL-1+Cy), mice were given an i.p. injection of Cy (20 mg/kg) one day prior to i.p. injection of IL-1 (360 ng, once a day for 5 days). In group E (B16-Neo), mice were immunized i.p. with 2×10^6 inactivated B16-Neo cells that had been treated with mitomycin C (80 ug/5×10^6 cells/ml for 1 h at 37 °C in RPMI 1640 medium). In group F (B16-IL-2), mice were immunized i.p. with 2×10^6 inactivated B16-IL-2 cells. In group G (B16-IL-2+IL-1), mice were immunized i.p. with 2×10^6 inactivated B16-IL-2 cells and injected i.p. of IL-1 (360 ng, once a day for 5 days). In group H (B16-IL-2+Cy), mice were given i.p. injection of Cy (20 mg/kg) one day prior to immunized i.p. with 2×10^6 inactivated B16-IL-2 cells. In group I (B16-IL-2+IL-1+Cy), mice were administered 20 mg/kg Cy one day before the immunization with B16-IL-2 cells and injection of IL-1. To count the metastatic nodules on the surface of lung, the mice were killed 15 days after tumor inoculation.
[a]P<0.05 as compared to group A, B, C;
[b]P<0.01 as compared to group E;
[c]P<0.01 compared to group F

2 days) are administered against established metastatic melanoma (Harada et al., 1994). Secondly, low-dose IL-2 may contribute to reverse the tumor-induced immunosuppression. Low-dose IL-2 has been used clinically and was shown to be able to augment the antitumor immunity in cancer patients. Thirdly, as stated above, additive or synergistic therapeutic effects have been documented in the experimental tumor systems using IL-2 in combination with Cy (Ikemoto et al., 1992; Naito et al., 1988). Fourthly, it has been demonstrated that the combination of IL-2 and IL-4, IL-2 and TNF can activate antitumor immunity synergistically (Ohe et al., 1993; Ohira et al., 1994). A better antitumor effect could be observed when cancer patients were treated with IL-2 plus IL-3 (Lossoni et al., 1993). Therefore, IL-2 may cooperate with IL-3, IL-4 or TNF secreted from genetically modified tumor cells to induce immune response more potently.

IL-3 is best known as a multicolony-stimulating factor which affects the development and maturation of many hematopoietic cells, including macrophages, granulocytes, eosinophils, erythroid cells and mast cells. The fact that activated T cells are one of the most potent sources of IL-3 indicates that IL-3 may play a role in the immune regulation. Some experiments have demonstrated that IL-3 has some immunomodulatory properties such as activation of macrophages and induction of other cytokines (Frendl, 1992; Thomassen et al., 1993). It has been shown that IL-3 gene-modified tumor cells exhibit decreased tumorigenicity. Unlike IL-2 gene-modified tumor cells which could bypass T cell help to induce antitumor response, IL-3 gene-modified tumor cells depend on CD4+ T cell to generate CTL (Pulaski et al., 1993). After gene transfection, limiting dilution and biological assay, we obtained a B16 F10 melanoma clone (B16-IL-3) which secreted 806 U/ml IL-3. The decreased tumorigenicity and pulmonary metastastic capability of B16-IL-3 cells were found in C57BL/6 mice and even in the nude mice to some extent. Then, we used the inactivated tumor vaccine prepared from the B16-IL-3 cells to treat the tumor-bearing mice with preestablished pulmonary metastases. As shown in Table 8-3, treatment with B16-IL-3 vaccine alone was more effective in the reduction of pulmonary metastases compared with treatment with wild-type B16 melanoma cell vaccine. The combined use of B16-IL-3 vaccine with low-dose IL-2 (2000U, three times a day for 3 days) or low-dose Cy (20 mg/kg) could reduce pulmonary metastases more significantly, whereas low-dose IL-2 or low-dose Cy alone had no therapeutic effect. When B16-IL-3 vaccine was used in combination with low-dose Cy and low-dose IL-2, the best anti-metastatic effect was achieved and 62.5% (5/8) of tumor-bearing mice survived more than 90 days. Our results showed that the combined therapy augmented the splenic NK, CTL cytotoxic activities and enhanced the number and functions of peritoneal macrophages most significantly, indicating that low-dose Cy and low-dose IL-2 can improve the therapeutic efficacy of IL-3 gene-modified tumor vaccine through more efficient activation of specific and non-specific antitumor responses.

Table 8-3. The number of pulmonary metastases in tumor-bearing mice after treatment with inactivated, IL-3 gene-modified tumor cells in combination with low-dose Cy and low-dose IL-2

Groups	Number of metastases (mean ±SD)
Hanks	123.4 ±21.6
IL-2	128.4 ±13.5
Cy	121.2 ± 17.1
IL-2+Cy	85.6 ± 8.9[a]
B16-Neo	95.9 ± 11.3[a]
B16-IL-3	54.2 ± 5.4[b]
B16-IL-3+IL-2	28.5 ± 3.2[c]
B16-IL-3+Cy	24.5 ± 4.1[c]
B16-IL-3+IL-2+Cy	15.6 ± 3.6[c]

[a]$P<0.05$ compared to Hanks group
[b]$P<0.05$ compared to Hanks group
[c]$P<0.01$ compared to B16-IL-3 vaccine alone

The first cytokine gene to be transfected into tumor cells and to be used to treat established tumors was IL-4 (Tepper et al., 1989; Golumbek et al., 1991). Analysis of the inoculation site in murine tumor models showed that inflammatory infiltrates consist predominantly of macrophages and eosinophils and that a few lymphocytes infiltrate the tumor site only a few days after the initial inoculation. Thus, it is possible for IL-4 gene-modified tumor cells to induce more potent antitumor responses, if the lymphocyte response is amplified by the combined use of IL-2. It has been demonstrated that vaccination with the combination of IL-2 gene-and IL-4 gene-modified tumor cells can suppress tumor growth more effectively (Ohe et al., 1993). In our experiment, we used the inactivated cellular vaccine, prepared from the IL-4 gene-modified B16 F10 tumor cells to treat the experimental pulmonary metastastic tumor-bearing mice. After treatment with IL-4 gene-modified vaccine, the pulmonary metastases were reduced more markedly and the survival time of tumor-bearing mice lasted longer compared with control vaccine. Its therapeutic effect was better when low-dose IL-2 was also used. The best therapeutic effect was achieved when IL-4 gene-modified vaccine was used in combination with low-dose IL-2 and low-dose Cy (Cao et al., 1995a). The cytotoxicity of CTL and macrophages from the tumor-bearing mice receiving the combined therapy increased significantly, but the NK activity remained unchanged, indicating that the above combined therapy exerts antitumor effects mainly through the activation of specific immune responses.

TNF gene-modified tumor cells are sometimes rejected when injected into syngeneic mice (Blankenstein et al., 1991b). However, it has been reported that transfection of TNF gene into tumor cells does not generate enhanced systemic immunity or fails to inhibit the tumor growth (Karp et al., 1993). On the basis of the synergistic effect between IL-2 and TNF, Ohira et al. co-transfected IL-2 gene and TNF gene into the same tumor cells. They found that the immunization effect by the co-transfectant was superior to that of IL-2 and TNF transfectants alone (Ohira et al., 1994). In our murine tumor model with pulmonary metastases, we found that vaccination with TNF gene-modified B16 F10 melanoma cells had no anti-metastatic effect. But the significant therapeutic effect on the pulmonary metastases was obtained when the irradiated TNF gene-modified vaccine was used in combination with low-dose IL-2 and low-dose Cy, suggesting that low-dose IL-2 plus low-dose Cy can help TNF gene-modified melanoma cell vaccine, which has no effectiveness to induce potent antitumor response in the tumor-bearing mice.

8.5 Conclusion

Our studies have demonstrated that administration of low-dose Cy prior to the vaccination with cytokine gene-modified tumor vaccine can improve the therapeutic efficacy in the treatment of the established tumor. This is consistent with the previous observations in which low-dose Cy can enhance the antitumor effect of the standard tumor vaccine. When low-dose Cy is used together with low-dose IL-2, which has adjuvant activity but has no *in vivo* antitumor effect, the augmentation of antitumor responses, induced by cytokine gene-modified tumor vaccine, is more remarkable than that when only low-dose Cy is used. In our murine tumor model with pulmonary metastases, we showed that the combined therapy can activate antitumor immune response more effectively, although, distinct immune responses are generated by the vaccination with different cytokine gene-modified tumor vaccines. Though many questions such as the optimal dosage of Cy and optimal schedule of the combined therapy remain to be answered, our results raise the possibility that the combined use of cytokine gene-modified tumor vaccine, low-dose IL-1/IL-2 and low-dose Cy may be a promising approach to cancer treatment.

Acknowledgments

I wish to thank Dr. Thomas Blankenstein for critical review of the manuscript. This work was supported by grants (39470800, 39421009) from the National Natural Science Foundation of China.

References

Abrams, J. S., Eiseman, J. L., Melink, T. J., Sridhara, R., Hiponia, D. J., Bell, M. M., Belani, C. P., Adler, W. H. and Aisner, J. (1993) Immunomodulation of interleukin-2 by cyclophosphamide: A phase IB trial. *J. Immunother*. 14: 56–64.

Askenase, P. W., Hayeen, B. J., Gershon, R. K. (1975) Augmentation of delayed-type hypersensitivity by doses of cyclophosphamide which do not affect antibody response. *J. Exp. Med*. 151: 697–702.

Awward, M. and North RJ. (1988a) Immunologically mediated regression of a murine lymphoma after treatment with anti-L3T4 antibody. A consequence of revoving L3T4+ suppressor T cells from a host generating predominantly Lyt-2+ T cell-mediated immunity. *J. Exp. Med*. 50: 2228–2233.

Awward, M. and North, R. J. (1988b) cyclophosphamide (Cy)-facilitated adoptive immunotherapy of a Cy-resistant tumour. Evidence that Cy permits the expression of adoptive T-cell mediated immunity by revoving suppressor T cells rather than by reducing tumour burden. *Immunology* 65: 87–92.

Awward, M. and North, R. J. (1989) Cycolphosphamide-induced immunologically mediated regression of a cyclophosphamide-resistant murine tumor: A consequence of eliminating precursor L3T4+ suppressor T-cells. *Cancer Res*. 49: 1649–1654.

Bear, H. D. (1986) Tumor-specific suppressor T cells which inhibit the *in vitro* generation of cytolytic T-cells from immune and early tumor-bearing host spleens. *Cancer Res*. 46: 1805–1812.

Berd, D., Mastrangelo, M. J., Engstrom, P. E., Paul, A. and Maguire, H. (1982) Augmentation of the human immune response by cyclophosphamide. *Cancer Res*. 42: 4862–4866.

Berd, D., Maguire Jr. HC and Mastrangelo MJ (1984) Potentiation of human cell-mediated and humoral immunity by low-dose cyclophosphamide. *Cancer Res*. 44: 5439–5443.

Berd, D., Maguire Jr. HC and Mastrangelo MJ (1986) Induction of cell-mediated immunity to autologous melanoma cells and regression of metastases after treatment with a melanoma cell vaccine preceded by cyclophosphamide. *Cancer Res*. 46: 2572–2577.

Berd, D. and Mastragelo, M. J. (1987) Effect of low dose cyclophosphamide on the immune system of cancer patients: Reduction of T-suppressor function without depletion of the CD8+ subset. *Cancer Res*. 47: 3317–3321.

Berd, D. and Mastragelo, M. J. (1988) Effect of low dose cyclophosphamide on the immune system of cancer patients: Depletion of CD4+ suppressor-inducer T-cells. *Cancer Res*. 48: 1671–1675.

Blankenstein, T., Qin, Z., Uberla, K. (1991a) Tumor suppression after tumor cell-targeted tumor necrosis factor alpha gene transfer. *J. Exp. Med*. 173: 1047–1052.

Blankenstein, T., Rowley, D. A. and Schreiber, H. (1991b) Cytokines and cancer: Experimental systems. *Curr. Opin. Immunol*. 3: 694–698.

Cao, X., Yu, Y., Xu, Z., Zheng, L., Ye, T. (1995a) Treatment of pulmonary metastatic melanoma with cellular vaccine genetically engineered to secrete interleukin-4 and its immunological mechanisms. *Chin. J. Microbiol. Immunol*. 15: 53–57.

Cao, X., Zhang, W., Gu, S., Yu, Y., Tao, Q., Ye, T. (1995b) Induction of antitumor immunity and treatment of preestablished tumor by interleukin-6 gene-transfected melanoma cells combined with low-dose interleukin-2. *J. Cancer Res. Clin. Oncol*. 121: 721–728.

Cao, X., Zhang, W., Zheng, L., Yu, Y., Tao, Q., Chen, G., Ye, T. (1994) Immunotherapy of cancer by IL-2 gene-transfected tumor vaccine in combination with IL-1, low-dose cyclophosphamide and its immunological mechanisms. *Chin. J. Immunol.* 10: 289–293.

Cayeux, S., Beck, C., Aicher, A., Dorken, B. and Blankenstein, T. (1995) Tumor cells cotransfected with interleukin-7 and B7.1 genes induces CD25 and CD28 on tumor-infiltrating T lymphocytes and are strong vaccines. *Eur. J. Immunol.* 25: 2325–2331.

Ciolli, V., Gabriele, L., Sestili, P., Varano, F., Proietti, E., Gresser, I., Testa, U., Montesoro, E., Bulgarini, D., Mariani, G., Peschle, C. and Belardelli, F. (1991) Combined IL-1/IL-2 therapy of mice injected with highly metastatic Friend leukemia cells: Host antitumor mechanisms and marked effects on established metastases. *J. Exp. Med.* 173: 313–322.

Culo, F., Klapan, I. and Kolak, T. (1993) The influence of cyclophosphamide on antitumor immunity in mice bearing late-stage tumors. *Cancer Immunol. Immunother.* 36: 115–122.

Dranoff, G., Jaffee, E., Lazenby, A., Golumbek, P., Levitsky, H., Brose, K., Jackson, V., Hamada, H., Pardoll, D. and Mulligan, R. C. (1993) Vaccination with irradiated tumor cells engineered to secrete murine granulocyte-macrophage colony stimulating factor stimulates potent, specific and long-lasting antitumor immunity. *Proc. Natl. Acad. Sci. USA* 90: 3539–3543.

Dye, E. S. and North, R. J. (1981) T cell-mediated immunosuppression as an obstacle to adoptive immunotherapy of P815 mastocytoma and its metastases. *J. Exp. Med.* 154,1033–1042.

Frendl, G. (1992) Interleukin-3: From colony-stimulating factor to pluripotent immunoregulatory cytokine. *Int. J. Immunopharmac.* 14: 421–430.

Golumbek, P. T., Lazenby, A. J., Levitsky, H. I., Jaffee, L. M., Karasuyama, H., Baker, M. and Pardoll, D. M. (1991) Treatment of established renal cancer by tumor cells engineered to secrete interleukin-4. *Science* 254: 713–716.

Harada, M., Matsuzaki, G., Shinomiya, K., Kurasawa, S., Ito, O., Okamoto, T., Takenoyama, M., Sumitika, H., Nishimura, Y. and Nomoto, K. (1994) Generation of tumor-specific cytotoxic T lymphocytes *in vivo* by combined treatment with inactivated tumor cells and recombinant interleukin-2. *Cancer Immunol. Immunother.* 38: 332–338.

Hengst, J. C. D., Mokyr, M. B. and Dray, S. (1980) Cooperation between cyclophosphamide tumoricidal activity and host antitumor immunity in the cure of mice bearing large MOPC-315 tumors. *Cancer Res.* 41: 2163–2167.

Hoon, D. S. B., Foshag, L. J., Nizze, A. S., Bohman, R. and Morton, D. L. (1990 Suppressor cell activity in a randomized trial of patients receiving active specific immunotherapy with melanoma cell vaccine and low dosages of cyclophosphamide. *Cancer Res.* 50: 5358–5364.

Ikemoto, S., Nishio, S., Kamizuru, M., Kishimoto, T., Wada, S., Maekawa, M. and Hayahara, N. (1992) Combined effect of interleukin 2 and cyclophosphamide in therapy of mice with transitional cell carcinoma. *Urology* 40: 574–578.

Karp, S. E., Farber, A., Salo, J. C., Hwu, P., Jaffe, G., Asher, A. L., Shiloni, E., Restifo, N. P., Mule, J. J. and Rosenberg, S. A. (1993) Cytokine secretion by genetically modified nonimmunogenic murine fibrosarcoma. *J. Immunol.* 150: 896–908.

Lissoni, P., Barni, S., Tisi, E., Rovelli, F., Pittalis, S., Rescaldani, R., Vigore, L., Biondi, A., Ardizzoia, A. and Tancini, G. (1993) *In vivo* biological results of the association between interleukin-2 and interleukin-3 in the immunotherapy of cancer. *Eur. J. Cancer.* 29: 1127–1132.

Maguire, H. C. Jr. and Ettore VL (1967) Enhancement of dinitrochlorobenzene (DNCB) contact sensitization by cyclophosphamide in the guinea pig. *J. Invest. Dermatol.* 48: 39–43.

McCune, C. S. and Marquis, D. M. (1990) Interleukin 1 as an adjuvant for active specific immunotherapy in a murine tumor model. *Cancer Res.* 50: 1212–1215.

Meuer, S. C., Dumann, H., Meyer zum Buschenfelde, K. H. and Kohler, H. (1989) Low-dose interleukin-2 induces systemic immune responses against HBsAg in immunodeficient non responders to hepatitis B vaccination. *Lancet I* 15–17.

Mitchell, M. S. (1992) Chemotherapy in combination with biomodulation: a 5-year experience with cyclophosphamide and interleukin-2. *Semin. Oncol.* 19 (Suppl 4): 80–87.

Mitchell, M. S., Kempf, R. A., Harel, W., Shau, H., Boswell, W. D., Lind, S. and Bradley, E. C. (1998) Effectiveness and tolerability of low-dose cyclophosphamide and low-dose intravenous interleukin-2 disseminated melanoma. *J. Clin. Oncol.* 6: 409–424.

Mitsuoka, A., Baba, M. and Morikawa, S. (1976) Enhancement delayed hypersensitivity by depletion of suppressor T cells with cyclophosphamide in mice. *Nature* 262: 77–78.

Naito, K., Pellis, N. R. and Kahan, B. D. (1988) Effect of continuous administration of interleukin 2 on active specific chemoimmunotherapy with extracted tumor-specific transplantation antigen and cyclophosphamide. *Cancer Res.* 48,101–108.

Nomi, S., Pellis, N. S. and Kahan, B. D. (1985) Antigen specific therapy if experimental metastases. *Cancer* 55: 1296–1302.

North, R. J. (1982) Cyclophosphamide-facilitated adoptive immunotherapy of an established tumor depends on elimination of tumor-induced suppressor T cells. *J. Exp. Med.* 55: 1063–1074.

North, R. J. and Awwad, M. (1990) Elimination of cycling CD4+ suppressor T cells with an antimitotic drug releases non-cycling CD8+ T cells to cause regression of an advanced lymphoma. *Immunology* 71: 90–95.

North, R. J. and Dye, E. S. (1985) Ly-1+2-suppressor T cells down-regulate the generation of Ly-1-2+ effector T cells during progressive growth of the P815 mastocytoma. *Immunology* 54,47–56.

Ohe, Y., Podack, E. R., Olsen, K. J., Miyahara, Y., Ohira, T., Miura, K., Nishio, K. and Saijo, N. (1993) Combination effect of vaccination with IL-2 and IL-4 cDNA transfected cells on the induction of a therapeutic immune response against Lewis lung carcinoma cells. *Int. J. Cancer* 53: 432–437.

Ohira, T., Ohe, Y., Heike, Y., Podack, E. R., Olsen, K. J., Nishio, K., Nishio, M., Miyahara, Y., Funayama, Y., Ogasawara, H., Arioka, H., Kato, H. and Saijo, N. (1994) Gene therapy for Lewis lung carcinoma with tumor necrosis factor and interleukin-2 cDNA cotransfected subline. *Gene Ther.* 1: 269–275.

Porgador, A., Feldman, M. and Eisenbach, L. (1994) Immunotherapy of tumor metastasis via gene therapy. *Nat. Immun.* 13: 113–130.

Proietti, E., Tritarelli, E., Gabriele, L., Testa, U., Greco, G., Pelesi, E., Gabbianelli, M., Belanlelli, F. and Peschle, C. (1993) Combined interleukin 1 beta/interleukin-2 treatment in mice: synergistic myelostimulatory activity and protection against cyclophosphamide induced myelosuppression. *Cancer Res.* 53: 569–576.

Pulaski, B. A., McAdam, A. J., Hutter, E. K., Biggar, S., Lord, E. M. and Frelinger, J. G. (1993) Interleukin 3 enhances development of tumor-reactive cytotoxic T cells by a CD4-dependent mechanism. *Cancer Res.* 53: 2112–2117.

Quan, W. D. Jr., Dean GE, Lieskovsky, G., Mitchell, M. S. and Kempf, R. A. (1994) Phase II study of low dose cyclophosphamide and intravenous interleukin-2 in metastatic renal cancer. *Invest. New Drugs* 1291: 35–9.

371

Rakhmilevich, A. L., North, R. J. and Dye, E. S. (1993) Presence of CD4+ T suppressor cells in mice rendered unresponsive to tumor antigens by intravenous injection of irradiated tumor cells. *Int. J. Cancer* 55: 338–343.

Ray, P. K. and Raychaudhuri, S. (1981) Low-dose cyclophosphamide inhibition of transplantable fibrosarcoma growth by augmentation of the host immune response. *J. Nat. Cancer Inst.* 67: 1341–1345.

Rosenthal, F. M., Cronin, K., Bannerji, R., Golde, D. W. and Gansbacher, B. (1991) Augmentation of antitumor immunity by tumor cells transduced with a retroviral vector carry the interleukin-2 and interferon-gamma cDNAs. *Blood* 83: 1289–1298.

Sahasrabudhe, D. M., deKernion, J. B., Pontes, J. E., Ryan, D. M., O'Donnell, R. W., Marquis, D. M., Mudholkar, G. S., McCune C. S. (1986) Specific immunotherapy with suppressor function inhibition for metastatic renal carcinoma. *J. Biol. Response Mod.* 5: 581–594.

Tepper, R., Pattengale, P. and Leder, P. (1989) Murine interleukin-4 displays potent antitumor activity *in vivo*. *Cell* 57: 503–512.

Thomassen, M. J., Antal, J. M., Connors, M. J., McLain, D., Sandstrom, K., Meeker, D. P., Budd, T. G., Levitt, D. and Bukowski, R. M. (1993) Immunomodulatory effects of recombinant interleukin-3 treatment on human alveolar macrophages and monocytes. *J. Immunother.* 14: 43–50.

Tuttle, T. M., Fleming, M. D., Hogg, P. S., Inge, T. H. and Bear, H. D. (1994) Ability of low-dose cyclophosphamide to overcome metastasis-induced immunosuppression. *Ann. Surg. Oncol.* 1: 53–58.

Verdi, C. J., Taylor, C. W., Crohgan, M. K., Dalke, P., Meyskens, F. L. and Hersh, E. M. (1992) Phase I study of low-dose cyclophosphamide and recombinant interleukin-2 for the treatment of advanced cancer. *J. Immunother.* 11: 286–291.

Vieweg, J., Rosenthal, F. M., Bannerji, R., Heston, W. D. W., Fair, W. R., Gansbacher, B. and Gilboa, E. (1994) Immunotherapy of prostate cancer in the dunnin rat model: use of cytokine gene modified tumor vaccine. *Cancer Res.* 54: 1760–1765.

Wersall, J. P., Masucci, G., Hjelm, A. L., Ragnhammar, P., Fagerberg, J., Frodin, J. E., Merk, K., Lindemalm, C., Ericson, K., Kalin, B. (1993) Low dose cyclophosphamide, alpha-interferon and continuous infusions of interleukin-2 in advanced renal cell carcinoma. *Med. Oncol. Tumor Pharmacother.* 10: 103–111.

Wise, J. A., Mokyr, M. B. and Dray, S. (1988) Effect of low-dose cyclophosphamide therapy on specific and nonspecific T cell-dependent immune responses of spleen cells from mice bearing large MOPC-315 plasmacytomas. *Cancer Immunol. Immunother.* 27: 191–197.

Subject Index

MCBU
Molecular and Cell Biology Updates

Molecular Aspects of Cancer and its Therapy

Mackiewicz, A.,
University School of Medical Sciences, Poland /
Sehgal, P.B.,
New York Medical College, Valhalla, NY (Ed.)

This book highlights recent progress in the molecular, cellular and immunological mechanisms that contribute to the pathophysiology of cancer and the design of therapeutic modalities based upon these molecular insights. Areas of particular emphasis include cancer immunology and the immunotherapy of cancer, the role of cytokines in modulating the social behaviour of cancer cells, the genetic alterations that characterize human cancer and metastasis, and a consideration of the more experimental approaches to cancer therapy, including gene therapy using expression vectors for cytokines and their receptors, antisense RNA therapy, and anti-idiotypic antibody immunization.

more research-oriented approach in discussions of specific research topics provides a stimulating and forward-looking volume which serves to update selected aspects of cancer research today. This combination will be useful to both the beginner as well as the more advanced biomedical scientist.

This volume serves to introduce the general reader as well as the cancer specialist to personalized perspectives of particular topics in cancer research by leading research groups in the field. The combination of a "reviews"-approach with a

MCBU – Molecular and Cell Biology Updates
Mackiewicz, A., et al. (Ed.)
Molecular Aspects of Cancer and its Therapy
1998. 240 pages. Hardcover
ISBN 3-7643-5724-X

BioSciences with Birkhäuser

(Prices are subject to change without notice. 10/98)

For orders originating from all over the world except USA and Canada:

For orders originating in the USA and Canada:

Birkhäuser Verlag AG
P.O. Box 133
CH-4010 Basel / Switzerland
Fax: +41 / 61 / 205 07 92
e-mail: orders@birkhauser.ch

Birkhäuser Boston, Inc.
333 Meadowland Parkway
USA-Secaucus, NJ 07094-2491
Fax: +1 / 201 348 4033
e-mail: orders@birkhauser.com

Birkhäuser